Y0-CAT-164

THE COMPLETE HANDBOOK OF
SOLAR AIR
HEATING SYSTEMS

To the homeowners in Colorado's San Luis Valley who have built their own solar heating system and who have demonstrated beyond a doubt that cost-effective solar heating is within the grasp of the average person.

About the Authors

Steve Kornher is a solar designer-contractor from southern Colorado with over 14 years' experience in the construction trades, 7 of which have been devoted to work in the solar field. He has worked closely with J. K. Ramstetter, a pioneer in air-heater construction, and has himself made innovations in collector and system design.

With over seven years of work in solar heating, research associate Andy Zaugg has also been involved in the development of air-heater systems. He is currently president of Hot Stuff Controls Inc., which manufactures dampers and other controls for air heaters.

THE COMPLETE HANDBOOK OF
SOLAR AIR
HEATING SYSTEMS

How to Design and Build Efficient, Economical Systems for Heating Your Home

BY STEVE KORNHER
WITH ANDY ZAUGG

Illustrations developed by Austin Canon

Rodale Press, Emmaus, Pa.

Copyright ©1984 by Steve Kornher

All rights reserved. No part of this publication may be reproduced or transmitted in any form or by any means, electronic or mechanical, including photocopy, recording, or any information storage and retrieval system without the written permission of the publisher.

Printed in the United States of America on recycled paper containing a high percentage of de-inked fiber.

Book design by Jerry O'Brien
Cover design by Daniel M. Guest
Photography by Steve Kornher, except:

   Photo 2-1 (p. 12), courtesy of Bill Doubleday
   Photo 7-2 (p. 137), photo 7-3 (p. 139) and photo 12-5 (p. 294), courtesy of
      Silas Kopf
   Photo 7-4 (p. 142), courtesy of Energy Training & Education Center, Northampton, Mass.
   Photo 7-5 (p. 145), Robert O. Ware
   Photo 11-2 (p. 270), and photo 11-3 (p. 278), courtesy of Jim Gunning

Creative development of artwork by Austin Canon
Camera-ready artwork by John Carlance
Cover photograph by Mitchell T. Mandel

Library of Congress Cataloging in Publication Data

Kornher, Steve.
   The complete handbook of solar air heating systems.

   Includes index.
   1. Solar heating—Handbooks, manuals, etc.
2. Dwellings—Heating and ventilation—Handbooks,
manuals, etc. I. Zaugg, Andy. II. Title.
TH7413.K66   1983   697'.78   83-11039

ISBN 0-87857-442-5   hardcover

2 4 6 8 10 9 7 5 3 1   hardcover

CONTENTS

This book is based on the author's experience with systems that he has worked with over the years. The systems described in this book have been refined and proven to be reliable when properly designed and installed. The ultimate benefits of solar improvements will of course depend on many factors, including the condition of the existing house, construction materials, hardware, and site and climate conditions. It is important to note that due to the variability of personal skills neither the author nor Rodale Press assumes any responsibility for any injuries or damages or other losses incurred that result from the material presented herein. All instructions and plans should be carefully studied and clearly understood before beginning any project.

ACKNOWLEDGMENTS

Many individuals contributed their time and efforts to create this book. First and foremost among these is my good friend, building partner and technical advisor Andy Zaugg. This book would not have been possible without his page-by-page advice and criticism. Andy developed the wiring schematics that appear in Appendix 2, and his knowledge and intuition about solar collector systems are evident throughout.

The hard work of Austin Canon in the creative development of the artwork for this book speaks for itself. I'd like to thank him for his patience and enthusiasm throughout this project.

J. K. Ramstetter, who has been building collectors for over 50 years, was a constant source of inspiration and advice on active systems. Most of the good ideas in this manual originated with J.K. The theory for the thermosiphoning air collectors came from W. Scott Morris, who has done extensive testing and design work on these systems. Scott, who sold me on mesh absorbers, was a tremendous resource. Bill Doubleday provided a lot of good information on the nuts-and-bolts building techniques of TAPs.

Technical advice and endless encouragement came from architect Akira Kawanabe. Solar contractor Terry McClaughry built many of the active systems featured in the book; and his practical experience, constructive criticism and help with photographs were invaluable. Roy Moore of Las Cruces, New Mexico, was the source of inspiration and reviewer for the section on black-roof collectors. Greg Peterson provided advice on glazing techniques for site-built collectors, and my brother, Gordon Kornher, outlined the section on dealing with tempered glass.

Silas Kopf, B. J. Rodgers and The Mark Jones Corporation all provided valuable material and insights for the convective collector sections. Kathy Kana reviewed the economics chapter, and Monty Holmes and Jim Gunning were responsible for the rock bin construction photographs. Steve Baer provided background material for the work Andy and I did on air-to-water heat exchangers. Lorraine Miller and Jim Toms did the majority of the photo finishing from my often poor-quality negatives.

Both the San Luis Valley Solar Energy Association and the Solar Energy Association of Northeast Colorado (especially member Vern Tyron) provided valuable assistance during the completion of this work.

I'd like to extend a special thanks to the individuals who spent many hours reading rough drafts and who offered their suggestions and comments: my father, John Kornher; long-time friend Jed Davis; solar contractor Mark Randall; hyrdoelectric enthusiasts Doug and Sue Bishop; solar consultant Bob Dunsmore; low-cost solar builders Arnie and Mario Valdez; and my business partner, Rick Roen.

A special thanks goes also to Nancy Wilkie, who made sense out of the second draft, and to Pat Todaro, who spent many late nights typing the majority of the manuscript.

The staff of Rodale Press has been instrumental in making this a well-organized, accurate and readable book, and to Rodale's hard-working editors I'd like to extend my thanks: to Joe Carter for his technical expertise and great job of editing and to Margaret J. Balitas for her organizational abilities and patience with a first-time author.

INTRODUCTION

All of us are experiencing the problem that this book can help to ease: the rising cost of energy. The fuel bills for heating our homes and businesses are steadily rising every winter with no end in sight; and more and more of our income is required to maintain the level of comfort to which we have become accustomed. Weatherizing your house by caulking, weather-stripping and adding insulation is a sure way to reduce your energy use and make your house more comfortable. Another approach is to replace an inefficient furnace or water heater with a more efficient one. You can also change to a cheaper fuel. Presently, gas and wood are typically cheaper fuels than oil and electricity. Sacrifice is yet another way to lower heating costs and is the approach many low- or fixed-income families must choose. Wearing sweaters and coats indoors and washing in cold water are reality for many people who are faced with a heat-or-eat situation.

Yet no matter how much you conserve or sacrifice, if you live where it is cold in the winter, some heat must be added to your house for it to be reasonably comfortable. We can look to the sun to produce this heat at a reasonable cost.

The purpose of this book is to help you avoid the high cost and unnecessary complexity of commercially manufactured solar air heaters by showing you how to build your own attractive, durable, low-cost collectors that deliver just as much heat as commercial systems costing over three times as much. Solar air heaters are easy to build, and hundreds of owner-built collectors have been used successfully nationwide, but some deliver a lot more heat than others. Most of the do-it-yourself installations we have inspected would work a lot better if only one or two things had been done differently. By presenting the basic principles involved and some of the mistakes made by others, we hope to help you build a trouble-free, efficient and cost-effective solar heater.

Site-Built Collectors in the San Luis Valley

Most of the information in this book is the result of the practical experience of do-it-yourself collector builders in Colorado's San Luis Valley. This high mountain valley has more owner-built collectors per capita than anywhere else in the world and is often described as the most solarized place in the United States. The San Luis Valley is a natural place for a

Photo 0-1: This vertical collector, one of the first built in the San Luis Valley, is shown in the late spring partially shaded by eaves. After ten years of use, the fiberglass glazing has yellowed and is in poor shape. The wooden trim is also falling apart, but the collector still delivers a lot of usable heat to an insulated crawl space.

grass-roots solar boom since the region is very cold and very sunny in the winter. Low income levels, high fuel costs and several weeks of subzero weather forced many families to seek an alternative to expensive propane heating. They found it with low-cost, site-built air-heating collectors.

In 1972 and 1973 the first of many systems were installed in the valley. They were designed by J. K. Ramstetter and built by his son-in-law, Bill North. The early "North" collectors were glazed with corrugated fiberglass, had wooden frames and baffles and used corrugated aluminum roofing for the absorber surface. Most of them were mounted vertically on south-facing walls and were used for space heating only. The heated air was blown into an insulated crawl space or directly into the house. Many systems costing under $300 reduced heating bills by 50 percent and paid for themselves within two winters. News of their success rapidly spread and farmers and ranchers throughout the valley began installing their own, easy-to-build systems onto homes, shops and outbuildings.

As with any new technology, innovations in collector design appeared everywhere. Changes in the original design were made; some were improvements, others mistakes. All of the systems built provided heat, but some gave much more than others.

Photo 0-2: People in the San Luis Valley began building site-built installations to reduce their fuel bills long before tax credits popularized collector construction. This older, owner-built system was constructed from corrugated aluminum roofing material and flat fiberglass glazing. A wall-mounted blower delivers air directly into the house.

In 1976 and 1977 the newly formed San Luis Valley Solar Energy Association began collecting the best design ideas and providing advice to people to help them build efficient, durable and cost-effective systems. Some of the design improvements included reducing the size of the airflow cavity, using a flat rather than a corrugated absorber plate and using metal for the collector frame and baffles. As some of the first North collectors began to age, it became obvious that even the best fiberglass wasn't a permanent glazing material. Nowadays, collectors are more often glazed with tempered glass. Older collectors also began to lose their seals and leak air, so improved sealing techniques were developed. These improvements increased the cost of construction slightly but resulted in much more durable and efficient collectors.

Lately, homeowners have been building bigger systems that provide a larger percentage of their heating needs, and systems with rock storage for nighttime use have become popular, especially in new construction. More do-it-yourselfers also use their space-heating systems for preheating hot water in the summer or design and build systems to heat water year-round. Nearly everyone in the San Luis Valley is enthusiastic about solar heating, and collector building is still booming. Several businesses now offer commercial systems and are doing a brisk business among the more affluent, but for those folks looking for thrifty heat and the satisfaction of doing their own installation, site-built air heaters are still the first choice.

The benefits of building your own collector are not restricted to rural residents. The city or suburban dweller can easily utilize the ideas presented in this book to design and build a system for his house.

Can You Build a Collector?

If you can finish out a room in the basement, weatherize your house or put in a patio, you can build some type of collector. Building the various collector designs presented in this book requires widely different levels of skill. The passive window box and TAP collectors are quite simple, but still require a dedication to detail. A large active air-heating system is much more complex, requiring no small amount of how-to savvy and a dedication not only to detail but also to what will probably be a long-term project. You'll find no exotic tools or construction techniques are required, and your tool box probably contains every tool you will need. Materials for the low-cost systems described in this book are all locally available or can be easily located.

Working with metal is a new experience for many do-it-yourselfers, but it need not be worrisome. All of the metal used in the construction of air collectors is lightweight and can be worked with simple hand tools. You may be surprised how easily the pieces go together.

Building your own solar system is a project that can be undertaken in the evenings and on weekends since most of the work can be done in stages. Depending on the type and size of system you're building, you could be looking at just a few days' work or many, possibly very many evenings and weekends. The bigger the project, the more planning becomes essential. It is, for example, a good idea to add more elaborate air distribution schemes, a hot water preheat system or rock thermal storage later, after the collector is already installed and providing heat.

The advantage of building your own system is that you can get just what you want. Nobody knows your house as well as you do, and with the ideas presented in this book, you, better than anyone else, can design a system to fit your needs. The authors of this book have retrofitted many different systems and have found that homeowners, with very little previous knowledge of solar heating, have come up with at least half of the good ideas that were incorporated into their own systems. With few exceptions, owner-builders have been very satisfied with their systems and are among the strongest proponents of solar heating.

The do-it-yourselfer must be willing to invest a considerable amount of time in building a system. About 100 hours are required to build a simple 100-square-foot collector system, but it can be enjoyable and satisfying work as well as economically rewarding. It is a true thrill to turn on the blower and feel that first blast of solar-heated air.

But Collectors Are So Ugly

It's true that a collector isn't cedar tongue-and-groove siding, a freshly painted white picket fence or a stained glass window. One greenhouse builder we know refers to collectors as "unsightly black squares" in his sales literature. But as energy prices continue to soar, collectors "look" more and more attractive. A collector can be absolutely beautiful on a winter day when it's sending all that heat into your house.

Using This Book

This book covers many different design options for air-heating collectors. Solar installations are very site specific; that is, the site where they are installed usually determines the size and type of system that is most desirable. There is a great deal of variation among different installations, and presenting one option as *the* system would be very misleading. So

Photo 0-3: Site-built collectors mounted at a slant are becoming more popular because they can be used for water heating in summer as well as for winter space heating.

by describing several options, we aim to help you choose system features that will suit your specific needs. Also, we allow options regarding the selection of building materials, since some of the materials we recommend may be hard to obtain in your area.

The use of solar air-heating collectors is not the complete answer to our energy dilemma, but they do provide an immediate, nonpolluting solution. We can use the sun for heating without the environmental problems involved in burning fuels or disposing of nuclear wastes. In 50 or 100 years our descendants will be living in energy-efficient homes heated by the sun, and small collectors will be used, as they should be, for supplemental heat or for specialized uses. But our existing houses and apartments will be inhabited for a long time to come, and retrofitting them with solar collectors is an appropriate way to heat them. As fuel prices rise, more of you will be clearing the way on the south side of your houses to let the sun shine in. So let's get on with it. We agree with our solar guru, J. K. Ramstetter, when he says, "We need a million site-built systems installed in the next few years." It's up to you to make the most of this most cost-effective approach to solar heating.

Photo 0-4: J. K. Ramstetter, the father of low-cost solar heating in the San Luis Valley, is presently exporting his know-how to the Denver suburbs. He designed and built this large, site-built system in Lakewood, Colorado.

1

KEEPING WARM: THE PRINCIPLES OF SOLAR HEATING

Nobody likes to be cold. That's why we have furnaces, boilers, woodstoves and solar heaters. In order to build an effective solar heating system, it's important to understand how our houses can make the best use of solar heat and how to minimize their loss of heat.

Whenever it is colder outside than inside, a house will lose some heat—no matter how well it's insulated and sealed. This heat transfer occurs in three ways: by conduction, convection and radiation.

Conduction is the movement of heat through a solid material, from the warmer side of an object to its cooler side. The rate of heat exchange depends upon the temperature difference between the two sides. The resistance that a material presents to conductive heat transfer is called its *R-value*. Wood, for example, has a higher R-value than metal. If you stir soup with a metal spoon, heat conducts rapidly through it while the opposite is true if you use a wooden spoon. In cold climates where there is a great difference between inside and outside temperatures, walls, floors and ceilings are insulated to a high R-value to resist potentially large conductive losses.

Convection is the transfer of heat by a moving stream of air (or water). As air in a room is warmed, it expands, becomes lighter and rises to the ceiling. It then gives off some of its heat to surrounding objects, becomes cooler and heavier and returns to a lower

level. This pattern of air movement, known as a *convective loop*, can also take place inside uninsulated walls. Air moves up the warm interior wall and falls down the exterior wall as it cools. In a poorly insulated house with lots of air leaks, a great deal of energy is wasted in heating the already too-hot area near the ceiling to maintain a comfortable air temperature near the floor. In such a case, insulation and weatherstripping are essential.

Radiation is the transfer of heat across an open space without changing the temperature of the air in the space. A warm object will radiate electromagnetic waves to any colder object that it "sees." Radiant heat is the heat that warms your front when you face a hot woodstove and warms your back when you turn around. On a clear, cold night the roof of a house radiates heat to the sky and can actually become colder than the outside air. Radiant heat losses are very site specific, which means that if the sky is cloudy where you live, your house will lose heat mostly by conduction, but if you live at a high altitude or where the sky is clear at night, your house's radiant losses will be increased.

Your first priority in keeping warm is to minimize heat losses from your house by first caulking and weather-stripping around doors, windows and attic openings to stop cold air infiltration and by insulating your walls and ceilings. Up to two-thirds of a house's heat loss can be through uninsulated walls and ceilings as heat travels through them by conduction, convection and radiation. Insulate your ceilings and walls. Single-pane windows can also cause major heat losses, and they should be double glazed in cold climates.

Adding Heat to Your House

A house can be warmed in two ways: by sensible heat transfer (convection and conduction) and radiant heat transfer (radiation). *Sensible heat* transfer is the kind that burns your finger when you put it in the flame of a match. A forced-air heating system heats your house by transferring sensible heat to a moving stream of air in the combustion chamber of your furnace. The warm air is then ducted throughout the house.

Radiant heat transfer occurs between any two surfaces that are of different temperatures and that can "see" one another. Because the sun's surface is so hot, it is the major source of radiant heat for the earth. A woodstove is a good example of a radiant heat source that many of us have in our houses.

Sensible and radiant heat environments are measured in different ways. Sensible heat is measured with a thermometer. The result is called *ambient temperature* and is expressed in degrees Fahrenheit (°F) or centigrade (°C). The radiant heat environment is described by the *mean radiant temperature,* which is the average temperature of the surfaces in the given environment (e.g., a room).

Both kinds of heat transfer can be used to heat our houses. We are just as comfortable in a room with cold air (sensible) and warm surfaces (radiant) as we are in a room with warm air and cold surfaces. However, ambient and mean radiant temperatures differ

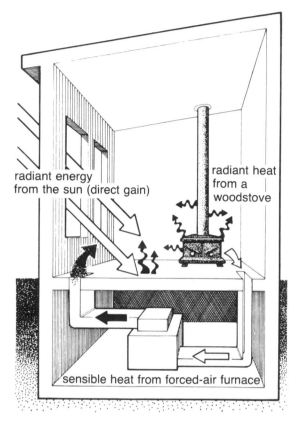

radiant energy from the sun (direct gain)

radiant heat from a woodstove

sensible heat from forced-air furnace

Figure 1-1: There are two ways a house can be warmed: by adding *sensible heat,* for example, heating air in a furnace and blowing it into the house, or by adding *radiant heat,* for example, using a woodstove or letting sunlight enter through a window. Radiant heat produces a comfortable environment at lower temperatures, so remember: it's fine to be sensible, but it's great to be radiant.

in their comfort values. For every degree Fahrenheit that the ambient temperature in a room is reduced, the mean radiant temperature need only be raised by 0.72°F to maintain the same level of comfort. Thus radiant heat sources are, in this sense, more efficient producers of comfort.

Your Solar Resource

Any discussion about the amount of heat that can be gained from the sun involves three factors: Btu's, insolation (not insulation) and heating degree-days.

A *British thermal unit* (Btu) is the basic unit of heat measurement; it is the amount of heat needed to raise one pound of water 1°F. A wooden match completely burned releases about one Btu. The hot water in a long shower contains 6,000 Btu, and it takes 400,000 Btu to heat a small house in Detroit for one day in the winter. *MBtu* refers to thousands of Btu's; *MMBtu* refers to millions of Btu's.

Insolation, incident solar radiation, which is solar energy striking a surface, is measured in Btu's per square foot per hour, day, month or year. The maximum amount of insolation available on a very clear day is between 280 and 310 Btu per square foot per hour.

A *heating degree-day* (HDD) is a unit of measurement representing a 1°F difference between a base temperature of 65°F and the average outside temperature for a one-day period. If the average outside temperature for 24 hours is 20°F, that day has 45 degree-days (65 − 20 = 45). Duluth, Minnesota, averages 9,250 heating degree-days a winter, while Los Angeles averages 2,060.

TABLE 1-1 Daily Insolation at 42 Degrees North Latitude		Clear Day (Btu/ft²/day)
January 21	vertical walls	1,664
	pitched roof	1,585
	(latitude − 10°)	
March 21	vertical walls	1,521
	pitched roof	2,283
	(latitude − 10°)	

Sunshine contains a lot of usable energy. On nearly every day of the year the heating requirement of your house is exceeded by the energy striking it in the form of sunshine. For an example let's look at how much sunshine strikes a 1,000-square-foot house in Des Moines, Iowa, at 42 degrees north latitude. The house is 25 feet by 40 feet with the long side facing south (the area of the south wall is 400 square feet; the area of the roof, 560 square feet).

On a typical, partly cloudy January day in Des Moines, the sun will deliver 792 Btu to each square foot of the roof (50 percent of available, clear-day solar energy from table 1-1). A total of nearly 444,000 Btu will strike the entire roof through the day (560 ft² of roof x 792 Btu/ft²/day/443,520). The south-facing wall will receive nearly 333,000 Btu. If this house is well insulated, it will have a heat loss of 8 Btu per square foot of floor area per heating degree-day. If it is 10°F outside all day (24 hours), there will be 55 heating degree-days (65° − 10° = 55°), and the house's heating need will be 440,000 Btu (55 HDD x 8 Btu/ft²/ HDD x 1,000 ft²). Enough heat from the sun strikes the roof alone in the course of a day to equal its losses on even a cold, partly cloudy day in January.

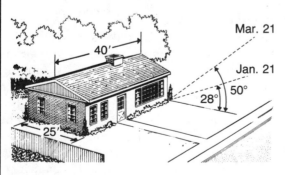

Figure 1-2

Solar Systems: Passive and Active

There are two main types of solar heating systems: passive and active. *Passive* systems operate without blowers or dampers to regulate the movement of heat. They work by the direct absorption of the sun's energy by a thermal storage mass located in the living space or by an exterior collector that works by *natural convection* (instead of *forced convection*, which is done with a fan or blower). Passive systems cost nothing to operate but can require manual operation of dampers and vents. *Active* systems, on the other hand, use electrical devices to move heat. Usually, they are more easily added on to an existing structure and can be completely automated. A third type of solar heating system is the *hybrid* system. Such a system includes active devices for moving heat but also relies on the natural convection heat flow that is vital in a passive system.

In any solar heating system, there are three basic processes: *collection*, changing sunlight to heat on an absorbing surface; *distribution*, moving the heat; and *storage*, holding the heat in thermal mass for later distribution.

The Greenhouse Effect

Sunlight is radiant energy, which is composed of short and long waves. Most of the sun's energy is in short-wave radiation, but it is the sun's long-wave radiation that warms us when we sit outside in the sunshine. Glass or any clear glazing material lets almost all of the short-wave solar radiation pass through it but very little of the long-wave heat radiation. Once the short waves pass through the glass and strike a solid, nontransparent surface, they turn into long-wave heat radiation and can't escape back out through the glass. A good example of this effect is that a car

parked in the sunshine for several hours will have an uncomfortably hot interior. The same thing happens in greenhouses, hence the term *greenhouse effect*.

In passive systems this trapped heat is often absorbed and stored by thermal mass behind the glazing, whereas in active systems this heat is carried out of the collector by a moving stream of air or liquid and may be stored elsewhere. Mass for active heat storage can take many forms. Air-heating collectors usually store the heat they gather in a rock-filled bin, and water-heating systems store it in a water tank.

Collectors

Collectors can operate both passively and actively. Passive collectors rely on natural convection for moving heat whereas active systems use mechanical devices and electricity for heat delivery. Both types capture heat from the sun in basically the same way. Sunlight passes through the glazing and heats up a lightweight, blackened metal absorber surface (aluminum, steel or copper). The heat is carried off by a flow of either air or liquid. To limit heat loss, the back of a collector is insulated, and often the front is double glazed.

Passive Collectors

Passive collectors, like passive solar systems, are self-operating, silent and, if properly built, quite efficient. Thermosiphoning air collectors operate by a natural flow of warmed air and must therefore be located below or at the same height as their point of use. Sunlight entering the collector becomes heat on the absorber surface. This heat is transferred to the air in the collector, which rises and enters the living area, pulling cooler air from the room back into the bottom of the collector. The more intense the sunlight, the hotter the collector becomes and the more forcefully the convective loop moves. At night a reversal of this loop must be prevented. The two major types of convective heaters, the *thermosiphoning air panel* (TAP) and the *window box collector,* accomplish this by different means. TAPs prevent nighttime reverse flow with lightweight backdraft dampers, and window box collectors use a heat trap in which the cold air settles to the bottom of the collector and is unable to rise into the room. Both TAPs and window box collectors need large, unrestricted airflow openings into the room being heated.

In small retrofitted systems it is not cost-effective to transfer the heat produced by passive collectors into thermal mass. In new construction large thermosiphoning collectors can be incorporated into the design, and the heat storage component can be installed as the house goes up.

Active Collectors

An active collector system requires a distribution system with electrical controls to move solar-heated air or liquid to its point of use. Air systems use ductwork for distribution while liquid systems use pipes. A thermostat turns on a blower (or pump) when the collector is hot, and dampers (or valves) control the flow of solar heat. Larger systems also have a separate heat storage area for

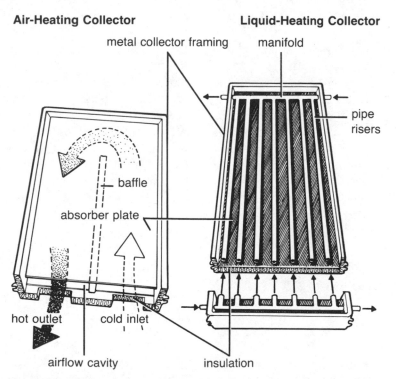

Figure 1-3: Both air- and liquid-medium active collectors work in basically the same way. Sunlight passes through glazing and is converted to heat on a black absorber surface. This heat is carried off by a blower or pump in a stream of air or liquid.

holding surplus heat that can't be used immediately. The operating cost of most active systems is minimal (see chapter 6). Most active systems warm the air in the dwelling directly but can be designed to provide radiant heat to the living area, as is discussed in chapter 2.

Flat Plate Collectors

The simplest type of collector is the *flat plate* collector. As the name implies, it has a flat absorber plate and uses either air or liquid as a heat transfer medium. Flat plate collectors deliver heat in the range of 100 to 180°F. Their main advantages are their ease of construction and installation and their relatively low cost. Well-built units operate with quite acceptable efficiency (30 to 50 percent efficiency).

We feel that the flat plate collectors discussed in this book represent the state of the art. They perform efficiently at a reasonable cost. It is very unlikely that in the coming years there will be any tremendous breakthrough in collector design that will enable a collector to deliver a great deal more heat without considerably increasing its cost.

Matrix Collectors

Another type of collector that is similar to a flat plate is the *matrix* collector. The absorber plate in this collector isn't flat but consists of a material that the air must pass through and/or around. Absorbers for matrix collectors can be black fiberglass fibers or expanded metal lath. Convective (passive) air heaters using a matrix absorber of expanded metal lath outperform similar collectors with flat plate absorbers and are therefore our choice for convective collectors. Since using a matrix absorber in active air collectors makes them more difficult to design and build and

does not substantially increase their output, we prefer flat plate absorbers for these systems.

Collectors Work Best for Retrofits

Passive systems such as direct-gain systems, greenhouses and Trombe walls, while good choices for new construction, can be difficult to retrofit because they can involve major changes to the house. The design and construction of a small solar air collector, on the other hand, is a project that is easily undertaken by the average home handyman. Modular convective air heaters are easily added on to any suitable south-facing wall to deliver heat to the adjoining rooms. Larger and more complex active collectors can be mounted in a variety of locations (on walls or roof), and the solar-heated air can be blown to a more remote point of use or to a remote heat storage box. The collectors are typically lightweight and usually require no extra structural support. These larger systems, as one would expect, represent a somewhat more challenging undertaking. People contemplating covering a wall or a roof with collectors and then installing the rest of the system should have some experience in a variety of home improvement skills.

There are other personal considerations in choosing the right solar system. For an owner-builder interested in horticulture, a solar greenhouse may be the first choice, but it is important to realize that a $2,000 collector will usually provide much more heat than a $2,000 greenhouse. The collector also qualifies for more tax credits, and its payback in fuel savings will be much shorter.

The success of a solar retrofit also depends on the type of house you own. The addition of a Trombe wall, rather than an air-heating system, may be the best choice if your house has a massive (brick, block or adobe), uninsulated south-facing wall. The thermal mass for your system is already in place, and you need only to glaze the wall to start enjoying the benefits of solar heating.

If you are planning a new house or a large addition to your present house, take a look at incorporating passive features in your design. However, a collector is probably your most cost-effective choice to heat an existing wood-frame house or to add supplemental heat to a new one.

Air versus Liquid Collectors

Air and liquid systems each have their advantages, but from a practical standpoint air systems are more foolproof since they cannot freeze and burst pipes. The skills required to build and maintain an air system are easily within the capabilities of the home handyman, whereas active liquid systems are more difficult. Air systems are also more versatile. Liquid systems need either storage or an expensive fan-coil unit to deliver heat, but in small-sized air systems the heated air from the collector is used directly to warm a house. Controls and hardware can thus be kept to a minimum and the cost per Btu delivered can be kept low.

Air systems do have their drawbacks, though, the main one being the large ductwork or vents that are necessary to move the air into the house. A 1-inch water pipe can carry as many Btu's from a liquid collector as a 16-inch air duct can from an air collector. Accommodating large ductwork can be difficult in retrofits. Table 1-2 lists other advantages and disadvantages of air and liquid systems.

When all is considered, it is our opinion that active liquid systems should be built only by somebody with plenty of plumbing ex-

perience. Freeze-up or leaking problems are *likely* unless you know what you are doing. If your roof-mounted air collector leaks, you get poorer performance but experience no major problems. If your roof-mounted liquid collector freezes and bursts or the pipes corrode and leak, you have a big headache in store. There are, however, many successful owner-built liquid collectors in use around the country. If you go with a do-it-yourself

TABLE 1-2	
Design Criteria Comparison	
Active Air Systems	**Active Liquid Systems**
Large ductwork	Small pipes
Can be less technically complex and less expensive	Can be more complex and more expensive
Little maintenance	More maintenance
More ductwork heat loss	Low pipe heat loss
Leaks degrade performance	Leaks shut down system, can cause damage
Larger storage volume required	Smaller storage volume
Fluid from collector (air) can heat house directly	Heat from collector fluid usually must be exchanged (liquid to air) before heat delivered to house
If tilted collectors are installed, they must be used or power vented in summer	Needs freeze protection, boiling protection and pressure relief

liquid system, get a good set of plans, take your time and pay attention to detail. The same can of course be said for an air system. As you'll see in subsequent chapters, planning and detailing are of utmost importance.

Site-Built Air Collectors

When all factors are considered, air-heating collectors are especially appropriate for space-heating retrofits. You can build a small convective unit yourself, but if you opt for a large active system, you must decide whether to build the system on site, buy commercially manufactured panels and install them yourself or have a solar contractor do the whole job for you.

Buying commercially manufactured panels and installing them yourself can save a considerable amount of money compared to hiring a contractor to do all the work, but for several reasons we favor site-built collectors over prebuilt panels:

• Collectors built on site can be designed to fit the available space. Generally, site-built collectors will make better use of southern exposure, especially when it is of an irregular shape (such as a gable end).

• Site-built collectors can often be sealed better than commercially manufactured panels. Commercial panels should be well sealed within themselves, but they are usually installed in an array, and sealing panels to each other or to an elaborate manifold can be a frustrating chore. Even using the manufacturer's method for joining collectors doesn't guarantee a permanent seal that can withstand the tremendous heat fluctuations present in all collectors. Site-built collectors have only one perimeter that needs sealing, and because of that they also lose less heat.

• There is more flexibility with the design of a site-built system, and it is easier to design the system to operate at peak efficiency (see chapter 4).

• Site-built systems are lower in cost than commercially manufactured panels. There is less total labor involved because you build and install the collector at the same time. You also don't have to pay for the panel manufacturer's overhead, profit or shipping charges.

• Site-built collectors can be as efficient, durable and attractive as commercially manufactured panels. The monitoring of site-built systems has shown that well-built ones can operate at the same efficiency as most commercial flat plate collectors. If care is taken in construction and good-quality materials are used, site-built systems will perform as well or better, last as long and look as good as a comparable commercial system but at a much lower installed cost.

Actual collector construction isn't difficult. In fact, it is often easier than doing the ductwork portion of the installation, but if after reading this book you decide you don't want to do it all yourself don't count out a site-built system. In many parts of the country there are solar contractors who do site-built systems, and if you can locate one, you may be able to save yourself a considerable amount of money by not buying an expensive, factory-built, modular system.

2

A SURVEY OF AIR-HEATER OPTIONS

Solar-heated air can be used in a wide variety of applications, but it is a limited resource. Only so much sunshine strikes each square foot of the earth, and a well-built collector system will typically deliver about half of this energy to a house in the form of heat.

The way this heat is used often means the difference between a useful, cost-effective system and one that never seems to perform properly. Effective use of collected solar heat is especially important in retrofitted systems which, because of space limitations, are often undersized for the houses they are serving. Yet even a small collector used as a supplemental heat source can make a substantial difference in a house's fuel consumption if the heat from it is used effectively. There are also limits to cost-effective collector sizing, which are discussed in chapter 3. Even where there is room for a large collector, it is usually not cost-effective to build one so large that it provides 100 percent of a house's space- or water-heating needs. So before diving into the actual design of your system, it is important to look at the best ways of making full use of the heat a collector delivers. There are a number of factors involved in determining the best ways to "manage" your solar heat, including the temperature of the heated air, the type and size of the collector and the size and layout of your house. These are discussed in the following pages.

Technically speaking, the heat generated by a solar air heater is a relatively "low-grade" heat. That's not a disparaging term, for it simply refers to the relatively low temperatures (80 to 140°F) at which air heaters operate. If the airflow rate were reduced, a collector would heat air above this range, but in terms of Btu's delivered, there is a lot more heat in a strong blast of 90°F air than there is in a tiny trickle of 140°F air from the same collector. Why? A collector operating at 90°F has less heat loss than one operating at 140°F and therefore operates more efficiently. In a hotter-running collector, the increased heat losses through the glazing, sides and back of the collector mean that less heat is actually delivered to the living space. This is a very important concept to keep in mind in the design and operation of both small and large systems. This low-grade solar heat is indeed very usable, but it must be handled differently from the 140 to 160°F "high-grade" heat produced by a forced-air furnace. A strong blast of 150°F air from a furnace will feel warm to the occupants of a home, whereas a strong blast of 90°F air from an efficient collector or from rock storage can feel drafty to the occupants, even though it is heating the house. Thus for solar air to heat a living space without drafty discomfort, it must enter the living space slowly and continuously and from several different points. This rule doesn't hold when the collector is very small, in which case it would be impractical and unnecessary to create more than one outlet. We can look again at the different types of air heaters, this time in terms of their specific air handling requirements.

Convective Air Heaters

Maintaining a continuous flow of low-grade heat is the goal of any convective air-heater design. Since a properly built passive air heater will raise the temperature of the air moving through it by about 30 to 40°F,

the output temperature in these collectors will typically be close to 110°F. This low-grade heat enters the house very slowly through large openings, and the air movement is hardly noticeable, thereby heating the home in a very comfortable way.

Convective air heaters dump the heat they produce directly into the adjacent room and therefore don't require a heat distribution component. If the south-facing rooms in a house are often occupied by day and thus require a lot of daytime heating, a simple passive collector is an appropriate choice. If these rooms are seldom used and require little heat,

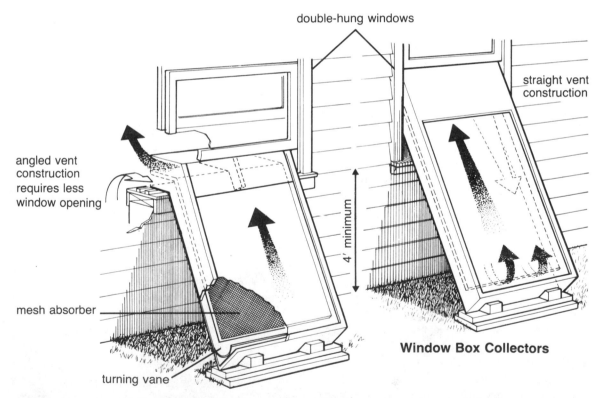

Figure 2-1: *Above and on opposite page.* Window box collectors and TAPs are easy-to-build collectors that deliver their heat by natural convection. They are best used to heat rooms that are frequently occupied and that require daytime heating. Window box collectors are self-damping at night. TAPs require lightweight backdraft dampers to prevent reverse thermosiphoning when the sun isn't shining.

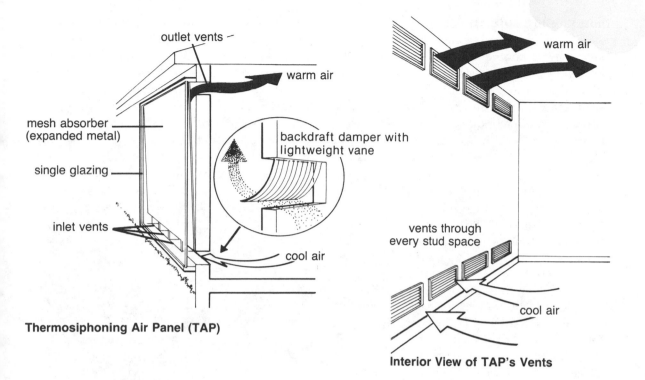

Thermosiphoning Air Panel (TAP)

Interior View of TAP's Vents

a more involved active collection and distribution system is needed to satisfy the daytime heating needs of nonsouth rooms. There's an in-between wrinkle, too: A large passive heater can provide too much heat to adjacent rooms, which necessitates additional passive (vents in walls or floors) or active (blowers and ducting) heat distribution.

Window Box Collectors

Window box collectors are one of the simplest solar heating devices you can build, but since they are quite small in relation to the size of the room they are heating, they don't provide a great percentage of a house's total heating needs unless several are used. These heaters provide a slow, continuous flow of heated air into the adjoining rooms. They are self-operating and, when properly built, have an advantage over other solar heating devices in that no dampers of any kind are needed to prevent nighttime heat losses. Their best application is on houses with wide, south-facing, double-hung windows that are 4 feet or more aboveground level. They can be installed on other types of windows, but the modifications required are more difficult and often expensive and unattractive. If the windows are less than 4 feet from the ground, the natural convection that moves air through them and into the house will be weak, and the collectors won't perform as well as they could. Since window box heaters are fairly small and act as a supplemental heat source, no heat distribution or storage is incorporated into their design. They don't involve any major modifications to the house and don't sacrifice any south-facing exposure that later may be desired for a larger collector installation.

Thermosiphoning Air Panels

Thermosiphoning air panels (TAPs) are also a good choice for buildings where the available south-facing wall area is small and where the rooms behind this wall can use supplemental daytime heat. Day-use spaces, such as small workshops and offices, are a good choice for TAPs because these collectors will deliver heat when the rooms are occupied.

Like window box systems, TAPs require a very free flow of inlet and outlet air, so the vent openings to the adjoining rooms need to be large (relative to vent requirements for active systems). When you install a TAP system on a frame structure, you can easily cut these vent openings between wall studs at the top and bottom of the wall. A Trombe wall, first cousin to a TAP, is usually a better retrofit choice for a masonry wall, especially since it incorporates the added feature of heat storage in the masonry.

Heat distribution in most TAP systems simply involves a convective flow of air (convective loop) inside the room behind the collector. This steady circulation of warmed air provides comfortable heat but can result in an overheated room if the TAP is large relative to the size of the room (where the collector area is more than 20 percent of the room's floor area).

In single-family dwellings it is often desirable to direct solar-heated air from a large TAP to rooms that aren't adjacent to the TAP. This can be accomplished with a small blower and ductwork, but if the retrofit project calls for extensive heat distribution, it is a better idea to design and build an active solar system to work with an active distribution system.

Even though passive air heaters are generally regarded and used as daytime heaters, heat storage can be incorporated into the design of thermosiphoning air panels on new houses. This involves building a large collector below the house or underneath a massive floor (concrete slab) so that solar-heated air rises by natural convection into and through the rock box. Storage in retrofitted convective air collectors is a tricky and expensive operation that is almost never justified unless the collector is very large and

Photo 2-1: This 8-by-14-foot TAP was retrofitted onto the Hitchcock Nature Center in Amherst, Massachusetts. It heats the second-floor office space and uses an open stairwell as the cool-air return. This collector experiences some morning shading, which indicates that a little tree trimming is in order at this site.

located well below the rooms to be heated. (Chapter 11 goes more deeply into heat storage.)

Both window box heaters and thermo-siphoning air panels are quite inexpensive and easy to build, which makes them good projects for first-time solar retrofitters. Since they operate passively, they deliver comfort-able heat without the need for wiring, blow-ers or thermostats. There are no operating costs as there are with active solar systems. Lack of heat distribution and methods of stor-ing heat are two possible disadvantages that can sometimes be solved by using a small fan. Actually, the transition from passive to active operation of these smaller collectors is not a major one, nor is it highly compli-cated. In the following section we'll look at different ways that active air heaters can give you more control over your solar Btu's.

Active Systems

Active solar air-heating systems are more versatile than convective systems because they allow you to direct the heat to rooms that aren't near the collector. Active systems can be more expensive to build than passive sys-tems, but they are easier to design because of their forced-air operation. With forced-air systems there is less concern about designing to maintain a delicate natural convection air-flow. Finally, active systems have often dem-onstrated better performance than their pas-sive counterparts, delivering more heat per square foot of collector.

A Collector-and-Crawl-Space Distribution System

A solar system that blows solar-heated air directly into a well-insulated crawl space or basement makes good use of low-grade so-lar heat. Because the flow of solar-heated air is isolated from the living area, 80 or 90°F air from a collector can be used without creating chilly drafts. This is significant because when a collector operates with a relatively high air-flow rate, it runs cooler and thus more effi-ciently than it would with a lower flow rate.

Because heat rises, this system heats the house in a nice way: by warming the floor during the day and early evening. It can cre-ate a comfortably warm floor, it eliminates forced-air drafts and, in cold climates, it can help prevent pipes from freezing. A crawl-space system is especially attractive in re-trofit applications where the collector is small relative to the size of the house and the added expense of including rock thermal storage or distribution ductwork probably isn't justi-fied. This system heats the house very subtly, and its effectiveness doesn't really show un-til there is a period of cloudy weather. When the collector doesn't operate for an extended period, the house and floor will be noticeably cooler and the back-up furnace will run more often.

A crawl-space system requires that the floor be uninsulated and that the crawl space (or basement), including foundation vents, be well sealed and insulated at the founda-tion or stemwall. It is also desirable, but not necessary, that the floors be uncarpeted to increase their ability to radiate heat to the living space. It is inevitable that there will be more heat loss with this system than with other delivery setups, but the higher oper-ating efficiency of the cooler-running collec-tor and the elimination of drafts in the house make up for that.

Ductwork for a crawl-space system should be kept to a minimum while still allowing for good heat distribution. The most desir-able method is to duct the hot air from the collector directly to the north side or coldest part of the house and let it find its way back to the collector return inlet located on the south side of the house. A "tee" fitting on the end of the outlet ductwork can help to distribute the air more evenly throughout the crawl space. Distribution ductwork in the crawl space can be left uninsulated since any heat losses from it help to heat the floor above.

Crawl-Space Controls

Controls and dampers for this system are simple. A standard differential thermostat (see chapter 6) turns on a blower whenever the collector plate is warmer than the crawl space and then shuts it off whenever the collector is cooler. A one-way backdraft damper or a motorized damper is mounted in the air return to prevent nighttime convective losses.

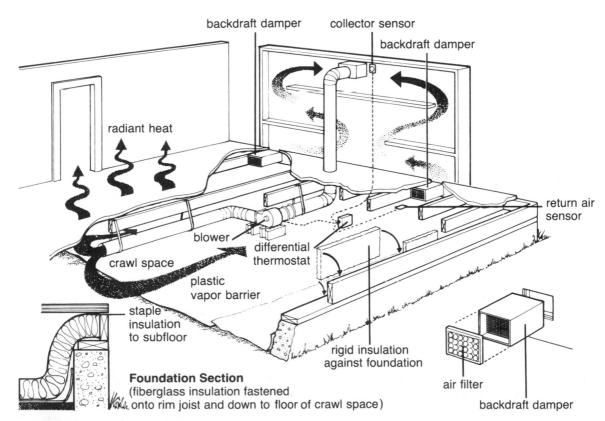

Figure 2-2: Blowing solar-heated air into a crawl space (or basement) is a good way to use the output from a small collector. Controls and wiring are simple. A differential thermostat turns on the blower when the collector is warmer than the crawl space, and a backdraft damper minimizes heat loss at night. During the day and into the evening hours, comfortable radiant heat enters the living area from the floor below. For this delivery system to work properly, the perimeter of the crawl space must be insulated and sealed, and a plastic vapor barrier should be laid over the dirt to minimize evaporative cooling.

The simplicity and effectiveness of this system, along with its low initial cost, make it a good choice for do-it-yourselfers, and many successful home-built installations incorporate this design. More elaborate air distribution schemes or a domestic water preheater can be added later.

Heat storage in most crawl-space systems is limited to the rise in temperature of the crawl-space dirt and the floor above it. The collector works well at keeping a house warm from late morning to bedtime, with back-up heat being needed in the early morning hours. Heat storage can be improved slightly by blowing the solar-heated air into a large,

---

Storage in a Crawl-Space System

Who says there's no storage in a crawl-space system? If you add up the floor joists, subfloor and flooring in a typical frame house, all of this wood has an equivalent thickness of between 2 and 2½ inches. Wood used for framing typically weighs about 32 pounds per cubic foot so a 1,000-square-foot, single-level house has 6,000 pounds of wood in the floor alone. The specific heat of wood (or its ability to hold heat) is 50 percent more than that of gravel (0.33 versus 0.21 Btu/lb/°F), so the floor in a 1,000-square-foot house can store as much heat as 9,000 pounds of gravel. Every 100 square feet of floor is therefore equivalent to a third of a cubic yard of gravel. Since the floor in a comfortable house can vary 15°F in temperature (60 to 75°F), the floor alone (not counting the walls and furnishings inside the house or the footing and crawl-space dirt under the house) can store all of the heat delivered by a collector that represents about 10 percent of the floor area of the house. The heat delivered from the floor is radiant heat so the warmth it provides is comfortable even at lower temperatures. If you can blow solar air into a basement, the amount of storage available there will easily handle a collector area that is 20 percent of the floor area.

---

black plastic bag in the crawl space. This bag is formed by running a 12-foot or 16-foot sheet of black 6-mil polyethylene down the center of the crawl space and weighing its edges down with rocks or bricks. It inflates when the collector is running and holds the solar air in closer contact with soil in the crawl space. The air then finds its way down the tunnel and out the end of the bag on the north side of the crawl space. At night, when the collector blower is off, the bag deflates and the crawl-space dirt radiates heat to the floor above. With the bag design the mass of dirt is heated more than the crawl-space air, providing more nighttime heat. The bag eliminates the need for distribution ductwork, which can be a major expense in low-cost systems. This approach won't work if you have a high water table, since groundwater will carry the heat out of the crawl-space dirt.

Direct-Use Systems

Heating your crawl space or basement with solar-heated air is an effective approach to solar utilization, but in some applications it is more desirable to blow warm air directly into the living space. There is often less heat loss with this method and less time lag in heating the building. This can be a big advantage in spaces that require heat only during the day.

Zone heating is the most successful way to utilize direct solar heating. Rather than trying to heat an entire house or office with an undersized collector, you are blowing the solar heat into the two or three rooms that need it the most. Zone heating is a good way to get more out of your collector than the heat it actually produces. We have all been in houses with poor heat distribution, where the living room had to be heated to 80°F to get the adjoining family room or back bedrooms to a usable temperature. If solar heat were added to the cooler rooms, the total amount of energy needed for comfort could be dramatically reduced.

Once again, since collectors produce lower temperatures than are usually produced by conventional heating systems, the solar-heated air must be delivered to the living space slowly or it will feel drafty. But at the same time, you want to move as much air as possible through the collector to get the most heat and the highest operating efficiency. Therefore, several branch ducts are needed to distribute the solar air from a large collector since it can't all be dumped into just one room. In direct-use systems a balance needs to be found between having enough duct work for good distribution and having too much ductwork, which entails excessive cost and complexity. If the ductwork is very extensive, it can lose a lot of heat even if it is well insulated (as it must be in this type of installation). This is especially important if the ductwork is located in the attic since any heat lost there is completely unusable and won't help heat the house. Long runs of ductwork also present resistance to airflow, thereby increasing the load on the solar blower, not to mention increasing costs.

Controls for direct-use systems are similar to those for the crawl-space system, but instead of using a differential thermostat, a *remote bulb thermostat* is used to control the collector blower. This thermostat is set to turn the blower on at a fairly high temperature (100 to 120°F) because warmer air is required to provide comfortable heat in a direct-use system. The cold air intake (collector inlet) has a backdraft damper installed in it and feeds the collector from the living area instead of from the crawl space.

Lack of control can be a drawback to an active direct-use system. Hot air enters the living or work area whenever the collector is hot, not necessarily when the space actually calls for heat, and unless the collector is undersized for the load of the space being heated, it's a good idea to have somewhere to dump surplus heat when the needs of the living space have been satisfied. The solution lies in building a system that has a second mode of operation.

Two-Mode Systems

Two-mode solar systems deliver heat to two different points of use and therefore require more electrical controls than do the direct-use or crawl-space systems. This added complexity and expense is usually only justified if the collector is large (more than 15 to 20 percent of the heated floor area) and delivers more heat than is needed in one part of the house. There are many different possibilities for a two-mode system. One option is to combine direct-use and crawl-space heating modes in the same system. Another combines direct-use space heating with domestic water heating, and a third option involves two direct-use heating systems serving different parts of the house.

Houses with central, forced-air furnace systems are a logical choice for retrofitting a two-mode system, since the ductwork is already run to all parts of the house and it can be used for distributing both the solar and the back-up (furnace) heat. Figure 2-3 shows a two-mode system that is tied into an existing forced-air furnace. Note that two thermostats are needed. One is a standard remote bulb thermostat (the *collector thermostat*) that operates the collector blower. It is located near the collector, with the sensing bulb itself mounted inside the collector. The other is a two-stage house thermostat that controls both the two-mode system and the conventional furnace. The collector thermostat should have an adjustment range of between 90 and 140°F for seasonal adjustment purposes (we'll talk more about this later). The collector thermostat and the solar blower operate independently from the rest of the system: When the collector is hot, the blower comes on.

The two-stage house thermostat operates independently from the collector thermostat and is mounted where a standard house thermostat would be located. The set-points of the two stages are different by about 3 degrees. When the house cools off and calls for heat, the first stage comes on, moving a motorized damper into a position that allows solar air from the collector to enter the furnace's ductwork, but only if the collector it-

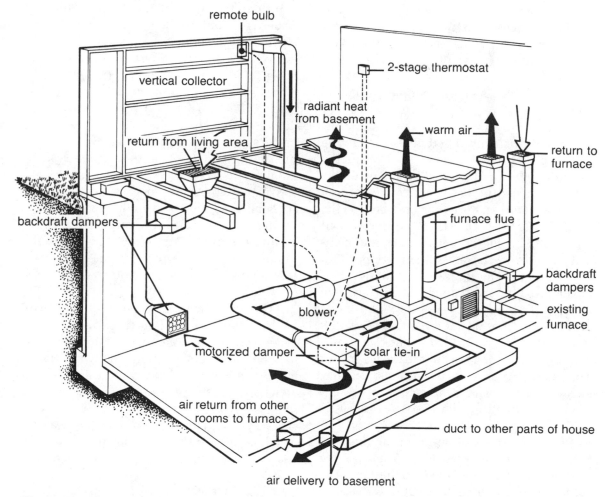

Figure 2-3: This two-mode system is simpler than it first appears. When the upper floor calls for heat and the collector is on, solar air is blown into the existing ductwork to warm the living area. When the needs of the upper floor have been satisfied, the motorized damper changes positions, and solar heat is delivered into the basement and stored in the concrete walls there. Note that two returns (inlets) to the collector, each with a backdraft damper, are required in this system.

self is "on." Solar air is then distributed throughout the furnace duct system. Since the collector blower is usually smaller than the furnace blower, the 100 to 120°F solar air enters the living space slowly and comfortably through several registers. If, however, the collector isn't hot enough, the collector blower won't be operating. In this case the house will cool another 3 degrees, and the second stage of the house thermostat will turn on the furnace in the normal fashion. Thus the solar system always gets first crack at space heating, but if it doesn't have enough heat to warrant delivery, the furnace takes over. When the solar blower is off, the damper will be in a position that prevents the furnace blower from circulating hot air through the collector.

All of this control activity occurs in the first mode of a two-mode system. Now for the second mode: Let's say that in this mode the collector heats the basement. At noon on a sunny winter day, the collector is hot and the solar blower is on, but the house doesn't call for heat. The motorized damper automatically goes to the basement-delivery position under orders from the two-stage house thermostat. As figure 2-3 shows, this is the same damper position as for the stage-two mode of the house thermostat except that now the independently controlled solar blower is running, delivering hot air into the basement. The basement acts as a temporary storage, and the house isn't overheated. In the unlikely event that this does cause overheating in the spring or fall, the setting on the collector thermostat can be raised, letting less total heat enter the basement. (You get more total Btu's when the collector thermostat is set at 100°F rather than at 140°F because of reduced collector heat loss.)

Notice that this system needs two air returns to the collector, one from the house and one from the basement, since at times each of these areas has air entering it. Both returns have backdraft dampers to prevent dense cold air from settling out of the collector into these spaces at night. Another backdraft damper installed on the furnace return grille or between the furnace and the furnace plenum prevents solar air from blowing into the furnace air return. Since air is taken from the top of the collector in our example, there will be little nighttime exchange of air in the hot air delivery (outlet) duct so a damper here is usually not needed.

When a two-mode system is tied into existing forced-air ductwork, there is usually good distribution of solar heat throughout the house. The furnace ductwork has been designed to accommodate a large flow of air from its blower, and the more slowly moving air from a collector may flow more readily through some branch ducts causing more air delivery at some registers, less at others. If not enough solar-heated air flows through the shorter ducts, the airflows can be balanced by closing down the adjustable floor registers (standard on most forced-air heating installations) that terminate the branch ducts from the furnace.

One of the nicest aspects of this two-mode system is that that furnace and solar heater operate independently while still using the same distribution ductwork. Two-mode systems are naturally more costly than crawl-space systems, and some do-it-yourselfers may prefer the simplicity of a crawl-space system. Distribution to the house can be added later if there is a desire for quicker heat delivery to the house in the early morning.

Tying into Other Heating Systems

Two-mode systems can be used with other types of heating systems, such as electric or hot water baseboard heating. When there is no existing duct system, hot air registers must be cut through the floor and connected to the collector with ductwork. Once again, it is important to have enough ducts to prevent having strong blasts of air coming from too few outlet registers. In systems like these a separate single-stage thermostat is mounted on the wall next to the thermostat that controls the baseboard system. This solar thermostat is set 2 or 3 degrees higher than the back-up thermostat so that they work together like the two-stage thermostat described earlier, in which solar energy gets the first chance to heat the house. An independent two-mode system such as this is often the best bet in retrofit situations where the existing heating system is electric baseboard since there is, of course, no way to tie the solar system into it.

Solar Water Heating

Heating domestic water often constitutes a large part of the household energy bill, and designing a solar air collector to help offset this expense is a viable option. Air-heating systems built for both space and domestic water heating are useful year-round and not just during the winter space-heating season. In some locations it is economically feasible to build a system that is used only to heat domestic water year-round, but the effectiveness of such a system should be considered next to a standard liquid collector domestic hot water system, which is the more common choice for heating domestic water (see chapter 10).

Collectors that are used for heating domestic water are always installed at a slant (typically at a tilt angle equal to your latitude plus or minus 5 degrees) rather than vertically, in order to take advantage of the high summer sun angles. They are also designed to operate at slightly higher temperatures than space-heating collectors because higher-grade heat is needed to heat water to a usable temperature of 100 to 120°F.

Systems that use air to heat water need an *air-to-water heat exchanger*. There are several options available for this exchanger, but the two most common are the *in-duct radiator type* (fan-coil or fin-tube) heat exchanger and the *finned-tank-type exchanger*. The radiator type fits inside the ductwork and takes heat from the moving stream of solar-heated air. A small pump is wired in parallel with the solar blower so that water will be circulated through the exchanger whenever the collector is operating. This heated water will then go to a tank that supplies preheated water to the existing water heater. The water heater will then bring the solar-heated water up to a usable temperature if it isn't hot enough. This type of system requires a very tightly sealing motorized damper in the solar ductwork to isolate the heat exchange coil from the collector at night if the coil is located below the collector inlet. Water won't be moving through the exchanger at night, so even a tiny trickle of cold air can freeze the stagnant water and burst the exchanger. The most appropriate design in this system would place the exchanger above the inlet so that cold air couldn't settle through the duct and reach the exchanger.

In the finned-tank exchanger the preheat tank acts as the exchanger. Fins attached to the outside of the tank pick up heat from the moving stream of air that is blown around the tank and transfer this heat to the water inside. This system operates as efficiently as in-duct exchangers. It is less expensive to build and simpler to operate since no pump is required. Freeze-up problems are very unlikely, but a tightly sealing damper is still a must.

A large, tilted space-heating collector can be used effectively for domestic water heating in the summer. A smaller summer-only blower or a variable or multi-speed blower can be installed to move air more slowly through the air-to-water heat exchanger when a larger blast of air isn't required for space heating. A system of this type will deliver very hot air to meet almost all of a family's summer hot water needs. Valves can be installed to completely bypass the conventional water heater so that all hot water comes from the solar preheat tank. If it is cloudy for a couple of days, the water won't be hot, but some folks are willing to put up with this inconvenience.

If a collector is going to be used for both water heating and space heating, a two-mode system is often a good choice. An adjustable thermostat on the hot water preheat tank allows more or less of a collector's air delivery to be used for water heating, depending upon the season. In winter this thermostat is set at a fairly low temperature, say 110°F, and after the water has been heated to this temperature, the hot airflow is used for space heating. In summer the thermostat is set much higher (180°F), and all of the solar air travels in a closed loop between the collector and the air-to-water heat exchanger. A two-mode water/space heating system, which is featured in chapter 12, is a good example of a two-mode design of this type.

Systems with Thermal Storage

A system with a separate heat storage component has definite advantages over the simpler systems we have been discussing up to now. Having storage means that a larger collector can be used because any surplus heat can be stored for nighttime use rather than wasted in daytime overheating of the living space. Storage systems add convenience and efficiency in large systems, but not without significant added cost for a rock storage bin and necessary additional controls (dampers, etc.). In fact, this cost can equal the cost of the collector itself.

The first thing to consider when planning for storage is whether or not the size of the collector justifies it. Building a separate heat storage area isn't usually cost-effective unless your collector is larger than 20 percent of the floor area to be heated (see chapter 3). The collector itself should also be larger than 200 square feet. Smaller collectors find their best use in heating a crawl space or in direct-use systems that heat two or three rooms in a house. Controls and dampers for these simple installations are straightforward and fairly inexpensive, and payback for these systems

Photo 2-2: Tank-type exchangers for heating water are easy to build and install in active solar air systems. Shown here is a 50-gallon finned exchanger, which will be used for summer water preheating in a small two-mode system.

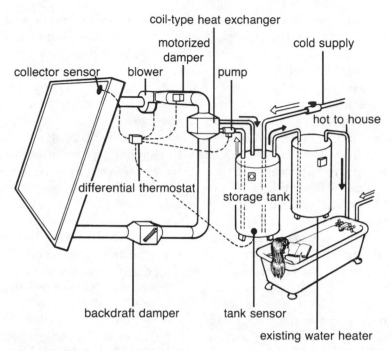

Figure 2-4: An air-to-water heat exchanger is required for heating domestic water with a solar air collector. A coil-type heat exchanger can be mounted inside a duct to take heat from solar air. Heated water is then circulated by a pump to a storage tank. Or a tank-type heat exchanger can transfer heat by blowing solar air directly over and around the storage tank.

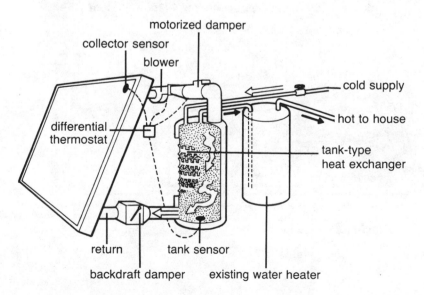

will be rapid. The added expense of a storage component won't pay for itself unless both collector and storage are large and supply a big percentage of the house's heating load.

Another factor to consider is whether or not you have room to accommodate a large rock bin. The bin is preferably located under, or even in, heated living space, not outside of it, so that any heat loss from storage is used. Some 7 to 8 feet of vertical clearance is needed for a standard bin that will also take up 25 to 35 square feet of floor area. Lower-profile bins with longer horizontal airflows have problems because the heat tends to rise to the top and thus isn't distributed properly through the rocks in the bin. Installing a rock bin in new construction is usually done with less difficulty and at lower cost, but that doesn't mean that retrofit applications are necessarily out of the question. They just have to be considered carefully. Heat delivery from storage is also more easily integrated into the back-up heating system of a new house since they can both use the same distribution ductwork.

The major advantage of having storage is that it enables you to build a larger collector than would normally be needed for daytime heating. Heat from storage can thus be used at night and during sunless days. Figure 2-5 shows a simple *four-mode system* with one blower that moves air for three solar modes: collector heats house, collector heats storage and storage heats house. A back-up furnace that uses the same distribution ductwork provides heat for the fourth mode when there is no solar heat available. Other system designs use two blowers or have a summer bypass for preheating domestic water when space heating isn't needed. All these options do get a little complicated, but they will be discussed in more detail in chapter 11.

Gravel Bed under a Slab

An attractive storage option for new construction uses a gravel bed located under a slab-on-grade floor. Solar-heated air is blown through the gravel, which then heats the slab to provide comfortable, radiant heat to the house. This system works like the radiant

System Components for 4-Mode System

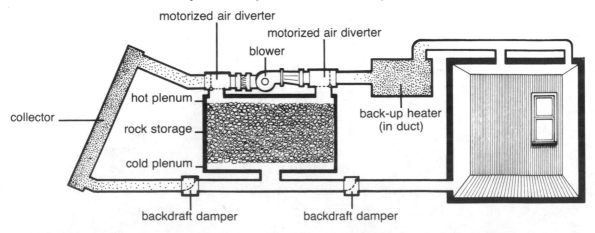

Figure 2-5: In this four-mode system the back-up furnace is integrated with the collector/storage ductwork.

floor heating systems that were popular in the 1950s in which hot water was pumped through pipes embedded in a slab. An air heat distribution system will never leak, however, like many of the hot water systems did. Since the solar loop is closed to the living space, low temperature air (70 to 80°F) can be used in this type of system to increase the output and efficiency of the collector without making the house drafty.

Controls for a gravel-bed system are fairly simple. A differential thermostat turns on the blower whenever the collector is warmer than the rocks, and a motorized damper wired in parallel with the blower prevents nighttime heat loss through the ductwork and collector.

This type of system requires a lot of storage, about five times more by weight than a conventional rock box. It has many advantages in new buildings, especially those that feature passive hybrid systems, but is very difficult to add on to an existing house.

Commercial Installations

Until now we have been discussing the use of solar heat in residential installations,

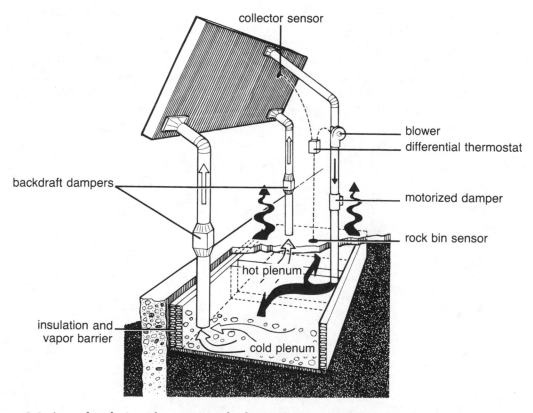

Figure 2-6: A good technique for storing solar heat in new construction involves blowing solar-heated air through an insulated gravel bed under a concrete slab floor. With this arrangement the collector operates very efficiently and the floor provides comfortable, radiant heat. Dampers prevent heat loss from storage at night.

but solar-heated air also has many commercial applications. Perhaps the best ones are in schools, hospitals, restaurants, and industrial shops—buildings that need a constant and extensive air change throughout the day. If the outside air being fed to the furnaces in these buildings can be preheated by a large, low-cost collector, the total heat consumption of the buildings can be greatly reduced. Air collectors operate best when they are fed cold, outside air and when they have a high airflow through them. Thus a great deal of preheated solar air can be delivered to the furnace quite economically from a relatively small collector, allowing the furnace to deliver an adequate amount of heat with a much lower consumption of fuel. Existing schools and post offices that have expensive, tracking, concentrating collectors mounted on them are often missing the boat when it comes to getting cost-effective solar heat. Larger, lower-cost air heaters can often be installed instead,

at one-quarter of the total cost while still providing the same contribution to the heating load.

Solar heaters are also useful around the farm. Portable solar dryers can be used for drying grain in the fall and then moved to the south side of a house or outbuilding for wintertime space heating. Grain drying usually requires a larger airflow through the collector than does space heating so, when used for these two different functions, the same collector will need either a variable-speed blower or two separate blowers.

Farmstead collectors are also useful in preheating air in applications where a constant air change is desired, such as in a farrowing house or in a dairy barn. In these applications they may work out even better than collectors mounted on residences because solar air of even lower temperatures can be useful.

3

THE ECONOMICS OF SOLAR HEATING

Building your own solar air-heating collector isn't difficult and can certainly be a very satisfying project, but the question is, Will it be financially worthwhile in terms of lower heating bills? Volumes have been written on the complicated subject of the economics of solar heating. Depending on what you have read, solar energy can look like a very good investment or a very impractical one.

This chapter examines a number of economic factors you should consider before installing any solar system. These factors include the length of time it will take to recover your investment; proper sizing of a collector to ensure an economical return; the cost of a do-it-yourself system compared to a contractor-built, site-built system or a commercial system; how much heat your house needs; and the amount of heat you can expect a collector to deliver.

When you spend the money to build and install a solar heater, you are paying today for heat that will be delivered in the future, which implies that any economic analysis of solar heating is subject to many variables. Assumptions must be made about such things as future interest rates and the rate of fuel price inflation. The investment potential of your solar heater is very sensitive to these factors, and in some economic calculations a 2 percent difference in projected fuel price increases over the next 15 years can change what looks like a questionable risk into a sound investment. Who knows what fuel will cost in 15 years? We can only safely predict

that it will be more than it is today. The purpose of this chapter is to give you solid information and not to lose you in a maze of numbers, assumptions and calculations. As you read this chapter, you can evaluate your house's solar potential without making extensive, complicated calculations. If you want to go further and make your own calculations, you'll find that the math is easy.

Each house needs to be considered individually in assessment of its solar potential. A system that represents a good investment for you may not work out nearly as well for your neighbor. Basically, solar heating is a good investment if your site receives enough sunshine, if your system is sized for your needs, if most of the heat it delivers can be effectively utilized and if the system doesn't cost too much to build and maintain.

It is important to realize that weatherizing and insulating your house are the first priorities when you consider any retrofit project. Insulation and weatherstripping will pay for themselves more rapidly than a solar installation. If your home is an energy hog, even a large, efficient collector won't make much of a dent in your fuel bill.

Your site is another consideration that influences the economics of using solar heat. The more directly your house faces south and the fewer obstructions of the sun it encounters, the better a solar heater will work for you (see chapter 4). If all of the available sites for your collector are shaded by the trees in your neighbor's yard, it is unlikely that you can have a cost-effective system.

Return on Investment

The two general theories behind fighting inflation are: 1) invest your money in short-term, highly liquid assets, such as money market funds, that pay out at higher rates than the general inflation rate, or 2) invest in long-term assets that keep pace with inflation. An investment in solar does the latter. As long as your solar unit is productive and maintains effective fuel savings each year, it will theoretically be worth more dollars each year during inflationary periods. Since a well-built collector has a long useful life, depreciation is only a small consideration. The only time it won't be a practical investment will be if its benefits are devalued during an inflationary period—in other words, if fuel oil or electricity, for example, become so cheap that you never reach a simple payback on your solar installation. As long as the simple payback looks reasonable (less than ten years), fuel continues to increase in price and the collector continues to deliver heat with little maintenance, it's hard to go wrong with solar heating.

You may want to compare current rates for money market funds (or other investments) with your savings per year divided by the cost of your solar investment to see what kind of effective interest rate you are earning. The effective interest rate is equal to slightly less than the inverse of simple payback so, if you save $200 per year on a $1,200 installation, a six-year payback will yield slightly more than a 15 percent effective interest rate.

If you have to borrow money to build a collector, the important rate to beat is the loan interest rate. It is hard to justify borrowing money to build a collector if you simply look at the first-year percentages. If, however, you consider the length of time it takes to break even between the dollars saved on fuel bills and the dollars spent on interest for a do-it-yourself collector, you should find that savings will outpace interest payments within about four years, even without claiming tax credits.

Payback

Once your house is properly weatherized and you have located an appropriate place to install the collector, your next consideration is *payback*, the length of time it will take to recover your investment in solar energy through fuel savings. The most straightforward calculation is for *simple payback*. If, for example, your system costs $2,000 to install, and you save $400 in heating bills its first year in use, the simple payback will be five years. Many people feel that simple payback is an overly simplified approach, but it can often tell you as much as more complex methods of economic analysis, which involve factors from table 3-1. If you live in a sunny, cold location, and you can claim a large tax credit on a small, low-cost collector, your payback period will be very short. If you live in a warm, cloudy area and have to borrow money to install a large, high-priced system, it may never pay for itself within its lifetime. Most people fall somewhere between these two extremes, where the choices aren't as clear. As you can see from all of the variables involved in table 3-1, any thorough economic calculation is complicated and involves many assumptions.

An important factor that can greatly influence payback has entered the picture recently—tax credits. The federal government offers a healthy 40 percent tax credit on active solar systems, and many states also have sizeable credits. (The federal credit expires at the end of 1985. If it isn't renewed, though, most state-level tax incentives are likely to be available for a long time.) When it is possible to recover nearly half or even more of the cost of your system in tax savings, it is hard to go wrong, but we are wary of tax credits. Like all good things from the government, the federal tax credit will probably end. So if you want to use tax credits to help finance your do-it-yourself collector, do it soon. On the bright side, the price of this book can be claimed as a tax credit if you actually build a system from it.

Low initial cost makes the most difference when considering payback for a particular system, and this is where do-it-yourself, site-built systems have an advantage. Since nobody knows what the energy or economic situation will be like in 10 or 15 years, a commercial system that has a 15- to 20-year payback (without tax credits) just doesn't make sense. A do-it-yourself installation that is properly designed and sized to utilize all the heat output from it should have a simple payback of 6 to 8 years (without tax credits). In our rapidly changing world, this payback period makes a lot more sense.

TABLE 3-1

Variables That Affect Payback

	Rapid Payback (2 yrs)	Moderate Payback (10 yrs)	Slow Payback (20 yrs or more)
Available sunshine (% of sunny days)	70	50	30
Average collector efficiency (%)	50	35	20
Climate (space-heating collector)	cold	cool	warm
Collector in use	all year (space and water heating)	year-round (water heating only)	only in winter (space heating only)
Cost of installation ($/ft²)	10	30	50
Fuel price inflation (%/yr)	15	5 to 10	no increase
Interest rate on money borrowed to build collector (%)	10	15	20
Size of system	undersized for house	properly sized for house	oversized for house
Tax credits (%)	70	40	no credits
Your yearly heating bill ($)	700	400	100

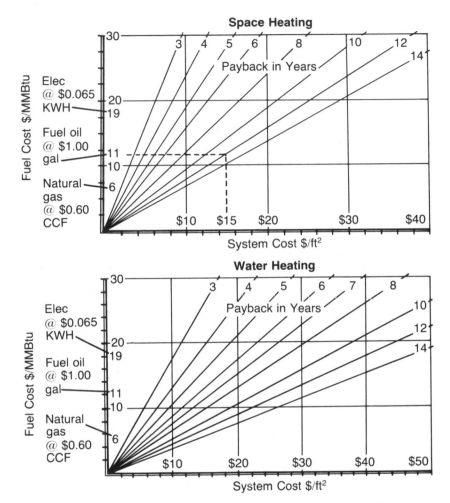

Figure 3-1: These charts show the simple payback period for two different collector systems, one a winter-only space heater and the other a hot water system used year-round. The length of time it takes a solar system to pay for itself is related to the present price for fuel and the installed cost of the system. The charts show that inexpensive solar installations that replace expensive fuel sources have rapid paybacks. The cost per square foot on the charts includes tax credits, so if you can claim a 50 percent credit on a $15-per-square-foot installation, its net cost will be $7.50 per square foot. The cost for fuel is a current cost for energy that will be replaced by solar energy. For example, let's assume you build a space-heating system for $15 per square foot and are presently heating with fuel oil. Your simple payback on this system will be 12 years. If fuel prices increase in subsequent years, the payback period will be shortened.

Certain assumptions were made regarding these charts.

The space heater:
- collector operates at 45 percent efficiency
- six-month heating season
- insolation = 1,300 Btu per square foot per day average
- heat delivery = 0.105 MBtu per square foot per heating season

The water heater:
- collector operates at 40 percent efficiency
- used all year
- insolation = 1,300 Btu per square foot per day average
- heat delivery = 0.190 MBtu per square foot per year

How Much Heat Should a System Provide?

When designing an air-heating collector, it is important to realize that it is uneconomical to design a system to provide 100 percent of your heat in midwinter. A collector that provides 30 percent of your heat in January can provide 95 percent in October and March and probably provide about 60 percent of your year-round heating needs. A collector of this size will pay for itself much more rapidly than one twice this size that produces 90 percent of your year-round heat.

Systems for preheating hot water are easier to size since the demand for hot water is fairly constant throughout the year, and the output from a properly oriented collector will be only slightly lower in the winter than in the summer in most locations. Properly sized water-heating systems will provide about 60 to 70 percent of your annual water heating needs. Efficient flat plate air heaters operate at fairly low temperatures (100 to 140°F) so it is often difficult to raise the temperature of your hot water tank much above 110°F

even on very sunny days. In a cost-effective system some form of supplemental heat will be required in order to provide the 120°F water we are accustomed to using and to heat water on cloudy days. It should be emphasized again, however, that a domestic-hot-water-only air system may not be as cost-effective as a liquid collector system, depending on many factors that would be specific to your site and climate, as well as the cost of solar hardware. For example, at a site where there was a great deal of cloudiness and limited space for collectors, the smaller area needed by a liquid collector system could make it more desirable than an air-heating system.

In many retrofit situations it is difficult to find a place suitable for building a collector large enough to provide 60 percent of your heating needs. Most retrofit systems are undersized and act as a supplemental heat source. There is nothing wrong with this, since undersized systems have a shorter payback, but don't expect 60 percent of your heat to come from a collector that is only equal in size to 10 percent of your heated floor area.

TABLE 3-2

Fuel Cost per Million Btu

Fuel	Heat Value of Fuel	Efficiency of Heater (%)	Cost per Unit of Fuel ($)*	Approximate Cost/Million Btu ($)
Coal	26,000,000 Btu/ton	75	75.00/ton	3.84
Electricity (resistance)	3,413 Btu/KWH	100	0.07/KWH	20.51
Fuel oil	138,000 Btu/gal	75	1.20/gal	11.60
Natural gas	94,000 Btu/CCF†	75	0.50/CCF	5.00
Propane	95,000 Btu/gal	75	0.80/gal	11.23
Wood	25,000,000 Btu/cord	50	90.00/cord	7.20

*Can vary with location.

†Can vary with altitude; CCF is 100 cubic feet and equals 1 therm.

TABLE 3-3

Cost of Heat from Solar Heaters

Installed Cost for Solar Space Heater ($/ft²)	Insolation (Btu/ft²/day)	Collector Efficiency (%)	Usable Life	Approximate Cost/Million Btu ($)
8*	1,300†	45	4,000 days or	3.41
15	1,300†	45	20 yrs × 200	6.41
45	1,300†	45	days	19.23
Installed Cost for Solar Water Heater ($/ft²)				
15	1,300†	40	7,300 days or	3.95
45	1,300†	40	20 yrs × 365 days	11.85

NOTE: The cost of heat delivered from a solar installation can be found by dividing the initial cost of the system by the amount of heat it delivers in its lifetime.

*After tax credits.

†1,300 Btu/ft²/day is a seasonal, clear-day average at a fairly sunny site.

TABLE 3-4

Cost of Heat Saved by Weatherization Measures

Measures	Cost for Weatherization ($)	Effective Life (yrs)	Approximate Cost/Million Btu for Fuel Saved ($)
Insulation: Increasing ceiling insulation from R-11 to R-30 by adding 6 inches of fiberglass	0.40/ft²	30	1.90
Weatherstripping: Stopping infiltration losses by reducing the air changes from 2 per hr to 1 per hr	100.00	6	0.61

NOTE: When considering weatherization measures, you need to look at the fuel cost savings that these improvements will realize in the future. The information in this table was generated by finding the difference in heat loss for a typical house (1,500 ft² in size, 5,000 heating degree-days and 5-month heating season) before and after weatherizing. The cost of instituting these measures was then divided by the Btu's saved.

How Much Do Solar Heating Systems Cost?

Asking the price of a solar heater is like asking what a car costs. You can buy a 1965, V-8 gas guzzler for $200, or you can pay $35,000 for a new Mercedes. Similarly, solar collectors have a very wide price range. Table 3-5 gives some typical costs for installed systems. The do-it-yourself column represents the cost of the medium-priced, site-built systems described in this book.

Site-built solar construction is like most other forms of home construction: The cost

TABLE 3-5

Typical Costs for Solar Installations (1983 prices)

Type of System	Do-It-Yourself Systems * ($)	Contractor, Site-Built Systems † ($)	Commercial Systems ‡ ($)
1. Window box heater	5 to 10/ft²		
2. TAP	4 to 8/ft²	9 to 15/ft²	
3. Vertically mounted collector (collector only)	6 to 8/ft²	12 to 18/ft²	20 to 25/ft²
4. Tilted collector (collector only)	7 to 9/ft²	13 to 19/ft²	22 to 27/ft²
5. Ductwork, blower and controls for crawl-space or direct-use system (for 100 ft² of collector)	150 to 200	300 to 400	400 to 500
6. Ductwork, blower and controls for a 2-mode system (for 120 ft² of collector)	400 to 600	700 to 900	700 to 900
7. Air-to-water heat exchanger (tank type) with additional ductwork, controls and plumbing (for 100 ft² of collector)	450	650	900§
8. 3-mode control system with ductwork, controls and blower	1,300 to 2,000	2,400 to 4,000	2,500 to 5,000
8a. Rock thermal storage bin (for 300 ft² of collector)	300 to 600	600 to 1,200	800 to 1,200

NOTE: To determine the cost of a direct-use space-heating system with a vertical collector, add the costs from items 3 and 5; to determine the cost of a 2-mode system (hot water and crawl space), add the costs from items 4, 6 and 7; to determine the cost of a hot water system, add the costs from items 4, 5 and 7; to determine the cost of a 3-mode system with storage, add the costs from items 4, 8 and 8a; to include domestic water preheating to the 3-mode system with storage, add the costs from items 4, 7, 8 and 8a.

* Cost is for new materials.

† Cost is for new materials and labor.

‡ Cost is for commercially manufactured systems plus installation costs.

§ In-duct heat exchanger, pump and storage tank.

of the materials is roughly equal to the cost of the labor. With installations using commercially assembled panels, the materials cost will be greater than the cost of labor. In contractor-built systems the builder's familiarity with solar will also affect the price. The first rock thermal storage bin a contractor installs is likely to be more expensive than ones he subsequently builds.

In retrofits the site can influence the cost of a system. If the roof to be retrofitted is very high and steep, or if structural changes to the building or extensive remodeling are necessary, the total price of the installation can be increased considerably. For example, long runs of ductwork can constitute a large percentage of the expense in low-cost installations.

What Are Your Heating Needs?

A look at last year's utility bills is the best way to assess your heating needs. If you didn't save them, your utility company can provide copies. Break down your energy consumption into four categories: space heating (56 percent of average total use), domestic water heating (15 percent), lighting, appliances and miscellaneous loads (25 percent), and air conditioning (4 percent). Costs for water heating, lighting and appliances will remain fairly constant throughout the year while expenditures for space heating and air conditioning will, of course, vary seasonally. Table 3-6 gives examples of typical yearly costs for the four categories for a family of four living in a well-insulated, 1,200-square-foot house in Denver, Colorado; a family of four in a similar house in Nashville, Tennessee; and an individual who lives alone in an 800-square-foot house in Los Angeles, California.

The family in Denver needs a solar system that reduces the $480 they spend on space heating and the $360 for heating water. The family in Nashville requires a collector that is used primarily for heating water, their largest expenditure, and that is also designed to provide supplemental space heat in the winter. The house in Los Angeles doesn't require much space heating, and the individual doesn't use much hot water, so solar heating is a more doubtful proposition in these particular circumstances. In this situation the installation of a demand-type water heater could probably reduce energy consumption.

Go Solar: Save at Least 20 Percent

The folks at the Small Farm Energy Project in Harrington, Nebraska, have been promoting low-cost, site-built solar systems for several years. They have also monitored existing systems and have reached some amazing conclusions. They found that, almost without exception, builders of do-it-yourself collectors save 20 percent on their fuel bills over and above the savings that can be traced to solar heat delivery. Even small systems that couldn't have provided the owners over 10 percent of their needs saved these folks 30 percent on their fuel bills.

A change in attitude and involvement was undoubtedly the source of this "free" heat. Once these homeowners built their own heating system, they realized that they weren't helpless in the face of skyrocketing fuel costs, and they became ever more energy conscious. They started doing small things like closing the door when calling the dog, turning down the thermostat every night or waiting until the solar heater had warmed the shop before starting work in the morning. All of these small things added up to dramatic savings. This just goes to show that careful economic analysis certainly isn't the only way to look at the effectiveness of solar heating.

The Economics of Domestic Water Heating

Let's look at the economic considerations of installing a solar system for preheating domestic water. Conservation is, of course, the top priority to ensure a cost-effective installation. If you can decrease your consumption of hot water by doing laundry in warm or cold water (not hot water), using low-flow shower heads and low-flow aerators

---

TABLE 3-6

Typical Residential Energy Consumption in Dollars

Family of Four in Denver, Colo.*

Space heating (natural gas)	4 mo @ $90/mo = $360
	3 mo @ 40/mo = 120
Water heating (electricity)	12 mo @ 30/mo = 360
Lights and appliances	12 mo @ 10/mo = 120
Air conditioning	3 mo @ 30/mo = 90

Jan. energy costs (approx) = $130
May energy costs   =   40
July energy costs   =   70

Family of Four in Nashville, Tenn.†

Space heating (natural gas)	4 mo @ $50/mo = $200
	2 mo @ 20/mo = 40
Water heating (electricity)	12 mo @ 30/mo = 360
Lights and appliances	12 mo @ 10/mo = 120
Air conditioning	4 mo @ 50/mo = 200

Jan. energy costs (approx) = $90
May energy costs   =   40
July energy costs   =   90

Individual in Los Angeles, Calif.‡

Space heating (natural gas)	4 mo @ $15/mo = $ 60
Water heating (electricity)	12 mo @ 15/mo = 180
Lights and appliances	12 mo @ 7/mo = 84
Air conditioning	4 mo @ 40/mo = 160
	2 mo @ 20/mo = 40

Jan. energy costs (approx) = $37
May energy costs   =   22
July energy costs   =   62

NOTE: The family of 4 in Denver lives in a well-insulated, 1,200-square-foot house; the family of 4 in Nashville lives in a similar house; the individual in Los Angeles lives in an 800-square-foot house.

* 5,524 heating degree-days.

† 3,578 heating degree-days.

‡ 2,061 heating degree-days.

TABLE 3-7

Sizing Water-Heating Systems

Btu/ft²/day for your site	Air Collector Sized to Provide 70% of Hot Water Needs (ft² of collector/gal of water)	Collector Size for 80-gal Storage Tank (ft²)
1,900	0.70	55
1,700	0.80	65
1,500	0.90	75
1,300	1.05	85
1,100	1.20	100

NOTE: The water-heating systems represented in this table produce a temperature rise of approximately 60°F (55 to 115°F) in 80 gallons of water at 35 percent system efficiency.

on faucets, insulating your hot water heater and plumbing pipes and putting a timer on your water heater, you can cut your hot water needs by one-third to one-half. This means that you can build a smaller solar system or one that can provide a larger percentage of your needs.

After conservation the second factor that influences the economics of a solar water-heating system is the amount of sunshine available at your site during the year. If you have lots of it, you can build a smaller collector and install a smaller storage tank than can someone who lives in a cloudy location. Average outdoor temperature is another, less important, consideration for heating water. You will have more heat loss from the collector if it is bitterly cold outside, so you will

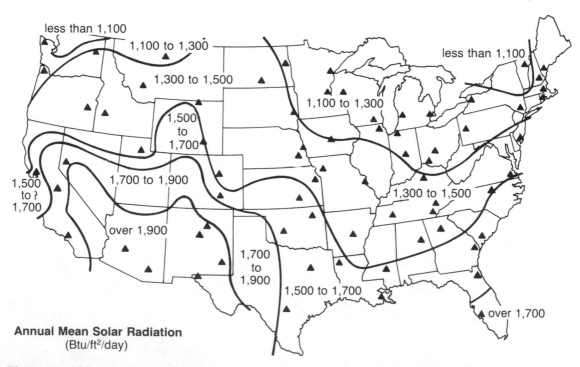

Annual Mean Solar Radiation
(Btu/ft²/day)

Figure 3-2: This map shows the annual mean solar radiation available in the United States in Btu's per square foot per day. If your site doesn't receive much sunshine, you will need to build a larger collector in order to meet your needs.

need to build a slightly larger collector. Table 3-7 indicates the size of collector an average family of four will need to provide 70 percent of their annual hot water needs. The table is based on the average American's per capita use of hot water (20 gallons per person per day), which actually is a fairly high level of use. If you have instituted conservation measures, your needs can be considerably reduced (by 30 to 50 percent).

Let's assume our typical families want to build only water-heating systems. Let's concentrate on the family in Nashville since water heating comprises the largest percentage of their fuel consumption. They can either build their own system with an 85-square-foot collector for about $1,305 (average cost from table 3-5) or have a commercial air system

TABLE 3-8
Recommended Insulation Levels for Five Heating Zones

Heating Zone*	Recommended R-Value for		
	Ceiling	Wall	Floor
A	R-26	R-13	R-11
B	R-26	R-19	R-13
C	R-30	R-19	R-19
D	R-33	R-19	R-22
E	R-38	R-19	R-22

NOTE: These R-values are the recommended amounts for a well-insulated house. Sources for R-value data include the Federal Housing Administration, the American Society of Heating, Refrigerating and Air-Conditioning Engineers (ASHRAE) and insulation manufacturers.

*Zones are indicated in figure 3-3.

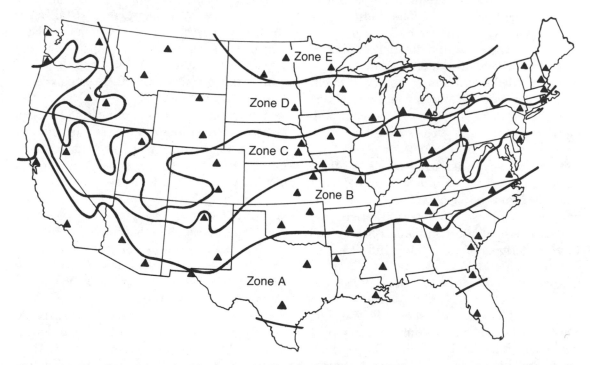

Figure 3-3: Use this map and table 3-8 to decide whether or not your house is insulated well enough to justify the addition of a solar heating system. Conservation improvements will pay you back faster than will a solar system. So if the insulation in your house doesn't meet or exceed the R-values in table 3-8, the most cost-effective step will be to add insulation.

installed for about $3,310. If these systems reduce the $360 per year they spend on water heating by 60 percent, they will save $216 per year. Simple payback (not including fuel price increases, available tax credits and other variables) will be just over 6 years for the home-built system ($1,305 ÷ 216) and over 15 years for the commercial system ($3,310 ÷ 216).

The Economics of Space Heating

There are many more variables involved in the economics of using a collector for space heating than in the economics of water heating. Your climate and your house are very important factors. Your house must be well insulated and weatherized if solar heating is to make economic sense. The question becomes, When do I stop weatherizing and start solarizing? The amount of insulation a house requires varies with climate. Table 3-8 lists recommended insulation values for different parts of the United States. If your house meets these guidelines, you are well on your way to having it properly weatherized.

Just having an adequate amount of insulation, however, isn't the whole weatherization story. For example, a well-insulated house in a cold zone such as Minnesota should have double-glazed windows with some type of movable insulation and an air-lock entryway or storm doors. Caulking and weatherstripping should plug any air leaks around windows and doors and in the building envelope. If your home is poorly insulated and weatherized, upgrading it should be your first consideration. Then, if your house is well insulated and your fuel bills still are high, you are ready to consider solar space heating.

Local climatic patterns greatly influence heating needs, and the amount of insolation obviously plays a part in determining the economics of a solar space-heating system. Figure 3-4 shows the generalized conditions for six zones in the United States regarding the amount of insolation and the need for space heating. The southern part of the United States (lined on the map) has a heating demand that is probably too low to justify a large expenditure for solar space heating, although solar water heating is an economical choice in this region. Zones 1 and 2 have lots of sunshine and a substantial need for space heating, so a properly sized system is a sensible choice. There is plenty of sunshine in zones 3 and 4, so it is hard to go wrong with a site-built system. Zones 5 and 6 are cloudy much of the time, and an owner-builder needs to take a critical look at his site, heating bills and budget before building a space-heating system. However, many successful solar space-heating systems have been built in New England (zone 6). In many cases an oversized water system that provides supplemental winter space heat will be the best choice for retrofits in this zone.

Your present source of heating fuel can also be a consideration. If your brother-in-law is a logger or if you live next to a coal mine, your low fuel costs may not justify a solar heater.

The size of the system you plan to install is also a factor to consider. Large systems, providing more than half of your heating needs, will need either elaborate ductwork and storage or several points of use for the solar heat. These added components will increase the per-square-foot cost of your system, and while such a system will provide more of your needs, the payback will be longer than that of a smaller, simpler system that provides only supplemental heat. Window box heaters built with recycled materials, for instance, can often pay for themselves in a single winter. This does not, however, mean that a smaller solar system is necessarily the better investment. There are several ways to judge the quality of an investment, and the one with the shortest payback is not necessarily the one that will turn out to be the best investment over the next 20 to 30 years.

Sizing for Heat Delivery

The size of the collector and your local climate will determine the most cost-effective delivery-and-storage arrangement in your system. A large collector in a sunny, mild climate will require storage for nighttime use because the large amount of heat delivered from the collector can't all be used during the day. In a cloudier, colder site less heat will be delivered from a collector, more heat will be required during the day, and storage won't be justified unless the collector is very large (see table 3-9).

For example, let's assume that you want to go whole hog and build an active air collector that will provide 60 percent of your annual heating needs. The size of the collector should be a certain percentage of the heated floor area of a house (see table 3-10). This percentage varies with climate and according to how well a house is insulated. For example, the size of a collector required to provide 60 percent of the heating needs of a moderately insulated, 2,000-square-foot house in Boston is 22 percent of the floor area of the house, or 440 square feet (0.22 x 2,000 = 440).

Table 3-10 clearly indicates the importance of proper weatherization. A well-insulated house needs about half the collector area that a poorly insulated one needs. Insulation is especially important at sites that are cold and receive little sun.

Let's look at the space-heating options available to our Denver family. Since Denver (zone 4) is a good area for solar space heating,

and the house is well insulated, we can expect that a solar installation will be economically favorable. Remember that this family spends $480 every year on space heating, and $360 is spent for water heating. For the sake of our example, let's assume they have unlimited space for a collector and good solar access and orientation, and that they are interested in building their own system.

Their first choice is a 156-square-foot, vertically mounted air-heating collector. According to table 3-10 a collector of this size will provide close to 60 percent of their heating needs. This collector is too large for a simple collector-and-crawl-space system but too small to justify the controls and storage required for a three- or four-mode system. They choose to build a direct-use-and-crawl-space, two-mode system costing approximately $1,592 (see table 3-5) and saving them $288 a year (0.60 x $480). Their simple payback is over five and a half years.

Their second choice is a 100-square-foot, vertically mounted air-heating collector. This collector should provide 35 percent of their yearly heat by warming a well-insulated crawl space or basement. The cost is approximately $875 and the simple payback just over five years.

Their third choice is a 250-square-foot, roof-mounted, slanted collector. This collector is slightly oversized for their house. It will require a rock thermal storage area because it will deliver more heat than is usable during the daylight hours. This system should provide 75 percent of their heating needs and cost approximately $4,100. The simple payback is 11 years for this three-mode system. If they add domestic water preheating to their third choice, the total cost will be $4,550, and it will also provide 80 percent of their hot water needs (it's greatly oversized for hot water) for seven months out of the year and

TABLE 3-9

Sizing Collector Area for Desired Percent of Heating Needs
(Collector Area Is Given as Percent of Heated Floor Area)

Desired Percentage of Heating Needs	Up to 30%	50%	Up to 70%
Recommended Delivery Setup	Crawl-Space Heater	2-Mode	3-Mode with Storage*
Zone 1[+]	up to 10%[‡]	10 to 15%	10 to 20%
Zone 2	up to 15%	15 to 25%	20 to 35%
Zone 3	up to 10%	10 to 20%	15 to 30%
Zone 4	up to 20%	15 to 25%	20 to 40%
Zone 5	up to 15%	15 to 25%	20 to 30%
Zone 6	up to 25%	25 to 35%	30% and up

NOTE: The figures in this table are based on a moderately insulated house. Since the climate within each zone can vary greatly according to site, this table contains generalized information.

*When one delivery mode heats water, subtract the square footage needed for hot water at the site from the total size of the collector before consulting this table.

+Zones are indicated in figure 3-4.

‡For example, a crawl-space heater can provide up to 30 percent of the heat desired if the size of the collector is equal to up to 10 percent of the heated floor area.

40 percent during the other months. The simple payback is nearly eight years for the combined system.

The fourth choice is a 120-square-foot, roof-mounted, slanted collector. This collector will be a two-mode system that heats an insulated crawl space and domestic hot water. It will save 40 percent on space heating and 60 percent on yearly water heating. The cost is approximately $1,910, and the simple payback is nearly five years.

The fifth choice is a 60-square-foot, vertically mounted thermosiphoning air panel (TAP). This fairly small convective unit should provide about 15 percent of the house's space-heating needs through the winter. It costs approximately $360, and the simple payback is five years.

Evaluating Your Site and Your Needs

Now it's your turn. As you weigh your solar options, site conditions and heating needs, keep in mind that in most locations you need to think big if you want solar heating to provide a large percentage of your heat. Building a small convective or active solar heater is a good introduction to solar energy that will rapidly pay for itself in fuel savings, but if you want to put a big dent in your fuel bill, you need to go for a large area of collector surface. Our advice to most do-it-yourselfers is to design and build a large two-mode system, which can provide a lot of heat without involving a lot of hassles. Building a larger system that includes rock storage and installing the accompanying controls can not only double the price of the installation but also double the amount of work you will need to do. Our favorite installation is the oversized two-mode hot water system that is used in the winter primarily for space heating. It is a realistic do-it-yourself project because it can be built in stages, and it represents a reason-able amount of work for anyone serious about solarizing his house (see the description of the Gonzales system in chapter 12). The pay-back period for this system is also usually short relative to other solar options.

As you have seen in this chapter, the economics of solar heating is a rather complex affair. When planning your own system,

[Continued on page 42]

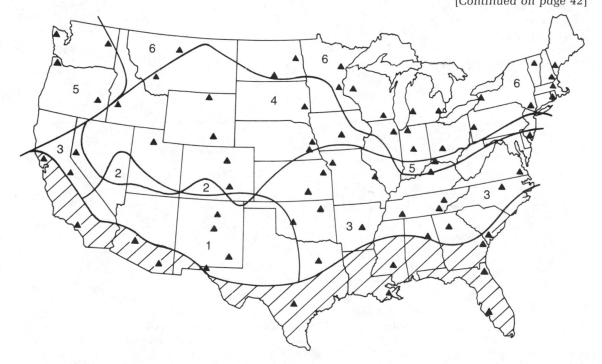

Figure 3-4: This map, which indicates the potential for solar energy use in six zones, considers both the need for heat and the amount of sunshine available at different locales and is helpful in evaluating the usefulness of solar space heating at your site.

Zone 1:   great sun, moderate heating need
Zone 2:   great sun, high heating need; in zones 1 and 2,
          you can't go wrong with a properly sized space-heating system
Zone 3:   good sun, moderate heating need
Zone 4:   good sun, high heating need; in zones 3 and 4,
          it's hard to go wrong with a space-heating system
Zone 5:   poor sun, moderate heating need
Zone 6:   poor sun, high heating need; in zones 5 and 6,
          you need to take a hard look at the economics involved

SOURCE: Adapted from TRW, *Solar Heating and Cooling of Buildings (Phase O)*, Vol. 3 (Washington, D.C.: National Science Foundation, 1974).

TABLE 3-10

Percentage of Collector Area to Floor Area for 60% of Space-Heating Needs

City	Zone*	Percentage of Collector Area for			Heating Degree-Days	Total Insolation†
		Well-Insulated House (%)	Moderately Insulated House (%)	Poorly Insulated House (%)		
Albuquerque, N.Mex.	1	8	10	15	4,348	330
Fresno, Calif.	3	7	10	14	2,611	215
Nashville, Tenn.	3	13	15	23	3,578	174
Denver, Colo.	4	13	16	23	5,524	269
Portland, Oreg.	5	15	19	28	4,109	187
Indianapolis, Ind.	5	21	26	38	5,699	190
Boston, Mass.	6	17	22	32	5,634	222
Chicago, Ill.	6	22	27	41	5,882	184

SOURCE: Edward Mazria, *The Passive Solar Energy Book* (Emmaus, Pa.: Rodale Press, 1979).

NOTE: A well-insulated house has a heat loss of 6.5 Btu/ft²/degree-day; a moderately insulated house has a heat loss of about 8 Btu/ft²/degree-day; a poorly insulated house has a heat loss of about 12 Btu/ft²/degree-day. In zones 1, 2 and 3 it is assumed that the collector operates @ 45% efficiency; in zones 4, 5 and 6 @ 40% efficiency.

*Zones are indicated in figure 3-4.

†Per heating season in MBtu/ft² on a horizontal surface.

TABLE 3-11

Amount of Heat Delivered from Collector Mounted at Latitude plus Ten Degrees

City	Average Daily Insolation in January (Btu/ft²)*	Insolation on Collector (Btu/ft²/day)	Heat Delivered from Collector (Btu/ft²/day)†	Average Daily Temperature (°F) in January
Albuquerque, N.Mex.	1,016	1,761	792	37
Fresno, Calif.	644	969	436	47
Nashville, Tenn.	579	819	368	43
Denver, Colo.	824	1,832	824	31
Seattle, Wash.	288	493	222	41
Indianapolis, Ind.	495	749	337	31
Boston, Mass.	475	784	353	31
Chicago, Ill.	506	854	384	25

SOURCE: Insolation values are from Sanford A. Klein, John A. Duffie and William A. Beckman, *Monthly Average Solar Radiation on Inclined Surfaces for 261 North American Cities*, Report no. 44-2 (Madison: University of Wisconsin, Solar Energy Lab, 1978).

*On horizontal surface.

† @ 45% efficiency.

How Much Heat to Expect from a Solar Space-Heating System

The following factors affect the amount of heat a collector will deliver:

- available sunshine at the site, percentage of cloud cover
- the collector's orientation and tilt, latitude of site
- the collector's glazing transmissivity, absorber performance characteristics
- the collector's efficiency and the total system efficiency, heat exchange losses
- outside temperature, wind speed and direction
- size of the collector
- the space-heating load of your house

Even simple calculations involving all of these variables can get very complicated. Let's look at a simple calculation that includes many of the possible climatic variations and assumes a set collector performance. For these calculations we will use the average daily insolation and the average daily temperatures at different locations in January, a cold month with little sunshine (see table 3-11). This will indicate the expected heat delivery from a solar heater under the harshest conditions.

For example, let's look at a 1,500-square-foot, well-insulated house in Indianapolis, Indiana. Assume that it has a heat loss of 7 Btu per square foot per degree day (see "Calculating Heat Loss by Using Your Fuel Bill" in this chapter). In January, the average outdoor temperature in Indianapolis is 31°F, so there are 34 heating degree-days (HDD) per day. The formula for calculating the number of Btu's needed to keep this house at 65°F on a typical January day is:

$$\text{area of house in ft}^2 \text{ x Btu/ft}^2\text{/HDD x HDD/day} = \text{Btu's/day}$$

or

$$1{,}500 \text{ ft}^2 \text{ x 7 Btu/ft}^2\text{/HDD x 34 HDD/day} = 357{,}000 \text{ Btu lost in one day}$$

To keep this house comfortable on a typical January day 357,000 Btu are needed. If we want the solar heater to offset 15 percent of these needs in January (or 50 to 60 percent year round) it will need to deliver the following number of Btu's per day:

$$357{,}000 \text{ Btu x } 0.15 = 53{,}550 \text{ Btu/day}$$

The heat we can expect from a collector at 45 percent efficiency on a typical day in Indianapolis is 337 Btu per square foot per day (see table 3-11), so in order to satisfy 15 percent of this house's needs the size of the collector is:

$$\frac{53{,}550 \text{ Btu/day needed}}{337 \text{ Btu/ft}^2\text{/day}} = 159 \text{ ft}^2 \text{ in size}$$

This is a fairly large installation, which emphasizes that big is beautiful when it comes to solar space-heating systems. There's also a lesson in this. The next time a fast-talking salesman tells you that two 4-by-8-foot panels will supply half your heating needs, get out your calculator. Unless you live in Phoenix, Arizona, or he is talking about the month of April, you will undoubtedly come up with other figures!

use common sense. Look at your site, your needs and your budget, and don't hesitate to get expert advice if you feel you need a more in-depth analysis. Your best bet might be to spend $50 for an economic analysis from a local solar consultant. They are often listed in the Yellow Pages and have computer software with programs for your particular area.

Buying Solar versus Doing It Yourself

If, after reading this book, you decide that building even a simple solar system is a larger project than you wish to undertake, your next step will be to shop for someone to install a system for you. We'd like to offer a few tips to help you obtain the best possible installation. First of all, hire a local builder or dealer. These contractors do business in your community, and they want a good name in the area. If something goes haywire, they are easy to call back. Look for someone with experience in the solar field. Don't be a "test house." Ask for references from your prospective contractor, or go see systems he has

---

Calculating Heat Loss by Using Your Fuel Bill

Calculating your house's heat loss by using your fuel bill is a good way to determine if it is ready for a solar space-heating retrofit or if you should first weatherize and insulate it. The following calculations tell you how to determine the number of Btu's each square foot of your house loses per degree-day. Any heat loss greater than 8 Btu per square foot of living space per degree-day should be considered too high to justify the addition of a solar heater. To do these simple calculations, you will need to know the heating degree-days (HDD) per month at your site. This information is available from your local fuel supplier, weather station or the National Climatic Center, which provides publications listing monthly heating degree-days for about 6,000 sites in the United States and daily heating degree-days for about 200 major locations.

Before you begin your calculations, you will need to determine the amount (not cost) of fuel your house uses for heating. For example, if your house is heated with electric resistance heating, look at your electric bills for months with very small heating and cooling loads. This will give you a good indication of how many kilowatt hours of electricity you are using for heating water and for operating appliances. Subtract this base amount from your consumption for a cold month, when you use a lot of electricity for space heating, to get the number of kilowatt hours consumed for heating. Similarly, if your house is heated with natural gas and you also have gas appliances, use the same procedure to find the number of cubic feet of gas consumed per month for space heating.

After you know your fuel consumption for space heating, the next step is to find the number of Btu's lost from your house in a month. The formula for calculating this is the following:

$$\begin{array}{c} \text{amount of fuel} \\ \text{used per month} \end{array} \times \begin{array}{c} \text{effective} \\ \text{heat value} \\ \text{for the fuel} \end{array} \times \begin{array}{c} \text{efficiency of} \\ \text{conversion} \end{array}$$
$$= \text{Btu's lost from house per month}$$

Electric Resistance Heaters

Let's look at electric resistance heating, which is the easiest to calculate— although not the easiest to pay for! You need to know that 1 kilowatt hour (KWH) equals 3,413 Btu and that electric resistance heaters operate at 100 percent efficiency.

For example, if your January bill indicates has

installed. Second, don't allow yourself to be talked into an elaborate system with storage unless your collector will be large enough to justify it. Don't buy a storage component instead of more collectors. Remember, rock storage doesn't produce any heat. Also, keep in mind the one-step-at-a-time approach. Many system components such as an air-to-water heat exchanger or more elaborate heat distribution can be added later when you know that your system can effectively use these components.

For Further Reference

Klein, Sanford A.; Duffie, John A.; and Beckman, William A. *Monthly Average Solar Radiation on Inclined Surfaces for 261 North American Cities*. Report no. 44-2. Madison: University of Wisconsin, Solar Energy Lab, 1978. Available for $10 from Engineering Publications Office, Engineering Library, University of Wisconsin, Madison, WI 53706.

---

that your house used 3,600 KWH for heating, and if this is your only heat source, the heat loss in Btu's from your house is:

$$3,600 \text{ KWH} \times 3,413 \text{ Btu/KWH} = 12,286,800 \text{ Btu}$$

To determine your heat loss per square foot per degree-day, use this calculation:

monthly heat loss in Btu's $\div$ ft^2 area of house $\div$ HDD/month = Btu/ft^2/HDD

For example, if your site has 1,000 degree-days in January and the size of your house is 1,200 square feet, your heat loss per square foot is:

$$12,286,800 \text{ Btu} \div 1,200 \text{ ft}^2 \div 1,000 \text{ HDD} = 10 \text{ Btu/ft}^2/\text{HDD}$$

If this is your house, nobody needs to tell you that it's time to insulate. Your January bill, at 6¢ per KWH, for heat alone is $216.

Natural Gas

To calculate heat loss for a house using a fuel other than electricity, you will need to know your furnace's efficiency. This figure can vary from 50 to 80 percent. If you heat with natural gas, you need to know the heat value for natural gas at your elevation, and this can also vary considerably. A phone call to your local utility can give you a pretty good idea of both of these figures. For our example let's assume we have a gas furnace with an efficiency of 75 percent and a heat value of 94,000 Btu per 100 cubic feet of natural gas (see table 3-2).

Consumption of natural gas is measured in CCF (100 cubic feet). If your monthly bill is for 83 CCF, your heat consumption for that month is:

$$83 \text{ CCF} \times 94,000 \text{ Btu/CCF} \times 0.75 \text{ (furnace efficiency)} = 5,851,500 \text{ Btu}$$

Fuel Oil

The heat value of fuel oil is 138,000 Btu per gallon, give or take one or two thousand Btu, depending on slight differences in fuels from different sources. As an example, assume a furnace efficiency of 75 percent. If your monthly consumption is 100 gallons, your heat load is the following: 138,000 Btu/gal $\times$ 0.75 (furnace efficiency) = 10,350,000 Btu

The heat values for other fuels are listed in table 3-2. Heat loss for houses heated by other fuels can be calculated in a manner similar to the examples shown here.

4

DESIGNING THE COLLECTOR

Before making a decision on the type of air-heating system that is best for your house, there are a number of points to consider. First, you must assess your site to determine if there is adequate solar exposure. Next, you must consider whether the system should be installed vertically or at a tilt, a decision that can be affected by the site, the structure of the house and whether a system is for space heating, water heating or both. Finally, you must consider the design features of a convective system versus an active system to determine which is the better choice for your situation. This chapter will help you make these decisions.

The Sun at Your Site

We all know that the sun is very large, very hot and 93 million miles away. We also know that the earth's axis of rotation is tilted with respect to the plane of its orbit around the sun. It is this tilt that causes seasonal changes as the earth makes its annual trip around the sun. In winter the Northern Hemisphere is tilted away from the sun, so the sun moves lower across the sky, and the days are shorter. Summer brings longer days and higher sun angles. Since we are interested in building a solar collector that will receive an adequate amount of sunlight to meet some of our heating needs, we need to take a more in-depth look at the sun's apparent motion.

The sun's *altitude* or position above the horizon changes throughout the year. It reaches its highest noontime position around the summer solstice (June 21). Its peak daily altitude gets lower and lower through the summer and fall until it reaches its lowest noontime altitude on the first day of winter (winter solstice, December 21). After the solstice the altitude and day length increase, once again peaking at the summer solstice. The sun rises and sets at about the same time every day and appears in approximately the same position in the sky for three or four months.

The total amount of sunshine striking the earth varies slightly in the course of the year because of the tilt of the earth's axis, which causes the seasonal changes. In the winter, not only are there fewer hours of sunlight but the rays from the low winter sun have to pass farther through the earth's atmosphere, which causes more of the direct radiation to be reflected. Winter sunlight is also less intense because it strikes the earth's surface at much less than a perpendicular angle. The farther a collector is from the equator, the more pronounced these effects become and the more important proper orientation towards the sun becomes.

Orientation

A collector receives the greatest amount of sunlight when it is oriented so that sunlight strikes it at a right angle. Whatever the angle, it's called the *angle of incidence*, and when the collector is perpendicular (90-degree angle) to the sun's rays, the angle of incidence is zero degrees (90 minus the angle at which solar radiation strikes a surface). If radiation reaches the collector surface at an angle of 60 degrees (relative to the collector surface), the angle of incidence is 90 degrees minus 60 degrees, or 30 degrees. *Tracking collectors*, which follow the sun through the sky, take the greatest advantage of this fact by staying always perpendicular to the sun, maintaining the lowest (and best) angle of incidence. However, the machinery required to accomplish this is very expensive, so *fixed flat plate collectors* are a more reasonable choice for residential space heating. Since the sun moves and the collector doesn't, it is important to locate the collector so that it intercepts as much sunlight as possible.

Proper orientation is more critical at northern latitudes, but in no case should a collector face more than 25 degrees away from true south (see figure 4-1 to determine true south). If it does, it won't gain enough heat to pay for itself within a reasonable length of time.

The optimal orientation with respect to south can be site-specific. A collector that faces slightly west of south often will gain more heat than a southeast-facing collector. It operates later into the afternoon, when outside air temperatures are higher, and thus there is less heat loss from it. A southeast-facing collector, on the other hand, could be an advantage if a building needs heat earlier in the day. Permanent obstructions to the sun can also influence the best orientation for a solar heater. If a building blocks out the afternoon winter sun after 3:00 P.M., an orientation slightly east of south is preferred.

The Solar Window

A good way to visualize the solar exposure available to the site for a collector is to take a bird's-eye view of the site. Imagine your house under a huge, opaque bowl with a window cut out of the south side of the bowl to let the sun shine in. This "solar window" is your access to the sun, and it should extend high enough to admit the summer sun (for domestic water heating) and, most importantly, low enough to let in the winter sun. Figure 4-2 shows a solar window for a site at 40 degrees north latitude. The opening for a site located further north would be lower on the bowl because the peak solar altitude is lower in summer and winter. A solar window for a site south of 40 degrees would be higher up on the bowl, and at the equator it would be directly overhead. The solar window will be 47 degrees in height for all locations because this is how much the sun changes its altitude through the course of a year. Since all of the sunlight that reaches your collector must come through this window, it must be as free as possible from obstructions such as trees and buildings if your collector is to operate properly.

It is difficult to find a collector site that is completely without obstructions. Most will have some shading, especially in residential areas. The critical hours for solar collection are between 9:00 A.M. and 3:00 P.M. (10:00 A.M. and 4:00 P.M. daylight savings time) since this is when the angle of the sun's rays is most perpendicular to the earth. For optimal performance a collector should be completely unshaded for this six-hour period. Shading from telephone poles won't be much of a problem, but shade from large trees, even bare trees in the winter, can be disastrous. Studies have shown that branches and tree trunks can easily block 20 to 50 percent of the available sunlight. A good rule of thumb to follow is that trees must be cut if they obstruct 20 percent or more of the sunlight during the main six-hour collection period. The 20 percent figure is too high if this occurs close to noon: For the three hours surrounding noon there should be no shade on your collector.

It's one thing to cut trees on your own property but quite another to cut them on your neighbor's. Some sites are completely unsuitable for this reason. Keep in mind, too, that trees grow. Your neighbor's 6-foot-tall, fast-growing evergreens, whose shadows now hardly reach your property line, may in ten years completely shade the whole south side of your house. If you have neighbors crowding you to the south it is a good idea to work out a *solar easement* with them before starting construction. There are laws in some states that guarantee your access to the sun, if it comes down to that (see Appendix 3).

Short, summer shadows grow to dark monsters in the winter and gobble up the sunshine for your solar heater. A thorough solar site analysis is very important and can spare you disappointment later. If, when considering a solar installation, you simply look south from the site and guess at the sun's possible location, you will probably be wrong. Even experienced solar installers have been fooled, and unfortunately their collectors have been shaded by large overhanging eaves, or by trees or buildings in the southeast or southwest portion of their solar window. A simple solar site analysis only takes five minutes. Do it!

Direct observation of the sun is the most straightforward approach. Since the sun lingers in its lowest position in the sky for two or three months, any observation between November 1 and February 1 will give you an accurate picture of your wintertime obstruction.

If you plan to do your installation in the summer, analyze the winter shading problems at your site by using a simple method developed by New York's Energy Task Force (see figure 4-3). Stand where the collector will be installed and face true south. Extend one arm and point at the horizon so that your finger and line of vision are horizontal. Place one fist on top of another in succession according to the number of fists listed in table 4-1 for your latitude. Repeat this process two more times pointing at 30 degrees east and west of true south. Any object above your top fist will shade the collector. Anything below your fists will be below the lowest path of the winter sun and of no concern. When checking for obstructions 30 degrees east or west of south, stand at the east edge of your proposed site and then at the west edge. If you plan to mount your collector on the ground, it will be necessary to lie on the ground to do your test or to stand 10 or 12 feet (the length of your winter shadow) south of the bottom of your proposed site.

Be conservative in your evaluation. If this simple site analysis leaves you with any doubts about potential shading problems, consult the sun charts in Appendix 1.

To Tilt or Not to Tilt

Properly tilting collectors will make them capture more of the available solar energy. The idea, when tilting a collector, is to have it point directly at the sun during the time of the year that you want the most heat output from it. A collector used for wintertime space heating should be tilted up more from the horizontal than a collector used for heating a swimming pool in the summer. Space-heating collectors can be mounted vertically with minimal losses in wintertime heat gain, whereas collectors for heating water must be tilted back in order to intercept the high summer sun and be useful year-round. The optimal tilt for a collector also depends on the

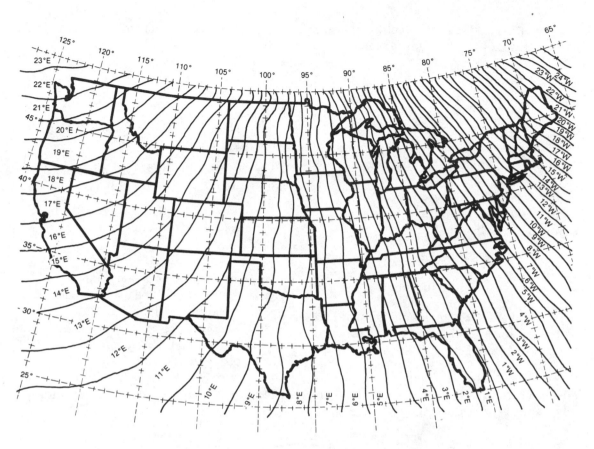

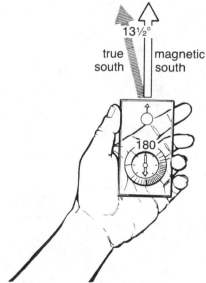

Figure 4-1: The first step in analyzing your solar site is to find true south. Looking at the position of the sun at noon (1:00 P.M. for daylight saving time) will give you a general idea, but for purposes of collector building we want a more accurate orientation and need to use a compass and a magnetic-variation map to find true north. It then becomes very easy to find true south. This isogonic map shows the approximate compass deviation from true south in the continental United States. For sites outside of this area, call the local airport for the deviation. For example, if you live in Denver, Colorado, your deviation is 13½ degrees east. That is, true south is 13½ degrees east of what your compass tells you is south. When using a compass, be sure to hold it away from all metal objects or the reading will be inaccurate.

latitude of your site and the configuration of the building to which it will be added.

Vertical Mounts

Most of the collectors built in the San Luis Valley are mounted vertically against existing walls and are used primarily for space heating. For this application, the vertical mount has some definite advantages:

• It is self-shading in the summer and therefore needs no summer venting system to prevent overheating.

• It requires little additional insulation on the back side since it is already mounted against an insulated wall.

• It also collects solar energy that is reflected off snow on the ground or a driveway located in front of it.

• It is close to the point of use (usually the crawl space or the living area) and therefore requires little ducting.

• It is often easier to build than a tilted collector. No tilt-up framing is necessary. Weatherizing the perimeter is easier than with roof mounts, and you can stand on the ground during construction.

• The vertical glazing stays cleaner than does tilted glazing.

• It doesn't alter a building's appearance as much as a tilted collector.

• It works better for space heating in northern latitudes because of low winter sun angles.

Vertical collectors have one disadvantage in that they can't be used for water heating in the summer. However, if all the cost factors are considered, vertical mounts are usually superior to tilted ones in terms of Btu's delivered for dollars spent (in space-heating applications), and they are especially attractive if you are considering building a small system.

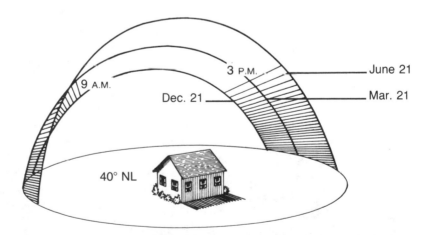

Figure 4-2: One way to visualize your solar exposure is to imagine a huge window in the sky south of your collector site. The top and bottom of the "solar window" at your site correspond to the summer and winter altitude angles. The sides of the window correspond to the sun's position at 9:00 A.M. and 3:00 P.M.

Figure 4-3: You can get a quick idea of potential shading problems at your site by using the fist method of site analysis. The woman in this illustration has stacked three fists to check for midday, wintertime obstructions at 36 degrees north latitude.

TABLE 4-1

Sighting Angles to Determine Winter Shading Problems

Your Latitude	True South	30° East and West of True South
28°N	4½ fists	3 fists
32°N	3½ fists	2½ fists
36°N	3 fists	2¼ fists
40°N	2½ fists	2 fists
44°N	2¼ fists	1½ fists
48°N	2 fists	1½ fists

SOURCE: Energy Task Force, *No Heat, No Rent: An Urban Solar & Energy Conservation Manual* (New York: Energy Task Force, 1977).

Tilted Collectors

A collector mounted at a certain angle (depending on latitude) off of a roof, directly on a roof or against a south-facing wall collects a greater percentage of the available solar energy in winter and can be used year-round for preheating the domestic hot water supply. A slanted collector that is roof-mounted has additional advantages:

• It can be made larger because the available roof space is larger.
• There are fewer problems with winter shading.
• The south wall can be used for a greenhouse or other solar addition.
• The collector doesn't interfere with south-facing, direct-gain windows.

A tilted collector works better in southern latitudes because of higher sun angles. But if it is unused during the summer months, a tilted collector must have a good heat release mechanism or be covered or shaded. (see the information on summer venting in chapter 6).

The optimal tilt angle for collectors that heat only water is equal to the local latitude. Thus the optimal tilt for a water heater in Columbus, Ohio (at 40 degrees north latitude) is 40 degrees from the horizontal. This tilt angle is lower than for space-heating applications, enabling the collector to take advantage of both the high summer sun and the low winter sun.

Collectors designed for space heating only should be aimed more directly at the low winter sun or at an angle from the horizontal equal to the latitude plus 15 degrees. (For Columbus, this is 40 degrees + 15 degrees = 55 degrees.) They will still collect a lot of heat during the summer but will have peak

[Continued on page 52]

Site Analysis

After trying the first method of site analysis, you have probably narrowed your list of obstructions down to a few shade makers, if any. Using a compass, sun charts and a large protractor with a weighted string, you can "shoot" these obstructions individually to determine precisely just how much shading they will create. Let's do an example at 40 degrees north latitude (see figure 4-4).

In the example there are three possible obstructions: two trees and a large building. First, use the compass to determine how many degrees from true south these obstructions are. In our example one tree is 35 degrees east of south, one tree is 15 degrees west of south, and the building is 30 degrees west of south.

Tie a knot in a 4-inch piece of string and push the string through a hole in the center of the protractor's straight edge. Tie a weight onto the dangling end. Sight down the straight edge of the protractor, and determine the altitude angle of the obstruction by where the string crosses the curved part of the protractor. In our example the top of the first tree is 25 degrees above the horizon, the second tree is 22 degrees, and the building is 20 degrees. Sketch these obstructions onto the sun chart for your latitude (see Appendix 1 for sun charts) as we have done for 40 degrees north latitude.

In the example the first tree will shade the collector from 9:00 A.M. to 10:00 A.M. in December and January. The second tree is no problem (unless it grows), and the building will shade the collector after 2:30 P.M. in December and January and after 3:30 P.M. in November and February. In December about 25 percent of the available sunlight is lost between the important period from 9:00 A.M. to 3:00 P.M. That makes this a marginal site for space heating unless the top of the eastern tree can be cut. Both of these trees could present future shading problems in the winter. However, our site is a good one for water heating since the collector is almost shade free from February through October.

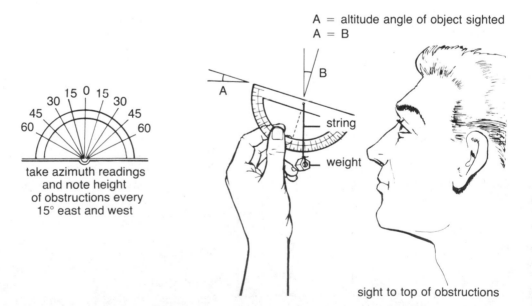

take azimuth readings
and note height
of obstructions every
15° east and west

A = altitude angle of object sighted
A = B

string

weight

sight to top of obstructions

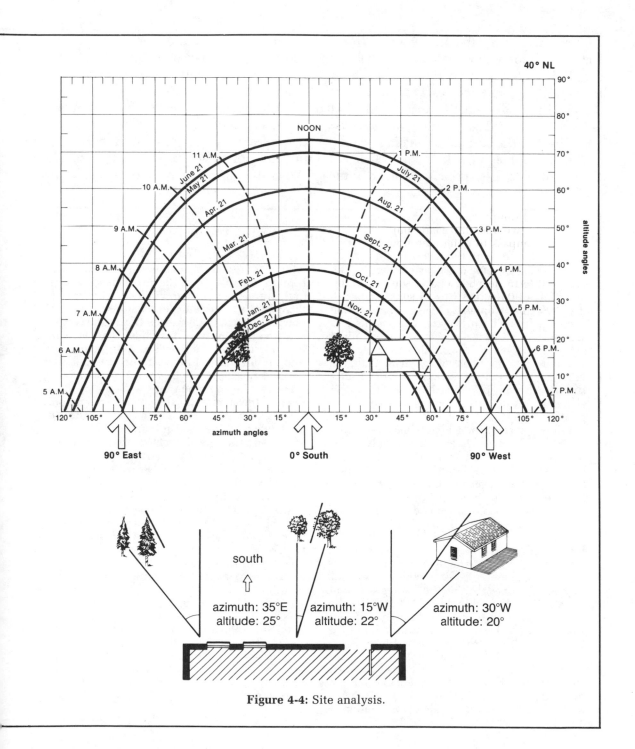

Figure 4-4: Site analysis.

performance during the heating season. Collectors used for both water and space heating should be tilted (at a compromise angle) to the latitude plus 10 degrees. Large collectors designed for space heating that are tilted at latitude plus 15 degrees may still gain too much heat in the summer even after heating water. Sometimes it is desirable to increase the tilt angle to lessen the heat gained in the summer so that venting the collector can be avoided. A variance of up to 15 degrees from the optimal tilt angles will make little difference in overall performance.

If your existing roof is close to the correct angle, it is usually more cost-effective to mount the collector directly on it rather than to build up the roof to accommodate the collector. When dealing with collector tilt, as well as

Moonshine Collector Siting

Moonshine (not the illegal beverage) can help you locate the best site for your collector. Most folks know that the full moon shines in the sky directly opposite the sun. That's what makes it full. The full moon rises as the sun sets during fall harvest because day and night are of equal length during this time of year. When day and night are of unequal length, the full moonrise will be slightly before or after sunset. What few people realize is that the moon also changes its altitude above the horizon in a pattern identical to that of the sun. The "lunar window" corresponds almost exactly to the "solar window" at a given site and only varies from it by a maximum of 4 degrees. The moon, however, completes its up and down motion through the sky every month, whereas it takes the sun a year to complete this cycle.

The changeable moon adds another twist to this phenomenon, as the path traveled by the full moon varies from month to month. The full moon is at its lowest in the summer night sky when the summer sun is at its highest. In fact, the full moon nearest the summer solstice travels the same path as the winter sun, and the shadows from this moon will indicate the obstructions that the winter sunlight for your solar heater will encounter. A full moon near Christmas shines from the same place in the sky as the summer sun and provides light for midnight skiing or skating as well as indicating summer sun angles. In the spring and fall the full moon and sun follow roughly the same path through the sky, and direct observation of

winter or summer sun angles using the moon is more difficult. Since the moon completes its cycle through your solar window every month, there are only three or four days each month in which it is in the same position as the winter sun, and figuring out which days these are can be rather complicated. Let's keep it simple and look at four times of the year during which the winter sun angles can be shown by the moon. They are the full moon on the date closest to June 21, the summer solstice; the quarter moon, waxing (getting fuller), on the date closest to September 21, the fall equinox; the new moon (or sun) on the date closest to December 21, the winter solstice; and the quarter moon, waning (decreasing in size), on the date closest to March 21, the spring equinox.

If you use the moon to check for sun angles, do it as close to the solstice or equinoxes as possible, *and* do it within two to three nights of the full or quarter moon. The transient moon travels the same path through the sky for only three or four nights, while the winter sun follows approximately the same path for three months.

While you are out looking at moon shadows, locate the Big Dipper and Polaris, the North Star. Since Polaris is located due north of all sites in the Northern Hemisphere, it becomes very easy to locate true south, and you don't have to use a compass and a magnetic variation chart to point your collector in the right direction.

with orientation, good is often better than best when structural, aesthetic and cost advantages are all considered.

Locating Your Collector

After determining that your site receives enough sunshine, you must decide where to place the collector. With a large system the problem usually lies in finding a spot large enough to accommodate the collector. The placement should also allow for the distribution of a large volume of solar-heated air into the house.

A window box heater, the simplest type of collector, works best when it has a long, vertical flow of air through it. The best application for this collector is on a house that has one or more wide, south-facing, double-hung windows that are 4 feet or more above ground level. If the window is narrow, the window box heater must also be narrow in order to allow a free flow of air through the window and into the room. If a window isn't double-hung (such as a casement, awning or hopper window), the modifications that are necessary may cost as much as the collector. Second-story windows are a good choice for window boxes mounted vertically because the long, vertical flow of convectively driven air will tend to push air into the building more easily.

Adequate venting is often the biggest problem when adding a TAP to an existing house. Again, there must be a very free flow of air into and out of the collector. This is more easily accomplished with a frame wall than with a masonry wall because the spaces between wall studs can be cut out for venting the collector. A TAP also works better with a long airflow path and is therefore a good choice for a building of more than one story. Cool air can be fed into the TAP at the first floor and exhausted into the upper floor. You may want to install a small blower with ductwork to help move warm air back down to the first floor to equalize temperatures between the floors. This is especially important if there is no easy way for the air to move between floors, such as through a large open stairway (see "Room Air Circulators" in chapter 7).

An active system relies on large ductwork to move air to and from its point of use. Accommodating these ducts is an important consideration when designing a blower-driven system. The best places for concealing ductwork are in the attic and in the crawl space, so these areas must be accessible and have enough working room. Proper orientation towards true south is especially important when the collector is mounted vertically on an existing wall; roof mounts can be mounted crosswise on a roof so orientation of the house towards south isn't always as critical with these systems.

The construction of the house itself can also influence the choice of a collector site. It is much easier to mount a collector onto a roof with asphalt shingles than one with wooden shakes or tile. Flat frame walls are the simplest place to mount vertical collectors since the collector backing can be secured to the wall studs. Masonry walls, especially uneven stone walls or those that have brick veneer half way up, present special problems. It may be better to plan for and build a slanted, ground-mounted collector when you don't have an adequately flat mounting surface. This decision depends on several factors, such as available space and the finished appearance of a slanted collector versus that of a vertical collector.

A final note: Don't cover existing windows with collectors. A south-facing window that is well sealed is an efficient heat collector as it is. If you already have large south-facing windows, open the drapes during the day, and mount your collector on the roof or elsewhere.

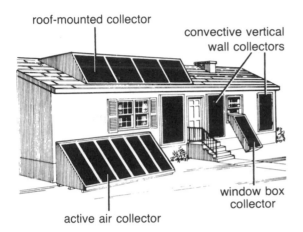

roof-mounted collector

convective vertical wall collectors

window box collector

active air collector

**Planning for Collectors
in New Construction**

Installing a solar heater on a building under construction is always easier and cheaper than adding one onto an existing structure. Air distribution, heat storage and interfacing with the back-up heating are all easier to accomplish.

As has been discussed, passive solar features are usually the best choice for new residential construction. In a well-designed building, the south-facing walls can include direct-gain windows, a greenhouse, a Trombe wall or the like. Collectors will be best mounted on the roof to provide additional heat to the north side of the house or for heating domestic water. A roof mount not only frees the south wall, it is also quite attractive in new construction because it looks

Photo 4-1 and Figure 4-5: This south-facing mobile home has tremendous retrofit possibilities. A window box collector and three vertically mounted TAPs could provide daytime heating to the living room and kitchen area. A tilted, roof-mounted collector could blow heat into a well-insulated space under the trailer or be used to heat water. A tilted, ground-mounted collector could be built on skids and have lightweight fiberglass glazing to allow for easy mobility or resale to the neighbors.

like it belongs and not like a "solar thumb." Solar roof trusses (which incorporate a 50- to 60-degree pitch for collector mounting) are becoming more popular, cost only slightly more than standard trusses and are a good way to go in conventional home construction.

In a new construction a large collector using a rock thermal storage bin can be built more economically than it can in a retrofit application. The storage can be placed under the house, where it belongs. The sides and bottom of the rock bin can be formed up and poured when the foundation is poured, and the crawl space can be designed so there is plenty of room to install necessary ductwork and air handling equipment. The new house can utilize an undersized forced-air heating system for back-up heating. It will be compatible with the air delivery from solar storage, and both systems can use the same distribution ductwork, further reducing costs. Large races (open spaces) can be designed into the backs of closets to accommodate the solar ductwork between roof-mounted collectors and rock storage, and all of the ductwork can be installed as the house goes up, not after. A space can also be provided next to the water heater for the domestic water preheating tank, thereby reducing plumbing costs and retrofit hassles.

Design Considerations: Convective Air Heaters

Convective solar air heaters, unlike active systems, can't be forced to work properly. Since these collectors work by the natural convection of heated air, and not with a blower, proper design is more critical than with active air systems. Heat is transferred to the air stream differently in convective collectors. Active collectors rely on resistance to airflow and on air turbulence within the collector to "scrub" heat off the absorber, whereas passive collectors need a very unrestricted airflow through them to maintain an adequate flow rate. The airflow cavity must be large, and all openings and changes in airflow direction must be very gradual and smooth, for the convective loop to operate efficiently.

Convective heaters, if properly designed and built, can perform nearly as efficiently as active collectors. Unfortunately, many poor construction plans have been published, and many convective heaters have been built that either deliver very little heat or, because of reverse thermosiphoning at night, lose most of the heat they gain during the daytime. In this section we will look at the design considerations of three different types of convective air heaters: the window box, the U-tube collector and the thermosiphoning air panel (TAP).

Window box collectors are typically small, removable units mounted at a slant below south-facing, double-hung windows. They require at least 4 feet of airflow up the absorber to be economically practical, and they are usually built to the same width as the window. U-tube collectors are vertically mounted window box collectors that are permanently attached to a south-facing wall (not to a window). They are typically placed below the room to be heated on a two-story building and vented at floor level into this room. Their height can be between 4 and 16 feet and their width is usually determined by available space, looks and wall-stud spacing.

U-tube collectors have limited applications, but they do have definite advantages. Like window box collectors they are self-damping at night (no reverse thermosiphoning), but they are not limited by the size of a window as are window box collectors. They require only one set of vents to be cut into the building while TAPs need two.

TAPs are mounted on vertical, south-facing walls of single-story buildings. They have two sets of vents cut into the adjoining room corresponding to the top and bottom of the collector. The height of the collector corresponds to the height of the adjoining room, less 6 inches, to keep the lower vent off the floor. Lightweight, tight-sealing backdraft dampers are installed in the lower vents to prevent reverse thermosiphoning at night.

The Chimney Effect

All convective heaters operate because warm air rises and cold air sinks. The push of solar-heated air that drives convective air heaters is also dependent on the length of the flow path. Chimneys that are built for wood-burning stoves and fireplaces illustrate both of these principles. When a fire is initially started, the draw (airflow) up the chimney is poor, and the fire is more likely to smoke because there is no difference in temperature

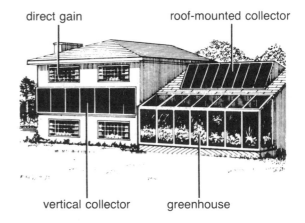

Photo 4-2 and Figure 4-6: This split-level is a natural for a large greenhouse or sunspace. A vertical active collector provides heat to the lower floor on the east end of the house, and a roof-mounted, two-mode system heats domestic water and/or the crawl space on the east end of the house.

between the air entering the stove and the air leaving the chimney. As the fire burns, the chimney warms up, and there is a greater temperature differential, creating a better draw up the chimney. The draw in fireplaces that tend to smoke can often be improved by increasing the length of the chimney. The warm air rising at the top of the chimney will pull cooler room air more readily into the fireplace. Chimneys (or stovepipes) that are large in cross-sectional area also draw better than narrow chimneys because the airflow through them is unrestricted. Convective air heaters operate in much the same way, and a fairly long, free flow of warmed air is the basis for their design. If the airflow path through the collector is too short or constricted, the flow rate of solar-heated air will be reduced. The hotter these collectors get, the greater the flow rate, but remember that hot collectors have a greater heat loss to the outside air. The challenge is to design a collector that gets hot enough to develop an adequate flow rate without losing heat to the outside.

Getting the most from a convective collector boils down to moving a large volume of air through it at a low velocity. The airflow cavity in the collector must be large, and the vent openings into the house must be as large as the cross section of this cavity. The airflow can't be choked down at any point, and a convective collector must never be baffled.

When properly designed, both the airflow and temperature rise through a convective collector will be about the same as for an active collector, about a 40 to 60°F temperature rise with an airflow of two cfm per square foot of collector (at sea level).

Laminar Airflow

Laminar airflow is another important design consideration in passive collectors. As

Photo 4-3 and Figure 4-7: It will take some special work to retrofit this home. A large tree to the southwest will need to be cut down or trimmed, and another tree to the southeast (on a neighbor's property) could also use some trimming. A roof-mounted collector will minimize these chores but will involve working up high. If the trees can be dealt with, TAPs or active collectors can be squeezed in between the entryway and the second-story windows, or a small, tilted, ground-mounted collector can be built.

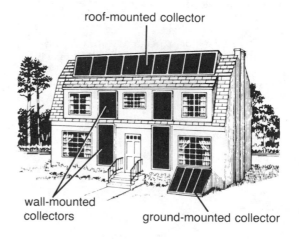

roof-mounted collector

wall-mounted collectors

ground-mounted collector

air rises through a collector, it tends to form layers of moving air that are of different temperatures. The layer near the glazing is cooler than the layer near the absorber. Active collectors rely on "scrubbing" the heat off the back side of the absorber and are designed to defeat laminar flow by forcing turbulent air rapidly through the collector. Since air in convectively driven collectors moves much more slowly, it is impossible to disrupt the laminar airflow. Ideally, the collector is designed to take advantage of this phenomenon in order to keep the glazing cool and to prevent excessive heat loss from the collector.

Optimizing Heat Transfer

There are several options available for the interior configuration of convective heat-

Photo 4-4 and Figure 4-8: Either a vertical or tilted collector could be built onto this south wall. Since the southern rooms on this house are bedrooms that aren't used during the day, a TAP isn't a very effective choice for this site. An active collector is needed to move solar-heated air to parts of the house that are occupied by day. A roof-mounted collector on this house would look like a "solar thumb," so it should be located on the back of the house and not be visible from the street.

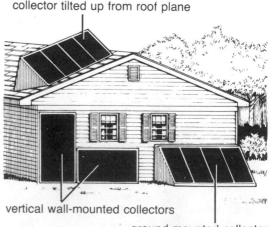

collector tilted up from roof plane

vertical wall-mounted collectors

ground-mounted collector

ers. In figure 4-9 collector design 1 shows a *backpass* design that looks much like a typical forced-air collector. Sunlight is converted to heat on the absorber surface and is carried off by a moving stream of air behind the absorber. Even though in a collector of this type there is a dead air space between the absorber and the glazing, double glazing is always recommended. A backpass collector of this type doesn't operate very efficiently because there is poor transfer of heat due to the low-velocity, nonturbulent flow. The absorber overheats and loses more heat out the front of the collector. A selective sur-

face absorber (in a collector of all-metal construction) greatly helps the performance of a backpass design but increases its cost.

In the second collector design, air flows on both sides of the absorber. The natural tendency of air to form layers is helpful in this design. The air in front of the absorber is cooler and moves more slowly than the air behind the absorber. Since the coolest air is in contact with the glazing, there is less heat loss from the collector, and double glazing isn't required.

Collector design 3 shows our preferred design for convective air heaters. Several lay-

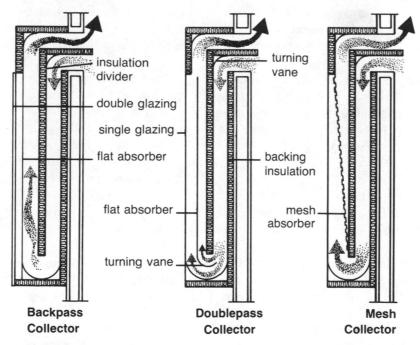

Figure 4-9: There are three different methods of transferring heat in convective collectors. Sealed backpass collectors are prone to excessive heat loss out the front of the collector, and they must be double glazed to minimize that loss. In the doublepass design, air moves both in front of and behind the absorber, which helps to keep the glazing cool. In mesh collectors, our preferred design, air is heated as it moves through an expanded-metal-lath absorber. [Illustration concept by W. Scott Morris.]

ers of expanded metal lath (mesh) are used as the absorber and are installed at a slant inside the collector. This design makes the best use of laminar airflow. The air in front of the mesh and in contact with the glass is cool, whereas the air behind the lath is warm. As in collector design 2 this layer of cool air helps to insulate the glazing and keep it cool, eliminating the need for double glazing.

W. Scott Morris of Santa Fe, New Mexico, who has tested these three designs, has found that a slanted-metal-lath absorber of this type will deliver 30 to 40 percent more heat than a flat plate absorber in a single-glazed, backpass convective collector. The lath offers a very large surface area for heat collection and transfer while encouraging, and taking the best advantage of, laminar airflow to reduce heat losses. In active systems, metal-lath absorbers work well but increase the cost of the collectors because, in most designs, double glazing is required to reduce heat loss from the glazing. Now let's look at some specific convective collector designs.

Window Box Collectors

Window box collectors perform best and are easiest to build when they are as wide as the windows in which they are mounted. A wide collector that funnels into a narrow window can be built if the vents through the window have the same cross-sectional area as the airflow cavities in the collector. Gradual transitions between the collector and the window penetration are a must, and square corners must be avoided.

When mesh is used as an absorber surface, the airflow cavities (top and bottom) should be sized so their depth is equal to about one-twentieth of the collector's length. Thus a collector with a height of 5 feet will

need two 3-inch air cavities; an 8-foot collector needs two 5-inch cavities. If the cavity depth is less than one-twentieth of the height, airflow will be restricted. If it is much wider than one-twentieth, air won't pass evenly through the entire absorber.

Proper design of the top of the collector will prevent reverse thermosiphoning at night. The collector is built so the top of the lower inlet vent is at the same level as, or above the top of, the glazing (see figure 4-10). It is wrong to extend the glazing above the inlet. A curved baffle known as a *turning vane* should be installed at the bottom of the collector to help the air find its way around the corner. Making the opening at the bottom of the collector slightly larger than the air cavity (1½ times) also helps ensure an unrestricted flow at this point.

The collector should be designed so that it can rest on bricks or blocks rather than directly on the ground to prevent the wooden framework from rotting. In high-wind areas it is also a good idea to stake down the lower corners to keep the collector from wiggling around inside the window frame.

The cost per square foot for a window box collector can be high if all of the materials are purchased new. Don't skimp on high-temperature insulation or good-quality caulking, but try to build the rest of your collector from low-cost, recycled materials in order to get a cost-effective installation.

A U-tube collector has the same basic design considerations as a window box collector but is vertically mounted directly onto the wall. Since it is a permanent installation, better-quality materials should be used. The vents into the house or building (best for multi-story buildings) must equal the area of the airflow cavity. Use angle braces at the bottom of the collector to support its weight.

TAPs

In order to perform effectively, the vent openings for a TAP must be sealed at night with a backdraft damper. The best-performing TAPs have vent openings that are slightly larger than the cross-sectional area of the airflow cavity. The best configuration is to have inlet and outlet vents that run continuously from one side of the TAP to the other, interrupted only by wall studs. An alternative is to use every other stud space, but the total vent area must still be larger than the airflow cavity. (Studies have shown, however, that with this vent design, compared to the continuous vent, heat output is reduced by 10 percent.) The lower vents should be located about 6 inches above the floor level to help keep dust out of the collector.

Preventing nighttime reverse thermosiphoning in a TAP is more difficult than with

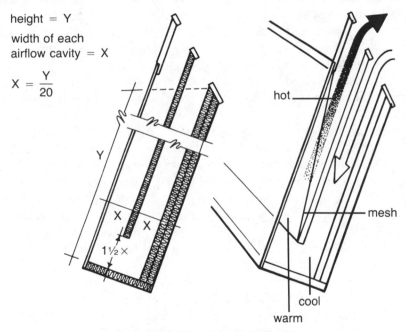

height = Y

width of each airflow cavity = X

$$X = \frac{Y}{20}$$

hot

mesh

cool

warm

Cross Section of Convective Collectors

Figure 4-10: The depth of the airflow cavity in a thermosiphoning collector should approximately equal one-twentieth the height (length) of the collector. The bottom opening in a window box or U-tube collector should be equal to 1½ times the depth of one air cavity. A turning vane is installed to help direct airflow around this corner. Mesh collectors take advantage of laminar airflows. The cooler air in the collector remains against the glazing while the faster-moving, hotter air stays behind the absorber. In window box and U-tube collectors, an insulated divider keeps the return air cool so that it will easily fall to the bottom of the collector. [Illustration concept by W. Scott Morris.]

the window box design because a heat trap isn't created. The backdraft dampers required to isolate the collector should have very thin, light flaps that open with extremely low pressure, and they must be tight sealing in order to be effective. A damper for each lower vent in a TAP is a must, and if the flapper vanes can be made light enough in weight, a damper in each top vent is a good idea to prevent a small amount of convective heat loss at the top of the collector at night. Manual registers can be installed instead, but they demand attention twice a day (morning and night). They are typically quite leaky and don't prevent reverse thermosiphoning if the sun is obscured by the clouds during the day. For these reasons we don't recommend their use. See chapter 7 for details on convective backdraft dampers.

TAPs are generally one-half as deep as a typical window box collector because there is just a single pass of air through them. The depth of the air cavity should, once again, be equal to approximately one-twentieth the length of flow for a collector with a mesh absorber.

Since there is no way to regulate the heat delivery from a TAP, overheating of the living space is a possibility, but this usually doesn't occur unless the collector is larger than 10 percent of the floor area of the living space behind the collector. If you decide to build a large TAP, you may need to install room air circulators (see chapter 7) to better distribute the heat throughout the house, or you may be better off building an active system with its accompanying distribution ductwork.

Adding a blower to a TAP is not a good idea and is usually only justified if you have built a collector with a poor design and aren't getting enough heat from it. Don't build a TAP that requires ducting. If you need distribution ductwork to the north side of your house, build an active system or install room air circulators. If you need to install a blower or an existing TAP, there are a few options available. The upper vents in a large TAP can be manifolded together, and air can be pulled from them with a single blower sized to deliver 2 to 3 cfm per square foot of collector. Restrictions will have to be placed in vents closest to the blower. This "fix" will entail an unattractive soffit and can involve placing a noisy blower in the living area.

Another option is to use a 100 cfm computer fan (such as Grainger's no. 4C549; see chapter 5 for the address) for each 30 to 40 square feet of collector. This involves completely sealing the upper vents so the blower pulls air from the collector and not from the room. An inexpensive snap disc thermostat (such as Grainger's no. 2E245) placed in an accessible spot may be used to control the blowers and will help keep costs down. Computer fans aren't too noisy if vibration is reduced by mounting them on a bed of silicone caulk, but several in operation at the same time can be distracting.

Keeping the Absorber Clean

There is a possibility that the mesh used for the absorber in a convective heater could act like a filter and collect dust and lint. Since the air will be moving very slowly through the collector, this usually isn't much of a problem, especially if the lower vents are located at least 6 inches off the floor. We have heard of only one installation where it was a problem, and this was a case where the owner was drying clothes in the room behind the collector, and lint collected on the mesh. Window box or U-tube collectors can easily be cleaned. A flexible vacuum cleaner hose can be inserted into the air cavity, pulling dust out as you tap on the absorber. TAPs should be designed so that the backdraft dampers are removable for cleaning the collector.

Performance Evaluation

The airflow rate in convective collectors is very difficult to measure. Large vents delivering the correct amount of air will have an airflow through them that is noticeable but gentle. Observing whether the glazing is cool and whether there is a proper temperature differential between the incoming and outgoing air is the best way to tell if the collector

---

TABLE 4-2

Temperature Differentials between the Incoming and Outgoing Air in a Convective Collector

Temperature Differential (°F)	Comments
30 to 40	Very good flow, high efficiency, collector requires a long flow path to achieve these temperatures
40 to 55	Good flow, good efficiency, flow is about right for a home-built retrofit
55 to 70	Flow is adequate, reasonable efficiency
70 to 85	Poor flow, poor efficiency, collector is getting too hot and losing too much heat
over 85	Back to the drawing board, or add a blower

NOTE: The temperature differentials are for a collector operating at noon on a sunny day.

---

is delivering an adequate amount of air and performing efficiently (see table 4-2). Cool glass and relatively low delivery temperatures are the signs of efficient operation. Restrictions in vent openings are the most common cause of poor airflow rates, high delivery temperatures and inefficient operation.

Design Considerations: Active Collectors

An active system has a blower to force air through a small airflow cavity and in most cases is easier to design than a passive system because a blower controls the direction of airflow and *makes* the collector work. This section discusses some of the basics of designing a single-glazed flat plate collector for residential use.

Proper airflow rate and airflow cavity size are vital to a collector's efficient operation. The collector design we feature has a fairly shallow (1-inch) cavity and a relatively high rate of flow through it, both of which help to take heat off the absorber plate. The length of flow is another consideration. The idea here is to move the air far enough through the collector for it to reach usable temperatures (between 100 and 140°F). Balancing these factors is the key to designing a cavity that will collect and deliver as much heat as possible.

Flow Rates

Within limits, the more air that flows through an active collector, the better its performance. The main consideration with active systems is that if too much air is moved through the collector, the air that is delivered is too cool to provide comfort. The desired temperature rise of air moving through a collector is the key to deciding upon the amount of air that needs to be moved. In most residential applications an airflow volume is chosen that will create a temperature rise of 50°F through the collector. For example, the incoming air will be 70°F, and the air leaving the collector will be 120°F. Calculations involving the specific heat of air (how much heat air can hold) and the insolation (the amount of heat available from the sun in Btu's per square foot) indicate that, to achieve this temperature differential, the collector must deliver between 2 and 2½ cubic feet per minute (cfm) for each square foot of glazing (at sea level). Thus a 100-square-foot collector needs an airflow of between 200 and 250 cfm. This gives an average temperature rise of about 50°F at noon and about 20°F early in the morning and late in the afternoon, which is about right for space heating. The fairly large temperature rise of 20°F in the morning and afternoon is desirable since it is high enough to keep the blower from cycling on and off as the airflow cools the collector. The outlet temperatures at this time are still high enough to be usable. If the flow rate is doubled, the temperature rise will be cut almost in half. In applications where low output temperatures (a small temperature rise) are usable, such as radiant storage floors, a higher airflow is desirable. This is because the higher-velocity airflow doesn't reach people directly. There are definite limits to the acceptable *velocity* (often measured in feet per minute, fpm, or feet per second, fps) when the flow enters the living space. An excessive velocity of warm air can have a cooling effect and can cause discomfort due to draftiness.

If less than 2 cfm of air per square foot of collector is delivered, the air will be hotter, but there will be less total heat delivered in the course of a day. A lower flow and a higher temperature rise can be useful in some applications, however, as when an oversized collector, primarily designed for space heat-

ing, is used for heating water in the summer.

Another factor to keep in mind is that the density of air varies with altitude. A cubic foot of air at sea level weighs considerably more than a cubic foot of air at 5,000 feet and can therefore hold more heat. The optimal airflow will vary with altitude as indicated in table 4-3.

Heat Transfer

The next consideration in designing an efficient active collector is transferring heat from the absorber to the stream of moving air. The best way to ensure a good transfer is to have a properly sized airflow cavity to maximize (to a point) the air turbulence. Turbulent air that is "rubbing" against the absorber plate will collect more heat than will a free flow of air. If there is no turbulence, layers of air (known as *boundary layers*) of different temperatures will form. This phenomenon can be used to advantage in passive collectors, but if layers form behind the absorber plate in an active system, they will insulate the plate and poor heat transfer will result.

To create the desired air turbulence, the airflow cavity must be long and shallow; in fact it should be as shallow as is practical.

TABLE 4-3		
Airflow for a 50°F Temperature Rise		
Elevation (ft)	Cfm/ft² of Collector	Cubic Feet of Air in 1 lb @ 120°F (ft³/lb)
Sea level	2 to 2½	14.6 (0.068)
2,500	2½ to 3	16.1 (0.062)
5,000	3 to 3½	17.6 (0.057)
7,500	3½ to 4	19.4 (0.052)

Limitations involving materials and construction skills make it very difficult to build a low-cost collector that has an airflow cavity less than 1 inch deep. This is a good figure to shoot for since it allows for the expansion and contraction that all absorber plates undergo as they are heated and cooled. If available materials dictate that the airflow cavity be as deep as 1½ inches, it is advisable to put airflow disrupters in the airflow cavity

---

Static Pressure and Water Gauge

As air moves through a solar heating system, it encounters a resistance that is called *static pressure*. Static pressure is measured by a device called a U-tube manometer, which is simply an open tube with the lower, U-shaped section partially filled with water (see figure 4-11). Since this is a water gauge, the unit for static pressure is measured in inches and indicated as *water gauge* or WG.

Straight runs of ductwork will present a certain amount of static pressure but not nearly as much as obstructions such as ductwork elbows, air filters, the collector itself or the rocks in a storage bin. To measure the static pressure presented by an obstruction, the two open ends of the manometer are inserted into the duct on opposite sides of the obstruction. Some of the airflow will try to "cheat" by flowing past the obstruction and through the manometer. The more restrictive the obstruction, the harder the air will try to force its way into the tube and the further it will separate the heights of the two columns of water in the U-tube. Figure 4-11 shows a static pressure of 0.5 inch WG.

Since the pressure is greater upstream from the obstruction than it is downstream, there is a pressure drop from one side to the other. That's why resistance to flow is often referred to as *static pressure drop*.

The static pressure presented by any obstruction will vary greatly depending upon the rate of airflow through it. Faster flows will produce higher static pressures.

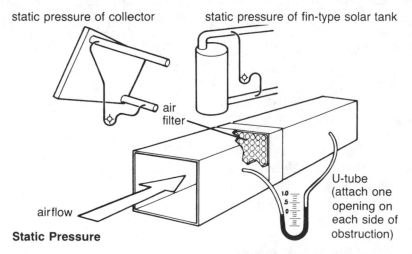

static pressure of collector

static pressure of fin-type solar tank

air filter

airflow

Static Pressure

U-tube (attach one opening on each side of obstruction)

Figure 4-11: Static pressure, the resistance to airflow through the collector and the distribution system, can be measured with a simple device called a U-tube manometer. The static pressure indicated in this illustration is 0.5 inch water gauge (WG).

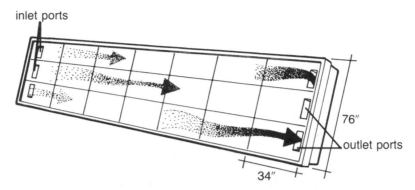

Figure 4-12: The length of airflow in this collector is 20 feet through a 1-inch-deep air cavity. The straight pass of air through the three baffle spaces behind the absorber is ideal.

to increase both the static pressure and the amount of heat transfer. This option is discussed in chapter 8, "Building Active Air Collectors." Collectors designed for a high airflow and small temperature rise can utilize a slightly deeper (1½ inches) airflow cavity without the need for disrupters.

In the most efficient active air-collector designs, the collector will present a *static pressure* (resistance to airflow) of between 0.20 and 0.35 inch water gauge. This static pressure is high enough to get good heat transfer but low enough so that an undesirable load isn't placed on the blower. Static pressure is directly affected by the length of flow, and in order to achieve the optimal static pressure in your collector, the air must be moved about 20 feet behind the absorber plate through a 1-inch cavity (see chapter 9 for a detailed discussion of the static pressure encountered inside a collector).

When a collector is designed this way, it has not only the proper static pressure but also a high airflow velocity through it. It makes sense to move air rapidly through a collector because rapidly moving air will get the heat

out of the collector and help prevent heat losses out of the front and back of the collector. The air velocity through properly designed collectors for residential use should be 600 to 800 feet per minute.

An example collector built to our guidelines (see the box "Airflow Rules of Thumb for Flat Plate Active Collectors") to accommodate seven panes of 34-by-76-inch tempered glass at sea level is shown in figure 4-12.

The airflow is:
  126 ft² of collector x 2.5 cfm/ft² = 315 cfm

The air cavity is:
  1" deep x 76" wide = 76 in²
  = 0.53 ft² in cross-sectional area

The air velocity is:
  315 cfm ÷ 0.53 ft² = 594 fpm

In many retrofit situations it is very difficult and not always cost-effective to build

Airflow Rules of Thumb for Flat Plate Active Collectors

- Airflow should be between 2 and 2½ cfm per square foot of collector (at sea level) for a temperature rise of about 50°F through the collector at noon.
- The air cavity depth should be 1 inch and in no case greater than 1½ inches. Airflow length should be close to 20 feet (between 16 and 24 feet).
- Airflow velocity should be between 600 and 800 fpm.

These rules of thumb will produce a collector with a static pressure between 0.20 and 0.35 inch WG. Static pressure is essentially a result, not a design goal. The primary design goals are achieving the appropriate velocity without excessive blower horsepower and achieving an appropriate, but not excessive, temperature rise through the collector.

a collector that meets all of these airflow rules precisely. Try to build your collector as close to these guidelines as possible, but don't be unduly concerned if your collector design varies slightly from them.

Baffling and Manifolding

Air must be directed in its path through a collector. If air is simply dumped into one end and taken out the other without regard to how it flows, the air will pick up some heat but not all that is available. Internal *baffles* direct the flow of the air inside the collector, and *manifolds* control the collector inlets and outlets. They both help to ensure even distribution of air inside the collector. Yet even in the most carefully baffled collectors there can be stagnant spots where there is insufficient airflow. "Hot spots" are the result. A long, straight flow without turns is

the best way to keep these hot spots to a minimum, but many collectors require more elaborate baffling schemes. The old method that many builders use with up-and-down serpentine baffles is not a good one because it tends to create stagnant spots. Baffles should be placed so that the airflow cavity between them is less than 24 inches wide. Try to design your baffles to accommodate the joints in the absorber plate that will lie on top of them, and/or put a "dummy" baffle under the joints (see figure 4-13).

When designing the baffles for your collector, take your time and imagine how the air will flow. Try to get the fewest turns possible that will still give you a 16-to-24-foot-

Photo 4-5: This simple baffle pattern is a reasonable choice for a collector made of five panes of 34-by-76-inch tempered glass, but turning vanes should also be installed whenever the airflow direction changes. This collector is a good candidate for a two-mode system: space heating in the winter and domestic water heating in the summer.

long airflow channel between the inlet and the outlet. If there are places where stagnation could develop (especially in square corners), install small turning vanes to help the air flow smoothly and without eddies through these spots. Turning vanes can reduce your collecting surface slightly, but proper airflow will more than make up for this small loss (see chapter 8 for more details on installing turning vanes).

Manifolds

Collector manifolds work like the manifolds on your car's engine in that they dis-

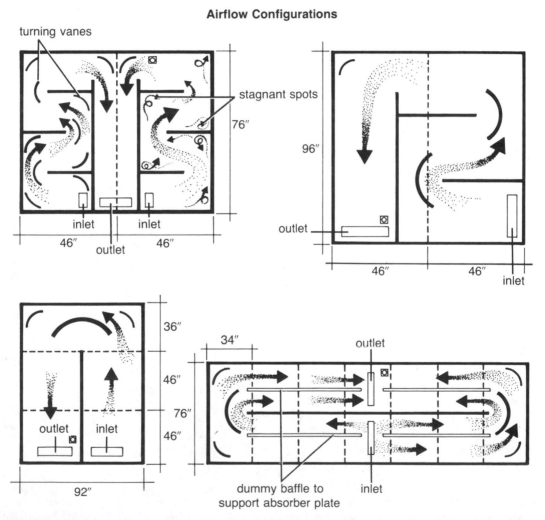

Figure 4-13: Smaller collectors often need internal baffles to move air through a 20-foot flow path through the airflow cavity. Collectors with convoluted airflow paths require turning vanes to prevent stagnant spots. Dummy baffles don't direct airflow but are used to support the edges of the absorber plate. Locations for the thermostat sensor are also indicated here.

tribute inlet air to more than one collector baffle space, then collect several streams of outlet air and consolidate them into one stream. Collectors that are very small do not require manifolding since the airflow will only take one or two very straightforward paths (see figure 4-14). Larger collectors, however, almost always need some form of manifolding in order to get an even distribution of air through the collector.

There are several methods for ensuring an even flow from the manifold into each baffle space. One approach is to design the manifold so that the sum of the cross-sectional areas of the inlets to the individually baffled spaces does not exceed the cross-sectional area of the feed manifold. In other words, the supply duct or manifold is made larger than the collector's total inlet area. The blower then slightly pressurizes the manifold, and since air is forced into each baffle space, each receives the same flow. Many mid-sized (around 100 ft²) collectors can be designed around this principle. They can utilize *internal manifolding* in which air flows through a supply manifold inside each end of the collector and is distributed into each baffle space. This type of manifolding is easy to build and seal since only two holes are cut into the collector. The holes feeding the individual baffle spaces in an internally manifolded collector must be sized correctly because once the absorber plate is installed, the amount of airflow down each space is difficult to regulate. If the holes are oversized, air will tend to short-circuit through the collector. Placing the inlet port diagonally across from the outlet port can also help ensure a good air distribution.

Collectors assembled from prebuilt panels and most larger, site-built collectors must rely on external manifolds with ports for proper airflow feeding. These manifolds are usually assembled with sheet-metal boots forming the ports. Round ductwork tees and elbows can be used to externally manifold two collector banks (see figure 4-14). *Butterfly dampers* are used in external manifolds in each inlet and outlet port to regulate the airflow to each baffle (see "Balancing Airflows" in chapter 9). When dampers are used, it is not absolutely necessary to pressurize the manifold although it is still a good idea.

Another option is to place the supply manifold directly behind the collector and cut holes into the airflow cavity to feed the collector. This is a desirable setup because all of the air makes one even sweep through the collector and stagnant spots are avoided. With this manifolding design, care must be taken to seal all of the ports from the manifold to the airflow cavity. Once again, it is important to under-size the inlet area relative to the size of the manifold or to install balancing dampers.

External manifolds are more exposed to heat loss than internal manifolds and must be completely airtight and well insulated to work effectively (see chapter 9 for specific information about building manifolds).

Blowers

Choosing the right blower for your system is not difficult. Since, at this point, you are still planning your system, let's look just at blower placement for now. There will be more discussion of blowers in chapter 6.

With crawl-space systems the blower can be placed in the inlet line, but it is usually a better idea to pull air out of the collector. In this way the blower will be located between the collector and the point of use. We use blowers with motors that are operated out of the stream of hot, outlet air. This prevents the possibility of hot air damaging the motor or bearings over time.

Structural Considerations

The single-glazed collectors discussed in this book are lightweight (6 pounds per square foot), so structural support for them isn't usually a problem. For roof-mounted collectors, consult a book on wood-frame construction that lists the loads that your rafters can bear. If you have an old roof or any doubts about your roof supporting a collector, talk to a lo-

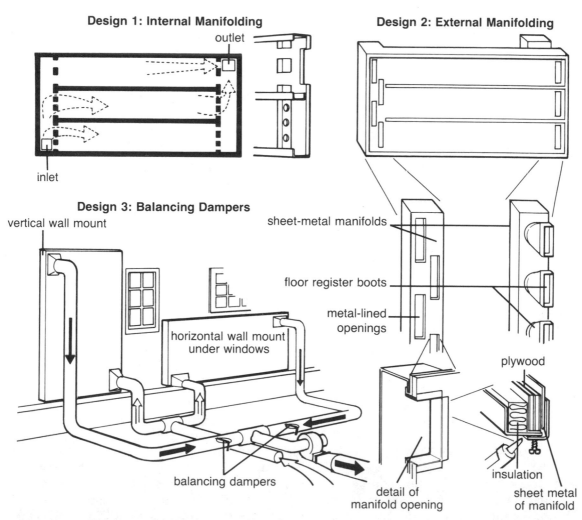

Figure 4-14: There are several options for manifolding larger collectors. Design 1 shows an internally manifolded collector where the area of the ports is smaller than the cross-sectional area of the manifolds themselves. Design 2 shows two options for building external manifolds. On the one side, the manifold is placed directly behind the collector. On the other side, floor register boots from the manifold feed the baffle spaces. Design 3 illustrates the need for balancing dampers when pulling air from two collectors with a single blower.

cal architect, structural engineer or building inspector. With roof-mounted collectors we prefer to build the collector separate from the roof rather than integrate it into the roof. This allows the collector to expand and contract freely and to be repaired independently of the existing roof. Vertically mounted collectors must be securely attached to the wall, and some require brace supports at the bottom.

Snow sheer is another factor to consider with roof mounts. Snow sliding off the collector can pile up below it. In a well-built collector the glazing is fairly cool even on a

Photo 4-6: The manifold for this collector is placed between roof rafters behind the collector. The inlet and outlet ports are cut through the plywood and litho-plate collector backing, lined with litho-plate angles and sealed with silicone.

sunny day, and if much snow accumulates against the bottom of the collector, it may not melt for several days. To avoid this problem, build the collector so that snow slides off and all the way to the ground, or build the collector a foot or more above any surface on which snow can collect. If snow piles up in large drifts on the sides of the collector, your roof load is increased, perhaps to more than the rafters can bear, and additional structural support may be necessary.

Performance Evaluation

After you have designed and built a collector, you will, of course, want to know how efficiently it is operating. One of the simplest tests is to feel the glazing near the hot outlet port. If the glazing here is relatively cool to

---

Reflective Gain

Any collector works best with a maximum amount of sunlight striking it. An easy way to increase solar gain is by reflection. Reflective devices are especially helpful for flat plate collectors, and often the collectors can be located north of an existing reflective surface such as a concrete driveway, a sloping tin roof or a white-shingled roof, which makes the addition of reflective panels unnecessary. With flat plate collectors the reflective surface is usually oriented so that the low-angle rays of the winter sun will be reflected onto the collector while high-angle rays from the summer sun will be bounced away to prevent collector overheating (see photo 4-7).

It is more difficult to use these "free" sources of reflective gain with tilted collectors, in which case special reflective panels would need to be installed. These panels can be adjustable in order to take advantage of the different sun angles throughout the year. They need to be of sturdy construction to keep high winds from damaging them, and often building and maintaining reflective panels for air-heating collectors is not a cost-effective undertaking.

Photo 4-7: Reflective gain from a permanently attached reflector panel significantly increases the output in this space-heating system. Manual dampers, a grille in the access door and a turbine roof vent on the roof pitch at the back of the collector (not shown) provide the small amount of convective venting this collector needs in the summer.

Internal Balancing Dampers

Internally manifolded collectors with two or three baffle spaces can often take advantage of adjustable dampers placed inside the airflow cavity. The best way to get an even airflow into each baffle space is to properly size the holes from the manifold, but in many installations this isn't a sufficiently reliable method. In Bill Gibson's installation his best access for ductwork to and from the collector was through the rim joist and into the crawl space. The length from one end of the collector to the other was 23 feet and very close to optimal.

The problem was how to get an even flow of air through the entire collector. Bill designed the elaborate baffling setup shown in figure 4-15, complete with several turning vanes, to help direct air evenly to all parts of the collector. He then installed an internal balancing damper in three baffle spaces to prevent air from short-circuiting through the collector and creating hot spots.

These adjustable dampers are simply short pieces of metal track that have been trimmed down so they can be moved inside the airflow cavity (by means of a bent coat hanger) after the absorber plate is in place. Bill installed all of the absorber (with a selective surface on the far east end) and turned on the blower. He then felt the surface of the absorber in different places near the outlet port. The lower baffle space was cooler (it had a higher airflow) so he damped down this line by moving damper 1. When both the upper and lower baffle spaces felt the same (you can sense a 3°F temperature differential with your fingertips), he knew he had the same flow through both of them. He next adjusted damper 2 to balance the lower lines, then damper 3 to balance the upper lines. After a small, final adjustment on damper 1, all lines were receiving the same flow. (See "Balancing Airflows" in chapter 9 for more details.) He secured the ends of the coat hangers to the outside perimeter of the collector, sealed the holes and started installing the glass—a nice trick.

A sliding type of damper is useful in installations where it isn't immediately obvious which path will have the highest flow. These dampers don't have to be very elaborate since you only have to adjust them once, but they must move freely underneath the absorber and be sturdy enough to stay put for years. Some air turbulence is created around the damper, of course, but this is a small price to pay for an even flow of air through the collector.

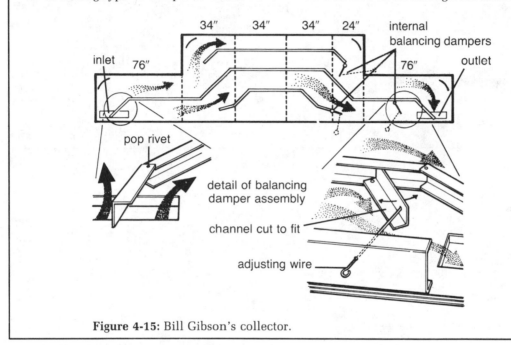

Figure 4-15: Bill Gibson's collector.

the touch (120°F or less), you know that the collector is removing and delivering most of the heat that it has collected. Another quick check is to measure the temperature differential between the inlet and the outlet air. As we have stated, the temperature rise should be about 50°F at noon on a sunny day (see table 4-4 for other temperature differentials and what they mean).

Airflow is another important consideration when evaluating your system, but this is more difficult to measure. For a detailed discussion on monitoring your active collector, consult Appendix 5. Also, Appendix 4 deals with collector efficiency. These discussions will be useful to those interested in designing a collector for a very specific application or those who are interested in taking a more in-depth look at collector design theory.

Active Matrix Collectors

In this book we don't go into details about building active collectors that use metal lath or another "matrix" material for the absorber surface. Many of the design considerations we have presented for flat plate collectors won't apply to matrix collectors. These col-

TABLE 4-4

Temperature Differential of Incoming and Outgoing Air in Active Collectors

Temperature Differential (°F)	(Temperature in °F)	Comments
20 to 30	(60 to 85)	Very high flow through collector, very good efficiency; limited usefulness, i.e., crawl-space heating
30 to 40	(65 to 100)	High flow, good efficiency; a good temperature range for radiant floor storage
40 to 60	(70 to 120)	Good flow, good efficiency; about right for direct-use space-heating applications
60 to 70	(90 to 155)	Reduced flow, poorer efficiency; higher temperatures may be useful for heating hot water
over 80		Poor flow, poor efficiency; you need a larger blower

Airflow Calculation

Let's look at a collector located at sea level, where it takes 14.6 cubic feet of air @ 120°F to equal 1 pound. Our example collector is designed to deliver 2.5 cfm per square foot of collector, or 150 cubic foot per hour per square foot of collector (2.5 cfm x 60 min).

$$\frac{\text{Air delivery in 1 hour}}{\text{Density of air at site}} = \frac{\text{lbs of air/hr/ft}^2}{\text{of collector}}$$

or

$$\frac{150 \text{ ft}^3/\text{hr/ft}^2}{14.6 \text{ ft}^3/\text{lb}} = \frac{10.3 \text{ lbs of air/hr/ft}^2}{\text{of collector}}$$

Since the specific heat of air is 0.25 Btu per degree Fahrenheit rise per pound, the amount of heat the 10.3 pounds can hold is:

weight of air x specific heat,
or

10.3 lbs x 0.25 Btu/°F/lb = 2.6 Btu/°F

or

2.6 Btu for each 1°F temperature rise

If the measured insolation (solar radiation) is 270 Btu per square foot per hour (a good, clear noontime average) on the collector's surface and the collector operates at a reasonable 45 percent efficiency, the collector can deliver 122 Btu per square foot of collector (270 x 0.45) in one hour. We need to put these 122 Btu into the air delivered in one hour, which can hold 2.6 Btu per degree Fahrenheit.

$$\frac{122 \text{ Btu}}{2.6 \text{ Btu/°F}} = \frac{47°\text{F temperature rise}}{\text{at 2.5 cfm/ft}^2}$$

Using the same calculation, the air will have a 59°F rise with a flow of 2.0 cfm per square foot.

lectors are usually built with a shorter airflow channel and with very different internal baffling. If you choose to go with a matrix collector, you will need to secure a good set of plans.

For Further Reference

Baer, Steve. *Sunspots*. 2d ed. Albuquerque: Zomeworks Corporation, 1977.

Energy Task Force. *No Heat, No Rent: An Urban Solar & Energy Conservation Manual*. New York: Energy Task Force, 1977.

Morris, W. Scott. "Convection Tested." *New Mexico Solar Energy Association Bulletin*, October 1977, pp. 9–13.

————."Natural Convection Collectors." *Solar Age*, September 1978, pp. 24–27.

————. "Natural Convection: No Moving Parts." *Solar Age*, January 1979, pp. 38–41.

————. "Natural Convection Solar Collectors." In *Proceedings of the 2nd National Passive Solar Conference: Passive Solar; State of the Art*, edited by Don Prowler. Newark, Del.: American Section of the International Solar Energy Society, 1978.

————. "Retrofitting with Natural Convection Collectors." *New Mexico Solar Energy Association Bulletin*, May 1980, pp. 9–17.

5

CONSTRUCTION MATERIALS AND USEFUL TOOLS — CHOOSING THE RIGHT STUFF

The proper selection of materials and tools is vital to the success of a site-built solar installation. Collectors need to be designed not only to fit into the best available space but also to be inexpensive and to use readily available materials. These materials are not new or exotic, and even though you may not have had experience with some of them, you shouldn't have trouble locating or working with the ones we recommend. If you are a do-it-yourselfer, your tool box probably has just about every tool you will need to build a solar system. If you need to purchase tools, it isn't necessary to spend a fortune, but there are some that will make the job easier. After materials are discussed, we'll talk a little about tools.

Collector Materials

Modern home building is centered around prefabricated, modular construction. Windows, doors, siding and kitchen cabinets all come in standard sizes that are easy to incorporate into designs and easy to install. All manufacturers of commercial solar collectors have followed this trend by offering modular units. A more cost-effective approach, the one emphasized in this book, is building collectors on site, and this of course involves locating the necessary materials. If you are the first collector builder in your neighborhood, this can be rather time consuming, but with a little looking and a little luck you may be able to get advice from other do-it-yourselfers or solar suppliers in your area. All of the materials used for site-built collectors are standard in various trades, though they may be new to you.

Use High-Temperature Materials

If a blower quits or the electricity goes off in an active system, the collector will *stagnate* and get very hot because there is nowhere for the solar heat to go. Although collectors usually operate between 90 and 150°F (and can be subjected to -30°F temperatures at night), well-built collectors under stagnation conditions can easily reach temperatures of 350°F. Collectors that were built with the wrong materials have actually ignited. All materials used in active collectors must be able to withstand temperatures of 350°F for extended periods.

Much more leeway is possible in material selection for convective collectors. They are, by nature, self-ventilating and should never reach temperatures over 180°F. Wood is a reasonable choice for the framework of a passive collector, but be sure to use fir and spruce rather than pine. Recycled materials can often be used for the absorber plate and glazing, which helps to keep costs low.

If you are considering using materials that are not discussed in this book, it is a good idea to test them beforehand. Place small samples of the materials in a covered, glass baking dish and bake them overnight at 350°F. If anything strange happens to them, or if they give off any gases, they are unsuitable for collector construction. It is better to stink up the kitchen for a night than have poor air quality from your collector for the next 20 years. A word of caution, though: Don't bake foam insulation because some types emit very toxic gases.

Air Quality

The quality of air coming out of your collector is critical—especially if you plan to blow solar-heated air directly into your living space. Good air quality cannot be overemphasized. We have seen collectors that were turned off and abandoned because they blew fiberglass fibers into the house or because the air coming out of them was so foul that a door had to be opened every time the collector came on. Proper material selection will help ensure good air quality. We perhaps go a little overboard in our designs because in the active collectors we build, the entire airflow cavity and all of the solar ductwork are bare metal. We don't paint the back of the absorber, don't use expanded metal lath in the airflow cavity and don't use fiberglass ductboard ducting for heat distribution. Foam insulation should not be used in any location that ever gets warmer than 150°F, and if litho plate is used anywhere in a collector, the "printing" side should be well sealed from the airflow cavity. Installing an air filter in the return ductwork is always a good idea and will not only keep the inside of the collector clean, it will prevent the blower from moving dust into your living space.

Locating Materials

You can probably locate all of the materials needed for your solar collector by looking through the Yellow Pages in the phone book and making seven or eight calls, but getting help from other sources will often save you time and money (see "Sources for Supplies" at the end of the chapter). Many communities have at least one solar supply store that caters to do-it-yourselfers. These outlets can furnish you with many of the odds and ends that you will need and can give you good advice for locating items they don't stock. These suppliers are often rather small and have a limited stock on hand, but they can order materials for you at reasonable prices.

Local solar energy associations (SEAs) are also good contacts for material suppliers. Your local SEA can often put you in touch with do-it-yourselfers who have built systems like the one you have planned. Other builders are a tremendous resource. You won't have any trouble getting them excited about your project, and they are, almost without exception, very helpful. Your local sheet-metal shop is another good contact for materials. The folks here are familiar with the materials mentioned in this book, and you may be able to get much of what you need at this one location. It's much easier to track down materials during weekly business hours than on Saturdays or in the evenings. If you need to place a special order from a supplier you have located, offer to prepay the order to get a better price.

Do-it-yourselfers are notorious scroungers. People build their own systems not only because they get satisfaction from it but also because they want to save money. Cutting costs by using recycled or bargain materials is a very understandable desire, but it isn't always a good idea when building a collector. Since active collectors can get very hot, materials must be chosen accordingly. Blowers and ductwork must also be properly sized, not "found," in order to deliver the proper airflow for efficient operation. Think twice before bringing home that bargain from the junkyard or auction, and make sure you are getting what you need in the way of materials and devices.

Glazing

The glazing material you choose should be durable, able to withstand high temperatures, relatively inexpensive and easy to work with. Available materials vary a great deal in cost, durability and ease of installation, but all have nearly the same ability to allow light to pass through. This is referred to as the *transmissivity* (or transmittance) of the material. Most glazing materials transmit between 83 and 92 percent of the available light. Most commercially manufactured collectors use tempered glass for an outside covering because it is a permanent, good-looking material. However, many do-it-yourselfers choose other materials because of their budget or because their installation represents a special application. For example, portable grain dryers usually have fiberglass glazing, and fiberglass is a sensible choice for collectors in areas where vandalism can be a problem.

Tempered Glass

All factors considered, tempered glass is the best glazing material for collectors. It is durable (it can withstand hail and high temperatures), it is fairly easy to work with, and it makes for an attractive, professional-looking installation. At $1.25 to $2.00 per square foot it is more expensive than certain other glazing materials, but it is permanent. Tempered glass panels, of course, can be broken, but they can withstand a pretty heavy impact without shattering. If tempered glass breaks,

it breaks into small lumps, not dangerous shards.

Many glass suppliers now offer low-iron, nonreflective tempered glass for collector construction at about the same price as regular tempered glass. One benefit of low-iron glass is that it allows a little more solar radiation to pass through it. Low-iron glass has a transmissivity of 92 percent versus 83 percent for regular glass. When viewed from the edge, regular glass has a definite green color whereas low-iron glass is colorless except for a very slight blue or green tint. The main advantage of nonglare glass is you can't see into it. It looks frosted, like the glass used in many bathroom windows, and is called nonglare because it doesn't create glaring reflections. Low-iron, nonglare glass makes a more attractive installation because it hides the absorber plate, screws, joints and silicone seams that are all part of collector construction. If low-iron, nonreflective glass is available for only slightly more (say 15 percent) than the best price you can get on standard tempered glass, it is a good choice. Solar glass usually has one smooth side and one side that is rough or textured. Always install the glass with the smooth side out so that the surface of the collector will stay cleaner. Be sure to call around town to get the best price. In any event, you shouldn't have to pay over $1.75 per square foot (1983 prices) for either of these materials.

The biggest drawback to using tempered glass panels for glazing is the size limitation they place on your collector since they are available in only a few standard sizes (typically 34 inches by 76 inches, 34 inches by 96 inches, 46 inches by 76 inches, 46 inches by 92 inches and 46 inches by 96 inches, though other sizes may be available in your area). Your collector must be designed to accommodate them. Nonstandard sizes would have to be cut to your order and then tempered, which makes them very expensive. Nontempered (annealed) glass can be used for collector construction, but it isn't recommended because it is more prone to breakage, and it is more dangerous to work with since it breaks into shards. Quarter-inch plate glass can be used to fill in odd-sized spaces on vertically mounted collectors but should never be used in slanted mounts because of the danger of breakage from hail. If you get a very good price on a quantity of nontempered glass panes, design a vertical collector to accommodate fairly small pieces (3 feet by 3 feet or smaller) since these are easier and safer to work with. Otherwise use tempered glass.

Weight and bulkiness are other drawbacks to using glass for collector glazing. Since large panes of glass are difficult to move and install, most collector builders prefer to use the smaller sizes, such as 34 inches by 76 inches.

Glass expands and contracts as it is heated and cooled through the course of a day so allow for this by leaving at least a ⅜-inch gap between panes when installing them in a collector. Also be sure that there is at least ⅛ inch of caulking or sealant between the pane and the support framework so the sealant won't shear off as the glass expands and contracts.

Fiber-Reinforced Plastic (FRP)

We feel that this type is not a great glazing material for air heaters. Although it is easy to work with and relatively inexpensive ($.80 to $1.10 per square foot), we don't know of any fiber-reinforced plastics that don't degrade rapidly at normal collector operating temperatures. FRP usually begins to yellow after about 4 or 5 years, even though it has a 20-year guarantee, and it often has to be replaced after 7 or 8 years. The solar transmit-

tance of FRP can be reduced by as much as 10 percent as it ages and yellows, reducing its effectiveness. Corrugated FRP can be hard to seal. "Flat" FRP isn't really flat when it's installed. It looks ripply unless it is installed in short, narrow sections. FRP also expands and contracts quite a bit as it heats and cools, which further complicates sealing chores.

Most FRP manufacturers no longer guarantee their products for collector applications because they don't hold up. Many do-it-yourselfers continue to use FRP, however, mainly because it is easy to work with (you can cut it with tinsnips or a utility knife). It is also unbreakable, which can be a consideration if your neighborhood is subject to vandalism or if you are building a portable collector. An FRP collector has some "give" to it and is easy to patch if you accidentally back the pickup truck into one while moving it from your grain drying operation over to the house for wintertime space heating.

If you choose FRP for your collector, get the best you can find. Some good brand names are Filon, Glasteel, Lascolite and Sun-Lite. A Tedlar (thin film) coating on the outside surface of the FRP is a must for prolonging its useful life.

Thin Film Glazings

Among the many types of thin films available for collector glazing, Mylar is a good choice. It withstands high temperatures, is inexpensive and is a good option for experimental or very low-cost collectors. Two-mil (0.002-inch) Mylar costs about 10¢ per square foot, and 7-mil is about 35¢ per square foot. Mylar is easy to work with and can be cut to any desired dimension. It is usually available on a roll that is 50 inches wide. Seven-mil Mylar can be used for the outside glazing, with 2-mil on the inside to create an inexpensive double glazing. Seven-mil Mylar will

[Continued on page 82]

More about Glass

Moving It Around

Large sheets of tempered glass are heavy, and moving them around can be awkward. Single panes should be carried vertically or on edge, never lying flat, by two people. The best way to transport glass in a truck is tilted with the long edge resting on wood that is covered with thick cardboard. It is easy to make a temporary A-rack for the back of a pickup to support glass in transit. Glass can also be carried flat in the back of a pickup or station wagon if the surface it is resting on is absolutely flat. When transporting it this way, move the glass into place holding it vertically and slowly lower it to lie flat.

Depending on how much glass you buy and the price you pay for it, most glass shops will deliver it free of charge. If charges are made for delivery, they shouldn't be any more than the shop's hourly rate ($15 to $20 per hour). Having the glass delivered may save you some hassles. Once you get the glass on site, store the panes against a wall at a slight tilt with flat boards and cardboard underneath. When storing sealed, double-pane units, be sure to support both glass edges of each unit. If it's stored outside, put a piece of plastic over the stack and tie a rope around it, making sure that rainwater can't get in. The presence of moisture between panes can lead to etching of the glass, which results in a not-so-nice "oil slick" appearance, a rainbow effect you don't want.

If your collector is roof mounted, moving the glass to the roof can be a little tricky. It is helpful to have four people around for the operation: two on the ground to pass it up and two up on the roof. Don't drag the glass over the shingles or you'll scratch it. A professional glazier can haul a pane of glass up a ladder without assistance. He carries one end in the crook of his arm (at the elbow) and balances the pane on his shoulder and head, leaving one arm free for climbing the ladder. If you're a maniac and decide to use this technique, try walking around the yard with the glass first, making turns and bending down until you get the feel of it. Be cautious but don't be afraid of tempered glass. If a pane breaks near you, the resulting small chunks can scratch you up quite a lot but won't cut you to ribbons. Nontempered glass is another story, however, and must be handled with leather gloves and much more care. If you want to spend the money, you can rent special level-actuated suction cups with handles from a glass dealer for moving the glass.

Regular annealed glass is stronger on the edge and weaker on the face. Tempered glass is stronger on the face than it is on the edge, and breakage of tempered glass usually results from nicking or scraping the edge on concrete or other rough surfaces.

Cleaning Glass

Everybody has a favorite method for cleaning glass, and whichever you use, you want to make sure that the interior surface is spotless. Streaks will show forever and cut down light transmittance. Windex or Glass Plus and paper towels work well, but we prefer to use dishwasher soap (a nonsudsing detergent) and a squeegee. After cleaning the glass, remove any streaks by buffing with a lint-free cloth that has been sprayed with a very dilute detergent-and-water solution. If the glass has sap, paint, caulk or other materials sticking to it, scrape the surface with a single-edged razor blade, keeping the blade flat to avoid scratching. Also keep it away from the edge of the glass, or the blade will be ruined in no time.

Nontempered Glass

Nontempered (annealed) glass is definitely more dangerous to work with than is tempered glass. Not only does it break into dangerous shards, but it also usually isn't "seamed" (sanded on the edges), so it will slice your hands if you let it slip. Wear gloves, and whenever you're working with nontempered glass, be extremely careful and have a reliable helper. *Never* install nontempered glass in large sheets on the roof or at a tilt. It can break from snow and wind loads and hail impacts.

Double-Pane Sealed Glass

Double-pane tempered glass units are expensive, but they are the best choice for double-glazed collectors. Don't try to make your own double-glazed units. They will inevitably fog up, come apart, leak or even explode with temperature cycling. Many commercial insulated units use a *single* polysulfide seal between the glass panes, but *double*-sealed units are a better choice for collector applications. In these units polyisobutylene is used as an inner, waterproof seal, and silicone is used for an outer structural seal. Double-sealed insulated units are more expensive (5 percent more) but worth the extra cost. Never use units sealed only with Butyl, or Thermopane-type units in which the edge seal is glass (i.e., the two panes are "welded" together).

Caulking

Silicone is the best for sealing glass to almost everything except wood. It will, however, pull away from wood in a couple of years unless the wood is dry and well sealed. Avoid Butyl caulks and any oil-, latex- or water-based caulks. They are unsuitable for collector construction in any case. Some caulks are not compatible with the sealants used on insulated glass units. If you are using double-pane units, check with your supplier to make sure the caulk you plan to use won't deteriorate the seal between the panes.

Photo 5-1: Moving glass onto a roof is not difficult, but it is helpful to have two people on the receiving end. The glass being passed up for the roof-mounted collector is low-iron solar glass. The ground-mounted collector in the foreground is glazed with standard, clear tempered glass.

probably have to be replaced after about three or four years. Tears and rips can be repaired with clear silicone caulk and clear Mylar "Band-Aids." Although collectors glazed with Mylar have a plastic-bag look to them, and the film tends to rattle in the wind, it is suitable material to use for very low-cost systems. J. K. Ramstetter of Denver, Colorado, has built several collectors glazed with Mylar (with litho-plate absorbers) that pay for themselves every month they are used, even if tax credits aren't claimed. A one-month payback is definitely worth giving this glazing material a second look, but if you use Mylar as an outer glazing material, consider designing your collector to accommodate tempered glass at a later time when you can afford it.

If a thin film (see table 5-1) is used in a collector, it's a good idea to mount it on a material that won't heat up too much. Many films will turn brittle after prolonged contact with hot materials. One problem that must be dealt with is the large amount of expansion (which causes sagging) that occurs when they're heated. Also, most thin films are not as opaque to long-wave heat reradiation as are glass and FRP and are therefore not as effective at keeping heat inside the collector. (See chapter 1 for a discussion of the greenhouse effect.)

Flexigard, Lexan and Teflon are suitable for a permanent inner glazing in a double-glazed collector or as a temporary outer glazing. Flexigard is a composite film made of polyester and acrylic. It comes in 10- and 15-mil thicknesses and usually retails for about 60¢ per square foot. It is sturdier than Mylar, more resistant to tearing and less likely to rattle in the wind. It also has a lower coefficient of expansion than most films, but since it is so thin it can't really be considered a permanent glazing material. Lexan is a po-

lycarbonate film that is easy to work with and considerably more durable than Mylar, but it tends to expand a great deal. Teflon film in 1-mil thickness is another choice for the inner glazing on a double-glazed collector. It has a high solar transmittance (96 percent) and a very high service temperature (400°F). It does, however, have the tendency to expand somewhat at high temperatures, which can cause it to sag toward the absorber. If you use Teflon, it must be supported at least every 2 feet. Also, this film is not opaque to long-wave radiation, so even though it will help prevent convective losses in a double-glazed collector, it won't help much in holding in heat by means of the greenhouse effect. There is no solar superfilm on the market. Most manufacturers come out with a new line every five or ten years because they haven't yet developed a film that works really well for solar collectors.

There are some glazing materials that you shouldn't use at all. Don't use any material that can't withstand continuous temperatures over 180°F. Acrylics, such as Plexiglas or Exolite double-wall panels, may be acceptable for Trombe walls or greenhouses, but they can't withstand the high temperatures encountered in an active collector. Acrylics also expand and contract a great deal with temperature changes, so sealing is difficult. Rigid plastics require sophisticated mounting systems so that they can expand without cracking. Polyethylene film is also completely unsuitable because it can literally melt at high collector temperatures.

Glazing Summary

• Tempered glass is the best glazing material. Use it whenever possible and design your collector to accommodate standard glass sizes.

- Corrugated fiberglass is a good second choice.

- If you glaze a collector with solar film, design it to accommodate glass at a later date.

Absorber Plates

A collector should have an absorber plate that can withstand high temperatures and that is easy to paint, install and seal. In a flat plate collector the absorber is usually a thin sheet of metal that allows for rapid transfer of heat

TABLE 5-1

Glazings for Collectors

Material Type	Collector Use	Thickness (in)	Solar Transmittance (%)	Maximum Temperature (°F)	Estimated Collector Life (yrs)	Estimated Retail Cost/ft² ($)
Fiberglass-reinforced plastic (FRP)	inner or outer glazing	0.025	0.025" = 87	200	7 to 10	0.025" = 0.78 to 1.03
		0.040	0.040" = 85			0.040" = 0.96 to 1.22
		0.060	0.060" = 72			0.060" = 1.29 to 1.55
Glass						
Low-iron solar glass	inner or outer glazing	⅛, 5/32, 3/16	90 to 91	400 to 600	50	up to 2.00
Tempered glass	inner or outer glazing	⅛, 5/32, 3/16	82 to 84	400 to 600	50	1.50
Solar films						
Flexigard* (laminated acrylic/ polyester)	outer glazing	0.011	89	275	3	0.65
	inner glazing	0.007	91	275	10	0.50
Lexan* (poly- carbonate)	outer glazing	0.010	89	190	3	0.45
	inner glazing	0.010	89	190	10	0.45
Mylar* (polyester)	outer glazing	0.007	82	300	3	0.35
	inner glazing	0.002	82	300	10	0.10
Teflon* (fluoro- carbon)	inner glazing	0.001	96	400	20	0.39 to 0.69

*Trade name

to the airflow cavity. Some of the materials that can be used include the following: aluminum litho plates, aluminum sheet stock, galvanized sheet metal and metal sheets that are coated with a selective surface for better absorption of solar radiation. Expanded metal lath ("diamond mesh") can be used in convective air heaters.

Litho Plates

The aluminum litho plates that are used in printing are a good, low-cost choice for site-built collectors. They are available for between 10¢ and 35¢ per square foot from your local newspaper, a large printer or even a savvy solar supplier. Since litho plates are flat, they are easy to install and seal (see the box on Pittsburgh joints). Litho plates are thin and have very good heat transfer capability.

Two standard sizes for litho plates are 32 inches by 40¾ inches and 44 inches by 50¼ inches, but they can vary from printer to printer. Commercial printers use heavier-gauge plates (between 0.015 inch and 0.018 inch), which are best for absorbers. Newspapers often use thinner plates that are less expensive to buy but more difficult to install as absorbers. A thinner plate does, however, make a good collector backing in the designs that call for it.

On most used litho plates the opposite sides are bent up at a slight angle so they fit in the presses. You can ask for plates that were spoiled before printing and don't have bent edges, or you can cut these edges off rather than trying to flatten them. Litho plates should be painted on the side that was used for printing, which, if clean, is already well primed to take paint. This will put the shiny side toward the airstream. When litho plates

are used for collector backing, the printing side should be placed against the back of the collector and well sealed. If the plates aren't clean, the ink can be removed easily with gasoline or paint thinner.

The major drawback to using litho plates in large collectors is that many relatively small pieces must be used to assemble the absorber plate. The usual procedure is to design internal baffles in the collector to support the overlaps of two plates. Unsupported joints are secured with Pittsburgh joints and the two layers are sealed with silicone (see figure 5-1). But if you don't have much help available when you assemble your collector, the relatively small size of the pieces may actually be an advantage because they are easy to handle. Litho-plate absorbers in large collectors tend to look wrinkled when they are assembled, so consider using the nonreflective low-iron glass for glazing.

Aluminum Plates

Aluminum siding for mobile homes is another good choice for absorber-plate material. Some mobile home manufacturers sell sheets of flat, 16-mil aluminum plate cut to your desired length (usually 76 inches). It is usually available in 3-foot and 4-foot widths and costs about 80¢ per square foot. Ask for unpainted, mill-finish plate rather than primed sheets because the primer may not be suitable for high-temperature use. Sheets of flat aluminum with Pittsburgh joints on one edge are very easy to install and seal.

Corrugated aluminum roofing material, such as Rainlock (manufactured by the Reynolds Aluminum Company), has been used for the absorber surface in many low-cost collectors. Rainlock is available in large sheets (4' x 8', 4' x 10' and 4' x 12'), which may or

Pittsburgh Joints for Absorbers

Cold air leaking into the airflow cavity definitely reduces collector performance. It is vital that the airflow cavity be completely sealed from both the outside air and the dead air space between the glazing and the absorber. In the past, corrugated, embossed aluminum roofing material has been used for absorber plates, but tests have shown that the closure gaskets and silicone caulk used to seal these plates hasn't produced a sufficiently tight seal. Corrugated sheets may be an effective shape for collecting solar energy, but this possible advantage is outweighed by the difficulty involved in making them airtight. Many site-built collectors now being built use flat, precut aluminum or sheet-metal plates that are joined on their edges with a *Pittsburgh joint* (see figure 5-1). It is made by a machine that rolls one edge of a flat sheet. The seam uses about an inch of material, reducing the width of the plate by this amount. Most sheet-metal shops have this machine. Two people can make over 500 linear feet of joint in an hour. The charges for making the joints for an average-sized collector (100 to 150 square feet) should be under $15.

At the time of installation of the plates, the groove in the Pittsburgh joint on one sheet is filled with silicone, and the next sheet is inserted into this groove, making a very good seal. Edges without Pittsburgh joints are caulked and screwed to metal track around the collector perimeter or to a baffle track. This method of plate attachment can also be used with some selective surface materials.

If you buy aluminum plates from a mobile-home manufacturer, they can often roll "hooks" on two edges of each sheet for you at a very slight charge. This is the method they use to join seams in mobile-home siding, and it is a good one for collectors if the hooks make a sharp bend. If the hooks are excessively open, they will need to be closed slightly by squashing them with a board pressed along their length. Be careful not to bend the metal to the breaking point. Hooks can also be formed on site using a folding tool (see "Useful Tools" later in this chapter).

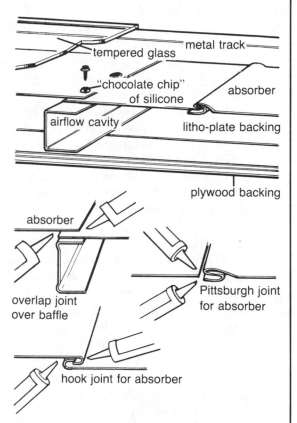

Figure 5-1: If possible, design your collector so that joints between absorber plates line up over underlying baffles. Joints that are unsupported from below should be Pittsburgh joints or hook joints that are filled with black silicone. After assembling the collector, run a continuous bead of silicone along the outside of all seams to ensure an airtight seal. Any screw that penetrates the absorber should be driven through a "chocolate chip" of black silicone.

may not be an advantage depending upon the layout of your collector and how much material waste you'll end up with. It costs about the same as siding for mobile homes but is much harder to seal when installed in the collector because it isn't flat. If aluminum roofing material is the only absorber material you can locate, buy the thinnest style available (usually 19 mils), and plan to spend a lot of time and to use a lot of caulk to seal the absorber. While it is easy to locate corrugated aluminum, it is often difficult to find the proper type of end-closure strips that must be used to seal the corrugated ends. These closure strips must be made from EPDM (ethylene-propylene-diene-monomer) material, with a maximum service temperature of 260°F, rather than neoprene with a maximum service temperature of 160°F.

Galvanized Metal

Galvanized sheet metal (zinc-coated steel) is another good choice for absorber-plate material. These plates are very easy to install and usually cost less than aluminum plates, but they transfer heat a bit more slowly. Always use the thinnest stock available, usually 30 gauge.

The biggest drawback to using galvanized absorbers is that it is much more difficult to get paint to adhere to them than to aluminum. Paint-Lock (a brand name) galvanized plates used for outdoor signs will hold paint much better than regular galvanized plates, but they are more expensive and not always available in thin gauges. Don't use preprimed galvanized plates.

Selective Surface Absorbers

Selective surface absorbers work great but cost a lot. These plates are in essence a one-way street: They take in solar radiation more readily than they reradiate heat. Heat tends to move through selective surfaces in only one direction. The back side gets very hot while the front remains relatively cool. These properties are referred to as *high absorptivity* and *low emissivity,* and they add considerably to a collector's overall performance. Metal plates that are simply painted black reradiate heat a lot more rapidly. The use of selective absorbers eliminates the need for double glazing in collectors. The glazing stays cooler and it loses less heat because the selective absorber isn't losing as much heat to the glass. If a single glazing can be used, more solar energy will reach the absorber, and therefore installing a selective surface is almost always more effective than double glazing.

When you choose a selective surface material, it is important to get one that isn't ruined by moisture in the air. Many of the first selective absorber plates manufactured had this problem, but almost all of the plates now being used are *black chrome* on copper, which isn't affected by the slight amount of moisture present in most collectors.

Selective surface absorbers are as simple to install as aluminum or steel plates in that they can be screwed, riveted and sealed just as easily. Care must be taken in handling selective surface material, however, because some types come with a fine ash coating that is easily smudged, leaving unsightly marks on the surface. Handle it by the edges as much as possible to avoid this. The surface can be touched up with a high-temperature paint, but this ruins its selective properties and doesn't improve the appearance all that much. It's better to use care and not smudge the surface in the first place. Some selective absorbers are covered with a thin plastic coating that keeps the plate from being scratched while in transit. Peel this coating off in the

Selective Surface Absorbers

Selective surface absorbers definitely increase the performance of site-built collectors, and we recommend using them in the "hot" one-third of an all-metal collector. When sunlight passes through the collector glazing, the selective coating absorbs up to 97 percent of it, converting it to heat. The heat is conducted to the plate, and since the coating also has a low emissivity, it transfers most of the heat to the air stream instead of radiating it back out through the glazing. Copper is an excellent conductor so it readily transfers heat to the airflow cavity. Selective surfaces typically have an absorptivity of 95 percent and an emissivity of 10 percent, while selective paints average 95 percent absorptivity and 30 to 50 percent emissivity.

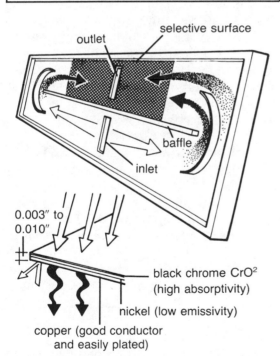

Figure 5-2: This illustration shows the possibility of using selective surface in the hot third of a collector. The detail is a cross section of selective surface material.

shade, not in the sun, or it can melt onto the plate. Selective surface plates can be run through a Pittsburgh joint machine (if they are at least 5 mils thick and not textured), which further simplifies installation. Selective absorbers get very hot during installation, and it is usually a good idea to wear at least one glove when handling them on a sunny day.

Selective surface materials are quite expensive, $2.50 to $3.75 per square foot, so it is a good idea to plan your installation to make the best use of them. The most cost-effective application is to place the selective surface in the "hot" half or third of a collector and use aluminum plate in the "cool" part. The cool half raises the air temperature, and the selective, hot half keeps the heat in, as well as further heating the air as it moves towards the outlet (see figure 5-2). This arrangement is particularly useful when you need hotter air from your collector, as in domestic water-heating applications. If a selective surface is used for the entire absorber surface, it will help raise temperatures somewhat in the cool end but probably not enough to be really cost-effective.

It is possible to create a semiselective surface by painting standard aluminum plate with a selective paint. This coating doesn't work as well as a manufactured selective surface, but is much less expensive and could be a good choice for the entire absorber or for the cool half when using a manufactured selective surface in the hot half. Proper application of selective paint is very critical, and our experience with the paints now on the market is that they don't significantly reduce heat loss from the collector. Selective paints are about twice the price of high-temperature enamels (40¢ per square foot versus 20¢), and since their use doesn't increase the total cost of the collector very much, using them in well-sealed collectors could be worthwhile (see "Paint and Painting Procedures" later in this chapter for application details).

Another possibility for creating a selective absorber is to use a thin, self-adhesive *selective surface foil*, which is bonded to a flat absorber on contact. This foil performs as well as selective plates and is nearly as expensive ($2 per square foot). Many manufacturers offer it only in narrow widths, so it takes time and patience to cover the absorber, but it is a good choice if manufactured selective surfaces are unavailable at a reasonable price.

A word of caution: If they are allowed to stagnate, selective surface collectors get much hotter than collectors that use black-painted absorbers. Selective material must *never* be used in wooden-frame collectors nor in those where all the other materials used aren't rated for temperatures over 300°F.

Expanded Metal Lath

In collectors that use expanded metal lath, heat transfer takes place as air travels through the layers of lath behind the glazing. This is a *matrix* type collector, and active systems of this design must usually be double glazed to create a dead air space, which helps to insulate the airflow cavity. Convective collectors using metal lath as an absorber can be designed so that double glazing is unnecessary. Metal-lath absorbers, in fact, deliver the best performance in passive collectors, and we recommend them.

Some collector builders place metal lath in waves (a corrugated configuration) in the airflow cavity behind a flat absorber in active collectors. This helps to increase output temperatures, since the lath gives the air more surface area from which to take heat, and helps to provide desirable turbulence in the airflow cavity. But on the other hand, lath painted with low-quality paint could be a possible source of noxious outgassing at the high temperatures present in stagnating active collectors. There is also a possibility that this paint could flake off, enter the airstream and flow into your living space. These factors aren't as much of a consideration in convective air heaters since they can't reach the high stagnation temperatures of an active system. They operate at much cooler temperatures and have a very slow flow of air through them. A mesh absorber plate in a convective collector also can be cleaned if necessary.

Generally, you shouldn't use anything for the absorber plate that isn't bare metal less than 20 mils thick unless the product is made specifically for collector applications, such as a selective surface. Also, don't use copper sheets. Copper, which transfers heat better than aluminum, is expensive, but you are wise to spend a little more and get a copper-based selective surface material.

Absorber Summary _____

• Use a flat aluminum absorber plate in the cool end of an active collector.

• If available, use a selective surface absorber plate in the hot one-third of the collector, but never use a selective surface in wooden collectors.

• Use metal lath for the absorber surface in a convective air heater.

_____

Metal Track

Metal track is the best choice for the supporting framework for active collector construction. It is commonly used for framing walls in commercial buildings and is readily available from larger building supply stores and drywall supply houses. It comes in standard sizes that correspond to standard lumber dimensions, 1⅝, 2½, 3⅝, 4 and 6 inches. Track, as it is known, is usually available in 10-foot lengths and is usually 28-gauge steel. Track with special dimensions is also available from most distributors. The uneven-leg track (also called unequal-leg track) used in the collector described in chapter 8 is a standard item. If a collector design requires a special size of track, be sure to check with a drywall supply house before going to a sheet-metal shop with a special order. Track custom made at the local sheet-metal shop will usually cost two or three times as much as standard track but can be considerably cheaper when ordered from a metal stud manufacturer. When buying track, be sure to get interior (0.020-inch-thick) rather than exterior (0.036-inch-thick) track. Interior track is easier to work with and completely adequate for collector construction. Also, for ease of assembly, be sure the track you purchase has long rather than short legs.

If you are a little wary of using metal track in construction, don't be. Installing metal track is something that's new to most do-it-yourselfers, but it is surprisingly easy to work with once you get the hang of it. Corners are easily formed by cutting the legs with snips and folding the track into an angle. To make a corner of this type (see figure 5-3), use a combination square and a felt-tipped marker to mark a square line on the back of the track. Scribe a right angle from this line down the legs, and cut out a small wedge section on both sides of the line with snips (aviation snips are very helpful here). The track is then easily bent into a 90-degree angle. A pop rivet can be placed through both sections after installation to further stiffen the corner. This

corner won't be perfectly flush but will be very sturdy. Some installations require that angles be very flush, and this can be accomplished by cutting a 90-degree angle from the track. Scribe a mark on the back of the track and down each leg. Then scribe a 45-degree angle on each side of the mark down each leg. Cut out the angles and bend the track. The back of the track can be tapped lightly with a hammer if there are slight bulges around the angle.

Seams are formed by overlapping two sections of track about 1½ to 2 inches. Cut a notch along the edge of one track so that the second track telescopes into it. Both legs of one track will be on top of both legs of the other track. Use screws on the bottom overlap to secure the two pieces together, and seal the seams with silicone.

Self-drilling screws usually screw easily into metal track, but if you encounter difficulty, use a hammer and nail (or punch) to start pilot holes. Before installation, set the track on the ground, put an appropriately sized piece of scrap wood inside the track (usually 1 inch by 4 inches or 2 inches by 4 inches), and punch a hole through the *outside* of the track. In most cases this is easier and quicker than drilling pilot holes. After just a little experience working with metal track, you will probably come to the same conclusion that most collector builders have: For building a collector framework and baffle, metal track is actually easier to work with than wood. It is also a lot more durable and can be used in a wider range of applications. Metal track is well worth the extra few pennies per foot it costs over wood.

Metal studs are another possibility for the supporting framework of a collector. They are usually a heavier-gauge metal and therefore sturdier than track, but harder to work with because the ends of the legs are turned in as shown in figure 5-3. Studs also have a lot of holes punched in them to accommodate wiring since they are used in the construction of buildings. These holes limit their usefulness in collector construction since they can't be used for the perimeter of a collector or anywhere the airflow must be isolated. In most installations it is better to use metal track rather than metal studs.

Metal Track Summary

• Metal track is the best material for the framework for active collectors.

• Metal studs can be used for the supporting framework, but the holes in them limit their suitability for use in collectors.

Caulking

Proper sealing makes the difference between a collector that works and one that works well. Owner-built collectors can be just as efficient as commercial ones if care is taken in designing them and in caulking all seams and cracks to make sure the collector is completely airtight.

Although there are several sealing materials that are suitable for collector construction, we recommend *silicone caulk* for most applications. Silicone is a high-temperature sealant that comes in clear, black, white and silver colors. Black silicone is the right choice for sealing the collector backing and the absorber plate and for installing tempered glass glazing. Clear is useful for attaching and sealing the metal trim to the collector. Acrylic, Butyl or latex caulks are completely unsuitable for use in active

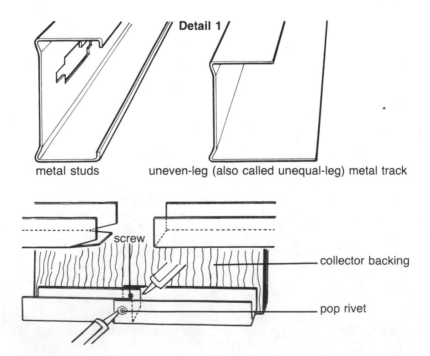

metal studs   uneven-leg (also called unequal-leg) metal track

screw

collector backing

pop rivet

collectors. Also, avoid elastomeric sealants and ethylene copolymer caulk because they can't handle the high temperatures present in these collectors.

Since you'll be using a lot of silicone (about one tube per 4 square feet of collector), it is important to buy it at a good price. Solar suppliers are usually good outlets for silicone,

and if you buy it by the case, you should be able to save around 30 percent.

Purchase a high-temperature silicone that is rated for use above 400°F, and don't buy silicone Sill Sealer, which is good to only 250°F. General Electric's SCS1200 and DOW CORNING 100% Silicone Rubber General Purpose Sealant are two good brands.

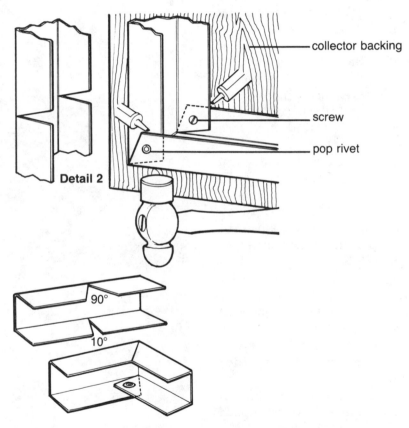

Figure 5-3: *Above and on opposite page.* Metal track (detail 1) is a better choice than metal studs for collector construction because it has flat legs and no holes in it. Overlap joints in track are made by snipping the corners of one piece and inserting it inside the other piece. Seal the joints with silicone, screw the back side of the overlap to the collector backing, and pop rivet the front overlap together. Right angle corners (detail 2) are made by notching the legs of the track and bending it. Lightly tap on the corners with a hammer to flatten bulges. Mitered corners are seldom needed but can be made by cutting out a 90-degree notch in the top leg.

A silicone seal should be used along the entire seam wherever you are joining two sheets of material. Special care should be taken to seal the perimeter of the collector (inside and out) and to seal between the different layers in your collector. Collector efficiencies drop rapidly if air from the dead air space can enter the airflow cavity, and it is disastrous if outside air enters the collector at any point. Try to make your collector ant proof: If ants can't get through, air can't either. If you ever wonder if you should caulk a seam, you probably should. Don't skimp on silicone.

No matter how much silicone you use, you probably still haven't used enough. Don't glob it everywhere, but fill and point all the cracks. At every collector workshop I have attended, several people have commented on the silicone frenzy going on, but there is no excuse for building an improperly sealed collector.

Silicone has its drawbacks, the main one being the speed with which it "skins over" and loses its surface stickiness, which takes five to ten minutes. When you are assembling the collector, make sure the parts fit before running a bead of caulk, and then work rapidly to attach the parts before the silicone starts to set. If the silicone skins over before the collector is assembled, scrape it off and run a new bead to ensure that the silicone sticks to both surfaces and that you have obtained a good seal. The only time this is a genuine problem is in assembling the absorber plates and installing the glazing. When you do these steps, plan ahead, have plenty of help and move rapidly. Most do-it-yourselfers are used to working at their own speed, but when using silicone to seal seams you need to work at the speed of silicone. A small timer is helpful in keeping track of how long the beads have been exposed to the air. Silicone cures by

taking on moisture from the air so it will tend to skin over more rapidly in humid environments. When rapid curing is a problem, the silicone must be applied to the

Photo 5-2: Silicone caulk is messy stuff, and it's hard to be neat when using it. Most collector builders eventually resort to wiping messy fingers on their pants after pointing beads of silicone. Here the author models his "twenty collector" silicone pants while caulking a glass seam. Silicone pants are very "in" among solar folks. Sunglasses are also quite fashionable as well as helpful during construction. The hat is optional, not fashionable.

materials to be joined in short 2-foot beads, which can be done with the procedures described in chapter 8.

Silicone is messy. Keep a rag handy for cleanups, release the pressure on your caulking gun after each use and wear old clothes when using it. Most collector builders like to finish the job by running a finger down the bead to fill up cracks and ensure a neat, smooth appearance. This is a messy but necessary step because an unsmeared bead of silicone doesn't make nearly as tight a seal. Silicone isn't toxic but the acetic acid it releases as it cures can be irritating to the

eyes and lips when it is used in an enclosed, unventilated area, and it supposedly can be slightly irritating to the skin.

To seal a partially used tube of caulk for later use, insert a wide nail or screw into the top of the tube and wipe a large glob of caulk around the tip. Depending upon how tightly the nail fits, this will keep the caulk in the nozzle from curing for between two weeks and several months.

Some collector builders use *polyurethane caulk* for parts that must be overlapped in active collectors. This durable caulk cures more slowly than silicone, but it doesn't have nearly as high a service temperature range. It is liable to break down or outgas at stagnation temperatures, especially in collectors with selective absorber plates. (Polyurethane doesn't pass the overnight bake test at 350°F.) Silicone doesn't adhere as well to porous materials such as wood as it does to metal or glass, while polyurethane does stick better to wood and is the best choice for sealing absorbers and glass to a wooden framework. Good-quality *Butyl caulk* can be used for sealing wood seams in convective air heaters if it is covered with insulation, but polyurethane is a better choice. Sears Best Caulk, available from Sears, Roebuck and Company, is a good-quality polyurethane caulk that is readily available.

Another material worth mentioning here is silicone rubber coating. This paintlike, thinned-down silicone sealant is typically used for a permanent seal on gutters and roof flashing, but unlike paint it remains flexible. In collector construction it is useful for applying a thin, noncorrosive barrier between dissimilar metals such as aluminum absorbers and galvanized metal track. A 1-quart can costs about $6 and is sufficient for even the largest site-built collectors. DOW CORNING

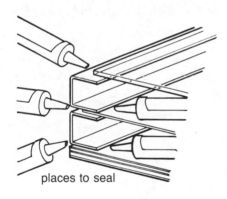

places to seal

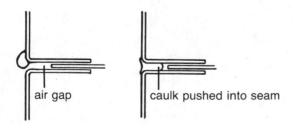

air gap   caulk pushed into seam

Figure 5-4: It is important to "point" beads of silicone to push them deep into seams. Beads that aren't smeared tend to leak air. A wetted finger is actually one of the best tools for this.

100% Silicone Rubber Protective Coating is a good brand.

Caulking Summary

• Use silicone caulk for sealing metal to metal, glass to metal, foil-faced insulation to metal, and metal to primed or painted wood.

• Use polyurethane caulk for sealing wood to wood, glass to wood, metal to wood, glass to metal in low-temperature applications, and metal to metal in low-temperature applications.

• Don't use acrylic, Butyl, latex, elastomeric or ethylene copolymer sealants. Avoid using silicone Sill Sealer in high-temperature applications (above 200°F).

Screws

No matter which materials you chose for building a collector, a lot of screws will be required to attach everything together. Nails of any type are unacceptable and should not be used because they will eventually work loose due to the tremendous heat fluctuations.

Blackened drywall screws are one of the best choices for attachments that won't be exposed to the weather. They are available in many different lengths (¾ inch through 3 inches), and they have "saber" points that make them self-starting (no need for predrilling) in most collector materials.

For the many applications where two sheets of metal are joined to each other, ⁷⁄₁₆-inch, saber-pointed, *binder-head* screws are a good choice. They are not available in most hardware stores but can be located easily at drywall supply stores or by looking under "Fasteners" in the Yellow Pages.

Cadmium-plated, *hex-head* screws are good for joining sections of ductwork and fastening control devices such as blowers and

dampers to the ductwork. They can also be used to attach metal to wood on the back side of a collector.

Stainless-steel screws with pan heads should be used in all applications that are exposed to the weather or where galvanic corrosion could be a problem (see table 5-2).

Collector Insulation

It is important to insulate the back and sides of a collector to at least R-5 to prevent excessive heat loss. Most foam insulations are unacceptable for active collector construction. White beadboard *expanded polystyrene* outgasses and comes apart at about 160°F, and polyurethane outgasses at about 220°F. If either of these materials reaches its upper temperature limit, it not only releases

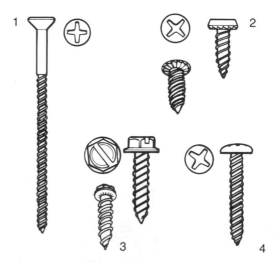

Figure 5-5: Illustrated here are screws commonly used for collector construction: 1) 2-inch #8 bugle-headed drywall screw; 2) ⁷⁄₁₆-inch #8 Parkerized screw with a binder head; 3) ¾-inch #8 cadmium-plated screw with a hex head and a saber point; and 4) 1-inch #8 cadmium-plated, self-tapping, pan-head screw.

some very nasty gases, it loses its insulation value and also shrinks, possibly pulling apart collector seals. There have even been cases where the polyurethane in collectors with selective absorber plates has actually ignited under stagnation temperatures. *Ductboard ducting*, a foil-faced, 1-inch rigid fiberglass material rated to 450°F, is a good choice when insulation is placed inside the collector. It's installed with the shiny side towards the airstream.

For lower-temperature tilted collectors (those without selective surfaces) another approach is to insulate behind the collector with *fiberglass batts*. If the collector is vertically mounted against an insulated building, little or no insulation is required behind it because there isn't much heat loss through the back of the collector. In both cases the back side of the airflow cavity should be covered with aluminum litho plates, screwed to the plywood and sealed with silicone. Install the litho plates with the printing side against the wood. Heavy-duty aluminum foil can also be used for a backing material but isn't recommended because it isn't durable enough. Do not use aluminized Mylar for a backing because it can't withstand collector temperatures.

There is one type of foil-faced insulation, *polyisocyanurate* (Thermax and Rmax are brand names), that can be used with confidence in convective air heaters. This insulation is rated to 250°F, and since convective collectors should never reach temperatures above 180°F, no problem should arise. Some manufacturers of active collectors use polyisocyanurate as an insulation, but we don't use it and don't recommend it.

Whenever you use high-temperature foams inside a collector, use only materials that are foil-faced on both sides, and seal exposed edges and joints with high-temperature aluminum duct tape. When applying

TABLE 5-2

Screws

Application	Length (inches)	Type of Screw	Point	Head	Approximate Cost/100 ($)
Ductwork and devices	½ or ¾	cadmium plated	saber (self-drilling)	hex	3.00 (½″)
Metal to metal	7/16	black (Parkerized)	saber (self-drilling)	Phillips (binder)	2.00
Metal to wood	¾	cadmium plated	self-tapping	Phillips or hex	1.50
Miscellaneous unexposed applications	¾ to 3	drywall (Parkerized)	saber (self-drilling)	Phillips (bugle)	3.00 (1″)
Trim and flashing	½ to 1½	stainless steel	self-tapping	Phillips (binder) or pan	6.50 (1″)

aluminum duct tape, be sure the surfaces are clean and seal all edges of the tape with a smear of silicone. Never use aluminum duct tape on the absorber.

Insulation Summary ───────────

• Ductboard ducting is rated to 450°F and is therefore suitable for all collector applications.

• Polyisocyanurate insulation, rated to 250°F, is acceptable for use in convective collectors.

• Polystyrene and polyurethane insulation should *never* be used in collectors.

Paint and Painting Procedures

Paint for the absorber plate and collector interior should be a high-temperature, flat, black, acrylic enamel rated to at least 500°F. Barbecue-black paint, available in spray cans from your local hardware store, is suitable, but we prefer to use DEM-KOTE, a flat black paint that doesn't need a primer (available from W. W. Grainger). A DEM-KOTE spray can comes with an adjustable valve that lets you change the spray pattern. One can covers about 20 square feet. Painting a large collector with spray cans is a bit time-consuming, so for larger collectors you can use a compressor and spray gun or one of the newer hand-held airless paint sprayers. Never apply paint to the absorber plate with a paint brush because the covering will inevitably be too thick.

Never use a paint or primer that isn't rated for high temperatures, or it will peel off or melt into a puddle at the bottom of the collector. Priming is not necessary if a good-quality paint is used and if the surface to be painted is well prepared. Paint should be applied at near room temperature. In the winter it may be necessary to face the plates toward the sun or to wait until they are installed in the collector before painting. Always follow the directions for the type of paint you are using.

For whatever absorber plate material you choose (except manufactured selective surfaces), it is necessary to wash the plate before painting. Use a detergent solution and a mop to clean off any dust, oil or printing ink that may be on the surface. Rinse well and allow the plate to dry. This is the only treatment that aluminum absorbers require, although it is a good idea to further clean and slightly etch the aluminum by mopping on a strong solution of baking soda and water. Rinse well with clean water.

Galvanized metal is much more difficult to paint than is aluminum. Use Paint-Lock galvanized or a similar product if possible. We have had good luck painting regular galvanized plate for absorbers by giving it a two-step treatment. First, Metalprep is mopped on to lightly etch the plate and give it a rough surface for holding the paint, then Galvaprep is applied to coat the plate. Follow directions on the bottles and rinse well with water after both steps. Absorbers treated this way will hold paint very well but appear slightly streaky after painting, so use textured solar glass for glazing if possible. Before painting the galvanized track used for the glass supports, we clean it with full-strength rubbing alcohol or a mild muriatic acid solution.

Proper application is very critical when using selective paints for the absorber. Good selective properties depend upon having a very thin coat of paint. The performance of this paint drops considerably if it is applied too thickly. Use aluminum rather than steel absorbers for the best bond.

Painting the absorber and track several days before installation is always a good idea. This way the paint can completely cure before you install the glazing so that any outgassing that occurs won't cloud the glazing. This isn't a big problem, and small touch-ups can be done immediately before glazing.

Paint Summary

• Flat, black, acrylic paint (rated to at least 500°F) in spray cans is recommended for painting absorber plates and interior track.

• Selective paint in spray cans can be used on aluminum absorbers.

Flashing

Unless your collector is hidden from view, it is best to have the flashing for it made at a local sheet-metal shop because homemade flashing usually looks *very* homemade. Twenty-eight-gauge galvanized sheet metal is the most economical choice, and although it does look tinny and shiny, it is satisfactory for most installations. Painted or colored flashing is often desirable, but paint won't stick for long to exposed galvanized metal unless you follow the painting directions given previously. A local auto-body shop can bake enamel onto galvanized metal, but it's an expensive undertaking. Some glass or sheet-metal shops offer galvanized sheet metal that is preprimed to accept paint or coated with a dark PVC (polyvinyl chloride) or fluoropolymer compound. Flat sheets can be cut and formed to your requirements without cracking the coating. Flashing of this type looks very slick yet can cost twice as much as regular sheet-metal flashing.

Avoid wooden flashing and cap strips on any collector installation. Over time, wooden flashing will warp, shrink, fall apart and look ugly no matter how it's treated or secured.

Flashing Summary

• Have the flashing for a collector custom made at a sheet-metal shop.

• Don't use wooden flashing.

Useful Tools

If do-it-yourselfers like materials, they *love* tools. There's always a reason for buying that new tool, especially when economy is superseded by desire. The prospect of a home-built air-heater project can be a perfect excuse for a flurry of tool buying, but if you're only building a system for yourself, you don't really need to buy a lot of specialized tools. The only truly indispensable tools are aviation shears (tinsnips) and an electric screwdriver. However, if you're going to be building several systems, you will find all the tools listed in this section to be useful and necessary. Most are readily available at local building supply or hardware stores. Some of the specialized sheet-metal tools can be gotten from a sheet-metal shop.

Tinsnips

Collector construction involves a lot of metal cutting, and you will need a good pair of tinsnips. Regular snips are useful for cutting large sheets of metal (as in absorber plates) and can be used to build an entire collector, but the best type for collector building are *aviation snips*. These snips have heavy-duty, nonslip blades that are slightly serrated. They will cut through just about anything and are especially handy for getting into the tight corners you will encounter when working with

Expansion and Corrosion of Collector Materials

We have built many large collectors using metal track and tempered glass, and we have yet to see silicone seals pull apart due to expansion and contraction. Absorber plates (especially aluminum) do bow in and out during the course of a day, but this isn't a problem as long as they are securely attached with screws.

The worst expansion problem we encountered was in the installation of flat FRP (fiberglass) on a large collector while the absorber was hot. At night the FRP contracted and tried to pull away from the wooden framework, which is a clear recommendation for installing this material when it's not so hot.

If you are concerned about expansion in a large collector, consider sealing the joints in the metal track with silicone, but don't screw down the overlapping pieces. We have tried using slip washers with enlarged holes in the track (covered with silicone) when attaching metal collector frameworks directly to masonry, but as nearly as we could tell, they weren't needed.

We have never seen corrosion take place inside solar air collectors, but the possibility may exist, at least in coastal areas where there is salt spray in the air. *Galvanic corrosion* is an *electrolytic* process that begins with two different metals that are in contact with each other. Also present is moisture that contains some mineral that can combine with one of the metals. The metals exchange electrons via the moisture "bridge," and one of the metals corrodes.

In collectors, there are often dissimilar metals touching one another, but any water that is present is usually condensation, which is really a distilled water and which doesn't contain minerals that would contribute to galvanic corrosion. But just in case, we recommend that all surfaces of dissimilar metals be separated by coatings of paint and/or silicone.

Below is a list of metals in order of *nobility*, the common term for resistance to corrosion. For any pair of metals on the list, the one appearing lower on the list will be the one that corrodes.

For example, copper absorber plates could corrode aluminum plates or cadmium-plated screws if the conditions inside the collector were just right, but this is unlikely in air-medium collectors. Galvanic corrosion is much more of a problem in liquid-medium collectors.

Metals in Order of Nobility

gold	2024-T3 aluminum alloy
chromium	7075-T6 aluminum alloy
nickel	cadmium
silver	clad 2024 aluminum alloy
stainless steel	pure aluminum (1100)
copper	clad 7075 aluminum alloy
tin	zinc
lead	magnesium
mild steel	

metal track. Aviation snips are available with three different blade configurations for making straight cuts, left-handed circular cuts and right-handed circular cuts. For the most convenience buy all three types, but know that you can get by quite well with a pair of straight cutters. At about $12 per pair, aviation snips can save you a lot of frustration and make working with metal much more pleasant.

Screwdrivers

As we've said previously, a lot of screws go into the making of a collector, too many to put in with a regular screwdriver. An electric or battery-driven screwdriver or a ratcheted, push-type screwdriver are must items. You will always be glad you have one of these because they are so handy for a variety of projects.

A variable-speed electric drill with a reverse gear is completely suitable for collector construction. If your drill isn't variable speed, an attachment that decreases the drill's speed while increasing its torque can be purchased for about $15. Attachments of this type also allow an easy switch over to reversible operation, so an inexpensive nonreversing drill can be made into a more versatile tool.

Cordless, rechargeable drills are also good for driving screws. The better and more expensive ones (about $120) have interchangeable battery packs, while lower-cost models have an internal battery that takes the drill out of service when it needs recharging. Plug-in drills that hold a charge for about an hour retail for about $50. An eight-hour drill costs about $70. Be sure to get one that is reversible and that has a high and low speed. Tool rental

and drywall supply stores will rent drills for about $4 a day, and that is a good way to get acquainted with different models before you make a purchase.

When buying screwdriver insert bits for your drill, spend a little extra money to get the best ones. Ask for magnetic bit holders, which magnetize the bits and make it much easier to position screws in awkward places. They cost about $9.

Pop Riveter

Pop rivets are expensive and time-consuming to install, but it is difficult to build a collector without using a few of them. Don't spend a lot of money on a deluxe pop riveter. Look for one with long handles to get better leverage and to avoid painful pinches as the handles snap together. An adequate pop riv-

Photo 5-3: The only tools you really need to build an all-metal collector are an electric drill and a pair of aviation snips. Some tools that will make the work easier are, *clockwise from the left*, a battery-powered drill with an interchangeable battery pack, a push-type screwdriver with a homemade magnetized bit insert, an electric drill, right- and left-cutting snips and standard tinsnips.

eter will cost about $10. In most applications ⅛-inch, medium-sized pop rivets are useful. Steel and aluminum rivets of this size cost about 2¢ each (boxes of 500). Stainless steel pop rivets cost about 4¢ each.

One-Handed Caulking Gun

A caulking gun loaded with silicone will be your constant companion while building your collector so you should buy a good one. Look for a heavy-duty gun with a pressure-release tab that can be operated by the same hand that is operating the gun. This will make caulking much easier and neater when you are leaning out over the collector and trying to run a bead while holding on with the other hand.

Bending Tools

There is no substitute for the large sheet-metal brakes that are used at sheet-metal shops for bending flashing and other parts of the system, but hand brakes are very handy for the bending jobs that come up during collector construction. Plier-type or Vise-Grip-type brakes are fairly inexpensive tools ($10) that can be used for small jobs. The flat bar-type folding tool ($6) is used in standard sheet-metal practice for bending attachment tabs on rectangular ductwork. It is also useful in collector construction for bending hooks on the edges of absorber plates.

Reciprocating Saw

Everybody who owns a good reciprocating saw is in love with it, and there isn't a sheet-metal worker who would work without one. When it comes time to cut holes for your ductwork, borrow or rent one for the afternoon. Plunge cuts are no problem, and wood/metal combination blades are available for cutting through flooring that has nails in it. Blades are expensive (up to $3 each),

but they usually only break if you are careless. If you are considering buying one, plan to spend about $150 or you will end up with a toy, not a tool.

Locking Pliers

Locking pliers such as Vise-Grips are very handy if your collector design involves a lot of drilling and pop riveting. These pliers will hold pieces tightly together to allow for correct positioning and snug fitting of metal track and absorber plates. They come in many different styles and sizes.

Duct Crimpers

A brief attempt at crimping ductwork with needle-nose pliers will convince you there must be an easier way to make male ends. There is! The right tool is a duct crimper. This handy, $12 item puts three neat crimps into ductwork with each squeeze of the handles. If you are running a lot of ductwork for your system, buy a pair. Otherwise, persevere with your needle-nose pliers since the only other use I can see for a duct crimper is in making Christmas tree decorations from tin cans.

Insulation Stapler

An insulation stapler can save a lot of wrestling with duct insulation in a crawl space. This one-purpose tool drives the ends of the staples outward, securely attaching overlaps in foil-faced duct insulation. This will save you the trouble of "stitching" and wiring insulation. Ask to see one at your local sheet-metal shop.

Sources for Supplies

Mail-Order Catalogs

There are two solar supply catalogs that are worth having on hand and are the best

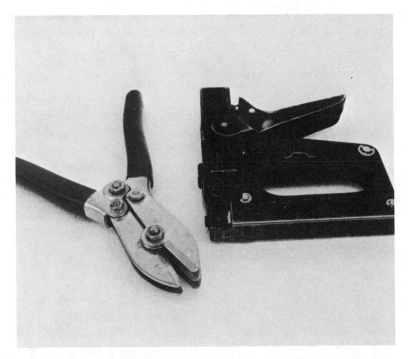

Photo 5-4: Two ductworking tools that aren't essential, but that you will find very handy, are duct crimpers, *left*, for making male ends on round ductwork and a special duct insulation stapler, *right*, that curls the staples outward as it shoots them.

sources for a few specific materials. You can often beat their prices, but you can't beat their availability. These catalogs are:

People's Solar Sourcebook
Solar Usage Now, Inc.
P.O. Box 306
420 E. Tiffin St.
Bascom, OH 44809
Phone: (800) 537-0985

Solar Components Catalog
Solar Components Corp.
P.O. Box 237
Manchester, NH 03105
Phone: (603) 668-8186

The first catalog has 354 pages and sells for $5 and the second one has 80 pages and sells for $3.

Ordering from *Grainger's*

W. W. Grainger, Inc., offers a lot of useful materials for the do-it-yourselfer in *Grainger's Wholesale Net Price Motorbook*. The company has a national distribution network, but it doesn't sell retail. You must place your order through your local solar supply dealer, sheet-metal shop or hardware store. To locate your local W. W. Grainger office, contact the following:

W. W. Grainger, Inc.
General Offices and Central Distribution
  Center
5959 W. Howard St.
Chicago, IL 60648
Phone: (312) 647-8900

Solar Magazines

Solar magazines have ads for numerous products and should not be overlooked as a source for locating supplies.

Solar Age
Solar Vision, Inc.
Church Hill
Harrisville, NH 03450
Phone: (603) 827-3347

Solar Engineering & Contracting
P.O. Box 3600
Troy, MI 48099
Phone: (313) 362-3700

Solar Heating & Cooling
Gordon Publications, Inc.
20 Community Place
CN 1952
Morristown, NJ 07960-1952
Phone: (201) 267-6040

Sun Up Energy News Digest
P.O. Drawer S
Yucca Valley, CA 92284
Phone: (714) 365-0604

Absorber-Plate Materials
Aluminum Plates

To locate mobile-home siding, look in the Yellow Pages under "Aluminum," "Aluminum Products" or "Mobile Home Manufacturers."

Galvanized Metal

Paint Lock is available at sheet-metal shops.

Litho Plates

Small local newspapers and large printers are good sources for litho plates. You can often get a very good price if you buy a large amount, but many businesses sell small quantities at reasonable prices. Look under "Printing" and "Printing Plates" in the Yellow Pages. Ask for sizes and thicknesses as well as prices. Litho plates are often sold by the pound. Try to get them for less than 45¢ per pound.

Caulk
Silicone and Polyurethane

Look under "Caulking" or "Plastics" in the Yellow Pages, and find a dealer who sells at retail. Ask for the price of a case (24 tubes). You may be able to save 50 percent over your local hardware store. All solar suppliers carry high-temperature silicone. Ask for DOW CORNING 100% Silicone Rubber General Purpose Sealant, DOW CORNING 100% Silicone Rubber Protective Coating or General Electric SCS1200.

Cadillac Plastic and Chemical Company (distributor) is a good source for cases of silicone. Contact their home office for a distributor near you, or look for their company under "Plastics" in the Yellow Pages.

Cadillac Plastic and Chemical Co.
1221 Bowers St.
Birmingham, MI 48012
Phone: (313) 646-5100

Polyurethane caulk is available from Sears, Roebuck and Company. Ask for Sears Best Caulk, catalog no. 30AY38211.

Flashing

Look in the Yellow Pages under "Glazing" or "Sheet Metal" to find a shop that offers preprimed or coated galvanized flashing that can be bent without cracking the coating.

Glazing

Fiberglass-Reinforced Plastic (FRP)

Good quality FRP is available at most large greenhouse supply stores. If you buy FRP from a local lumber store, make sure it is of top quality. Don't use cheap FRP. A good source for larger orders of flat FRP (Sun-Lite) is Solar Components Corporation (see previous section on mail-order catalogs in this chapter).

Solar Films

Look under "Plastics" in the Yellow Pages. Shop around because prices vary greatly with the source of supply and the amount ordered. Clear Mylar is readily available in any large metropolitan area.

Solar Glass

Look under "Solar Energy Equipment" and "Glass" in the Yellow Pages. You won't find "seconds" on solar glass.

Tempered Glass

Check the want ads in the newspaper for patio doors which have been replaced with windows. Auctions and glass companies also often sell these. Large glass companies sometimes have sliding glass door replacement "seconds." These panes usually have only a scratch or two on the surface, cost about half as much as regular panes and are completely suitable for collector construction. Some small, local glass shops mark up tempered glass 100 percent over their wholesale price, so be sure to shop around.

Insulation

Ductboard Ducting

The local sheet-metal shop or heating contractor has ductboard in stock or can order it for you.

Polyisocyanurate Foam Insulation

Foil-faced polyisocyanurate insulation is available at most lumber yards. Rmax and Thermax are common brand names. Be sure you are getting polyisocyanurate and not polyurethane. They look the same.

Metal Lath

Metal lath is available at any large lumber store. Ask for diamond mesh with 1/4-by-1/2-inch holes.

Paint

W. W. Grainger's (see address previously listed) DEM-KOTE (order no. 2X717) is a good-quality, flat, black, acrylic enamel. Rust-Oleum, barbecue black and other high-temperature acrylic enamels readily available in hardware stores are also suitable.

Screws

Large lumber and hardware stores carry boxes (from 50 to 1,000 screws per box) of drywall screws with Phillips heads in several sizes. If they don't have the size you need, they can order it for you. Your local sheet-metal shop or heating contractor is the best source for hex-headed, cadmium-plated, self-drilling screws. W. W. Grainger, Inc. (see address previously listed) offers a wide assortment of drywall and hex-headed screws. Stainless steel and 7/16-inch binder-head screws are more available at drywall supply houses, or look under "Fasteners, Industrial" in the Yellow Pages.

Selective Paint

Thurmalox 250, a selective black paint, is available from the following company (order the 12-ounce can):

Dampney Co.
85 Paris St.
Everett, MA 02149
Phone: (617) 389-2805

The solar catalogs previously listed in this chapter also carry selective paints.

The following company manufactures selective coatings:

Olympic Solar Corp.
208 15th St. SW
Canton, OH 44707
Phone: (216) 452-8856

Selective Surfaces

These can be tough to locate. A local solar supplier or collector manufacturer is your best bet. Look for ads in local solar newsletters that mention selective surface. You will need to place a large order when purchasing from the manufacturer, and many companies are swamped with orders so service is slow. The following company offers a variety of selective materials that have been used with good results:

Berry Solar Products
2850 Woodbridge Ave.
Edison, NJ 08837
Phone: (201) 549-0700

Solar Components Corporation (see the previous section on mail-order catalogs) handles sales of Berry Solar Products under 100 square feet.

Self-adhesive foil is available from the following company:

Ergenics
681 Lawlins Rd.
Wyckoff, NJ 07481
Phone: (201) 891-9103

Solar Usage Now, Inc. (see the previous section on mail-order catalogs) offers self-adhesive selective foils. In the next few years selective materials should be more locally available.

Studs and Track

Look in the Yellow Pages under "Drywall Supply" to locate a fabricator of metal studs and track. Local drywall contractors can also help you locate metal track. The following company manufactures 2-inch uneven-leg track in Colorado, Phoenix and Houston and can ship to anywhere in the United States:

Metal Stud Forming Corp.
3800 E. Sixty-eighth Ave.
P.O. Box 450
Commerce City, CO 80037
Phone: (800) 525-1962

6

THERMOSTATS, BLOWERS AND DAMPERS FOR SIMPLE ACTIVE SYSTEMS

All of the control systems discussed in this book are designed to be as simple, reliable and inexpensive as possible while still performing their necessary functions. The primary function of any control system is to deliver a sufficient airflow when the collector is producing heat. This is insured by choosing the right thermostat and blower and by arranging the ductwork and/or the dampers to deliver heat to the point or points of use and to prevent reverse thermosiphoning from the collector at night. These are the essential system controls.

This chapter discusses the basic electrical controls common to all active solar air systems. More elaborate systems with several modes of operation or storage will be presented in chapter 11.

Controls for active systems with one point of use (or mode) are very simple, but for systems with more than one mode they can be quite complex. It's a good idea for you to draw an airflow schematic of your system for each mode, with arrows indicating flow paths. This makes it easier to see where dampers must be placed in your system. In your planning keep in mind the need to stop nighttime reverse thermosiphoning, and look for an opportunity to use a *heat trap* (see figure 6-1) to keep cold air from settling down either the hot or cold ductwork from the collector.

Thermostats

Thermostats are a vital part of all active solar systems. They sense temperature and send messages to the blowers and dampers to control their operation. In selecting one, you need to know the temperature range and differential at which it operates and whether it is a heating or a cooling thermostat, or both.

Solar heating systems use a thermostat that cools the collector. As the temperature in a collector rises, the contacts in the thermostat close and complete the electrical circuit to the blower. The temperature at which this happens is called the *set-point*. The blower cools the collector down to another temperature, the *cut-off point*, and the contacts in the thermostat open once again, turning off the blower. The difference between these two points is called the *differential* of the thermostat. Thermostats are manufactured with a variety of set-points and differentials for various uses. For use in a collector, the differential is usually fairly large. A collector with a properly sized blower cools down rapidly, so a thermostat with too small a differential would make the blower cycle on and off too often. Most collector thermostats have an adjustable set-point and a temperature differential of 10 to 15 degrees, which is about right for efficient operation without a lot of cycling. For example, a direct-use system can be set to turn on at 105°F and turn off at 95°F.

There are four types of thermostats that the do-it-yourselfer should be acquainted with. *Remote bulb thermostats* are used primarily for simple direct-use systems and in two-mode space-heating systems. *Differential thermostats* are used to control the collector for heating domestic water or thermal mass storage and in simple crawl-space heating systems. *Two-stage thermostats* are used to control air delivery to the house from the collector or

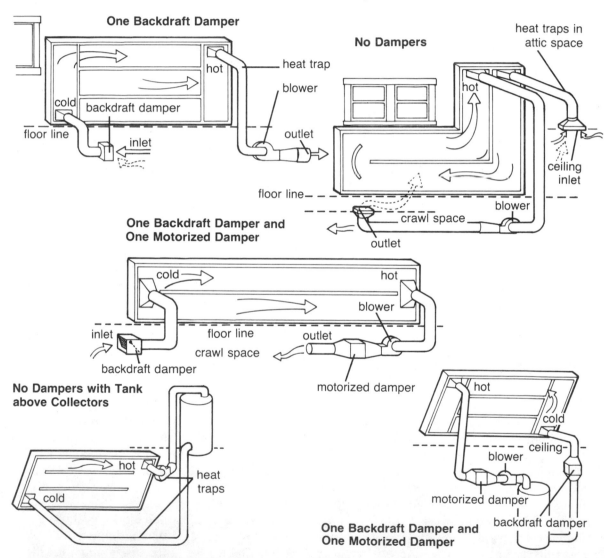

Figure 6-1: Preventing nighttime convective heat transfer from a cold collector is an important design consideration. This illustration shows damper placement to prevent this in three space heating and two water heating systems. Fewer dampers are required in systems where a heat trap is created in the ductwork.

storage. *Snap disc thermostats* can be used in simple, low-cost systems with small collector areas (less than 80 square feet).

Remote Bulb Thermostats

The remote bulb thermostat is commonly used in direct-use space-heating systems. It has a sensing bulb that is separate from the thermostat, hence its name. This bulb is connected to the thermostat by a very thin copper tube (about 3 to 6 feet long), and both the bulb and tube are filled with Freon. The Freon expands when the bulb is heated and pushes on a diaphragm inside the thermostat. The diaphragm pushes a switching mechanism, closing an electrical contact to complete a circuit and turn on the blower. As the Freon in the bulb cools and contracts, the diaphragm moves back, the switching mechanism opens, and the circuit is broken, turning off the blower.

The remote bulb is installed in the collector airflow cavity in a hot place about a foot from the outlet (hot) port to minimize blower cycling. There should, however, be some cycling in the morning and afternoon, when there isn't much solar radiation striking the collector.

The body of the thermostat is mounted by removing its front cover, revealing holes in the back mounting plate. Then it is screwed to a convenient location near the collector. You have to be careful not to kink or bend the fine copper tubing, or the Freon won't be able to move freely and the thermostat won't work. Don't mount the thermostat so that the tube is exposed to the weather.

When this thermostat is used to control a collector, power failure can expose the bulb to very high temperatures. There is one remote bulb unit that we highly recommend for collector use because of its wide set point range and high temperature tolerance: Honeywell no. L6008C1026. Another typical solar application for a remote bulb is sensing the temperature in a rock storage bin. For this use, several suitable thermostats are listed at the end of this chapter. A remote bulb thermostat costs from $30 to $50 retail. Almost all have adjustable set-points and differentials.

Differential Thermostats

Differential thermostats (DTs) operate according to a temperature difference. For example, let's look at a thermostat with a "cut-in" differential of 18°F and a "cut-off" differential of 8°F. It's installed in a collector crawl-space system. This thermostat has two temperature sensors: one mounted in the collector and one in the crawl space near the air return. Whenever the collector becomes 18 degrees warmer than the crawl space, the thermostat activates the blower. If the crawl space becomes heated to within 8 degrees of the collector temperature, or if the collector cools to within 8 degrees of the crawl space, the blower is turned off.

DTs are available with different differential settings, and some are user-adjustable. For air-heating systems, a thermostat with an 18-degree cut-in and an 8-degree cut-off differential is a good choice. Simple DTs retail for $50 to $70 and come with simple wiring directions.

A DT is a more precise control than a remote bulb thermostat. For example, in a crawl-space system, the DT-controlled blower comes on earlier in the morning, when the collector is warmer than the crawl space but not as warm as the 90°F set-point on a remote bulb unit. The blower never moves air into the crawl space that is cooler than the crawl space itself, which can sometimes happen with a remote bulb late in the afternoon. Domestic water-heating systems and rock storage systems also require DTs, as we will see

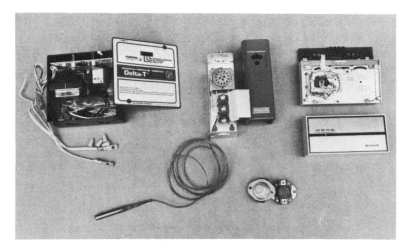

Photo 6-1: Thermostatic controls for active systems include, *left to right*: a differential thermostat with two sensing probes and Wire Nuts; a remote bulb thermostat; a two-stage house thermostat. Below the remote bulb thermostat are two snap disc thermostats.

later. Two-mode space-heating systems, on the other hand, require a remote bulb control because air is sometimes blown directly into the house, and this air must be at least 90 to 100°F (see the two-mode section in chapter 2).

The two sensors, or *thermistors*, for a DT are connected to the thermostat by thin, low-voltage wires (18- to 24-gauge). These sensors should have copper shields to prevent them from being heated or cooled too rapidly, causing unwanted cycling of the blower. Since long wires can be run from the thermostat to the sensors, DTs don't have to be mounted near the collector.

Two-Stage Thermostats

A two-stage thermostat interfaces a solar system with a back-up (existing) heating system. The first stage controls the solar heat, and the second controls the auxiliary or supplemental heating. In most two-stage thermostats the temperature difference between stages is about 3 degrees. If a house cools below the set-point, the contacts of the solar stage close. If solar heat is available, a blower is turned on to move warm air into the house. If no solar heat is available, the house cools another 3 degrees, and the second stage turns on the back-up heat. Thus the solar system gets a chance to heat the house before the back-up system is activated.

The unit is mounted in the living area in a location comparable to that of a thermostat for any central heating system. It should be 60 inches from the floor and near the place where air returns to the collector from the living space. Often a main hallway is a good location. Wiring a two-stage thermostat is not difficult but is a bit more involved than simpler types (see Appendix 2 for instructions).

If you already have a one-stage house thermostat in place, another one-stage thermostat can be mounted next to it, and they can work together as a two-stage control group. The solar thermostat is set 3 degrees higher than the existing thermostat. This is less expensive than replacing the existing one-stage with a two-stage, but the functioning of the two thermostats can be disrupted if the setpoint of the back-up thermostat is inadvertently set above that of the solar thermostat.

Snap Disc Thermostats

These simple, low-cost ($9) thermostats are used to control the collector blower. They operate by the "snap" action of a bimetal disc that responds to temperature. Their simplicity and economy does, however, include some drawbacks. They are difficult to wire in conformity with electrical codes, have too high a temperature differential between the setpoint and the cut-off point and are not as reliable as other types of thermostats. They can, however, be used in low-cost installations if they are mounted where they can be reached for servicing.

Manual On/Off Switches

It is always a good idea to install a switch to be able to turn off power to the solar system and override all other controls. Some DTs come equipped with this switch, and if the thermostat can be mounted in or near the living area, it becomes very easy to turn off the system.

Blowers

The best collector blowers are the squirrel cage type. Smaller models usually have a single inlet and a single outlet and are directly driven (instead of belt-driven) by a shaded-pole or a permanent-split-capacitor (PSC) motor. Shaded-pole blowers are less expensive and are available in more sizes, while the more efficient PSC blowers use less electrical current to deliver the same airflow and therefore are a better choice.

The following are 1983 prices for blowers that will deliver 800 cfm: a shaded-pole blower from $75 to $85; a permanent-split-capacitor blower from $85 to $95; a high-volume blower from $100 to $110.

Systems with large collectors generally use belt-driven blowers. They move a large volume of air efficiently, operate at lower rpm's and are quieter than smaller, direct-drive blowers. Belt-driven blowers offer the option of changing or adjusting pulleys to change the rate of air delivery to obtain optimal collector performance, but they are only available in fairly large sizes (above 600 cfm, which would be used with collectors larger than 150 square feet).

Whatever type of blower is chosen, make sure that the motor is mounted outside, rather than inside, the airstream to keep it from overheating, which reduces its service life. Along the same line, don't use the blower in an existing forced-air furnace to move solar air. It is undoubtedly the wrong size, and its motor will burn up in short order if it is fed with solar-heated air.

Various kinds of blades are available for squirrel cage blowers. Most blowers for small solar installations have blades that curve forward. These blowers are the quietest and can handle static pressures of up to 1 inch. *Forward-curve blowers*, along with other types that are more suitable for larger systems or for systems with high static pressure, are discussed in chapter 9.

Because of friction, air encounters resistance to its flow as it moves through the collector and the ductwork, so it is important to choose a blower that delivers the desired amount of air in spite of this resistance. All blowers are rated to deliver a certain volume of air over time (cubic feet per minute, or *cfm*) at a specific resistance to flow (static

Photo 6-2: Blowers come in a wide range of sizes and models. *Left to right* they are: a 265 cfm direct-drive, shaded-pole blower for a small, direct-use space-heating system; a 1,690 cfm belt-drive blower for a large system with rock storage; and a 595 cfm high-volume, direct-drive blower for a system with high static pressure.

pressure, measured in inches of water gauge, WG). A typical squirrel cage blower for delivering 300 cfm in "free air" (no static pressure) will deliver two-thirds as much if the static pressure is 0.5 inch WG. Most collectors and their accompanying ductwork will present about 0.4 to 0.8 inch WG, but this figure can vary greatly depending upon the system. High-volume blowers designed to move air through high resistance may be necessary if the static pressure in your system is above these figures.

Some systems benefit from having two-speed or variable-speed blowers that allow for changing the air delivery according to collector heat output or according to desired output temperatures. The speed of some blowers can be manually regulated using a standard light-dimmer switch, and there are differential thermostats that can raise and lower the blower speed automatically depending on temperature conditions. These controls, however, are very expensive and not justified in most installations. Calculating the static pressure in a system, choosing the best blower for a specific installation and regulating its air delivery are covered in chapter 9.

Dampers

All but the simplest of systems require at least one damper to prevent heat loss at night. Collectors lose heat to the night sky nearly as readily as they gain it during the day, and an absorber plate can be close to 0°F on a 20°F night. At night cold air can settle down through ductwork from the collector and be replaced by warm air. Any nighttime exchange of air between the house and the collector will considerably reduce your system's overall performance.

Backdraft (one-way) dampers are a simple, mechanical, nonmotorized means of preventing this reverse thermosiphoning because they allow airflow in only one direction. Nearly every solar system needs a backdraft damper on the inlet ductwork to the collector, and many systems need others. For example, if the solar system is tied into the furnace ductwork, another backdraft damper will be needed at that junction.

These dampers are fairly simple to build (see chapter 9 for building instructions) and/ or inexpensive to buy. Most consist of a lightweight flap mounted over a supporting screen inside the ductwork. This flap closes and restricts airflow whenever a blast of air from the blower doesn't hold it open. Heavy, cold air keeps the damper well sealed at night. Delivery systems can usually be designed so that only one backdraft damper is needed. Figure 6-1 (earlier in this chapter) shows several different system configurations and the dampers needed to control them.

If your ductwork is arranged so that a backdraft damper can't be used—if, for example, hot air is taken from the bottom of a collector and nighttime air can take the same path as solar air—you need to install a thermostatically controlled, motorized damper in the line. Motorized dampers, especially those with compression seals, seal much more tightly than backdraft dampers and must be used in any application where the nighttime airflow must be completely restricted. A system with an in-duct air-to-water heat exchange coil would need a motorized damper. Even a tiny trickle of very cold air could freeze the water in this exchanger at night and burst

the pipes. Motorized dampers are also used to control airflows in systems with storage or more than one mode.

These dampers use a small electric motor to move a metal vane or set of vanes inside a sheet-metal box. Some types "motor" in both directions (open and closed) while others motor to a closed position and are opened by a spring. Most motorized dampers either allow or restrict airflow through a single duct, but some dampers have their vanes arranged so that air can be directed to, or pulled from, two different duct lines. These *air diverters* operate like a three-way valve in a water line. There are three openings in the damper. Air flows in one of two possible directions while the third opening is completely sealed. In systems with more than one mode, air diverters simplify controls and keep costs down because one diverter can often perform the same function as two regular dampers.

No matter which type of damper you need for your system, remember that those with compression seals usually have less leakage than those that use wiper-type seals. Tightly sealing motorized dampers are quite difficult to build, so plans are not included in this book. Be sure to buy a low-leakage damper designed for solar installation. Most commercially manufactured dampers are used for restaurant exhaust vents. These are very leaky and almost worthless in a solar system. Dampers suitable for solar application specify the leakage as cfm (airflow) at a given pressure (in inches of water gauge). The tight-sealing dampers used in the heating and cooling trades specify their leakage as a percentage and should be avoided for solar system use (see chapter 9 for a simple test for leakage). A good-quality motorized damper approximately 14 inches by 14 inches should cost between $150 and $300, the same size back-

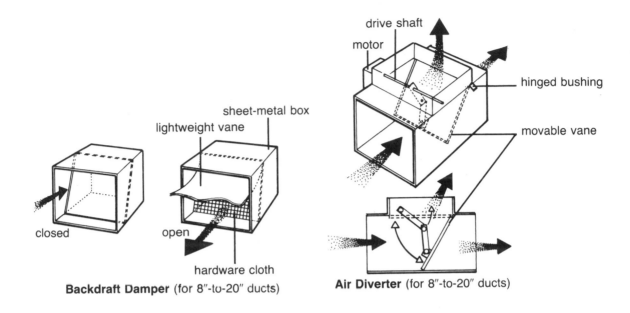

Backdraft Damper (for 8"-to-20" ducts)

Air Diverter (for 8"-to-20" ducts)

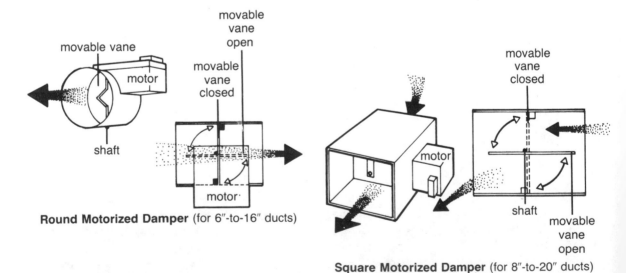

Round Motorized Damper (for 6"-to-16" ducts)

Square Motorized Damper (for 8"-to-20" ducts)

Figure 6-2: Most solar air systems require a backdraft damper at some point in the system. These simple, one-way dampers are self-operating and relatively easy to build. Many active systems require a tightly sealing motorized damper to control airflows. More complex systems can often utilize an air diverter, which functions much like two motorized dampers, to direct airflow along either of two routes.

draft damper between $30 and $60, and a 14-by-14-by-14-inch air diverter from $250 to $500.

Summer Venting

Summer venting can be an important consideration for systems with tilted collectors. The materials we recommend for the construction of all-metal collectors should be able to withstand temperatures of 350°F for extended periods, but it is a very good idea to keep all-metal collectors below 250°F to help protect the various seals through 20 or 30 years of use. Preventing wood and fiberglass collectors from overheating is much more important, and they should never be allowed to get hotter than 180°F. Vertically mounted collectors get warm in the summer, but they typically stay below temperatures that would present problems.

There are two options for cooling a collector in the summer: use the heat or vent it from the collector to the outside. Using a collector's output for preheating water is, of course, the most attractive prospect (see chapter 10). A two-speed or variable-speed blower can deliver a slower flow (1 to 2 cfm per square foot of collector) of hot air from a large tilted collector. This output will heat a large quantity of water and keep the collector adequately cool. If your collector is larger than 200 square feet, you can consider increasing its tilt to latitude plus 20 to 30 degrees if other system design factors allow it (collector location, appearance, cost, etc.). This won't significantly reduce winter output but will help keep the collector cool in summer while still providing plenty of heat for hot water needs.

If you can't use the heat your collector produces in summer, it must be vented. Venting by natural convection requires large inlet and outlet vents and a large airflow cavity to move enough air through the collector to keep it cool. Since a well-built active collector has a relatively shallow airflow cavity, there won't be enough natural convection airflow to keep it cool. A blower is needed to have adequate ventilation in the collector, but that is almost silly: running a blower to dump solar heat. Power venting represents poor system design, in our opinion, because of the solar heat that's wasted. It should instead be used to heat domestic water, a feature that would significantly boost the cost-effectiveness of the whole system.

Rather than using the large space-heating blower for venting, install a smaller blower that is controlled by the collector thermostat. As long as this blower pulls outside air through the collector at about 1 to 1½ cfm per square foot of collector (depending upon the altitude), it will provide adequate cooling. If the space-heating blower is a variable-speed or belt-driven unit, adjustments can be made to slow it down for summer use.

Many items on the market are advertised to provide emergency venting of a collector should the power go off or the controls malfunction. However, the reliability of such devices is questionable, in our opinion. The best idea is to build your collector with high-temperature materials so that it can withstand periods of stagnation. If your electrical service is unreliable, consider putting sealable exhaust vents in your ductwork to provide the collector with some natural-convection venting capability, but don't depend upon a setup like this for an extended period.

Wiring Simple Systems

In this section we'll take a look at how to wire a simple system with one point of use, such as a hot water preheat, crawl-space or direct-use system. All wiring is best done by a licensed electrician, but in most states it is legal for a homeowner to do electrical work on his own house. If you have ever installed or taken apart an electrical outlet box, you shouldn't have any problems wiring the blower and thermostat for these basic active systems. Systems with two delivery modes or with rock storage require relays and a control box and are more complicated (see Appendix 2 for information on wiring these systems). If you have any doubts about your

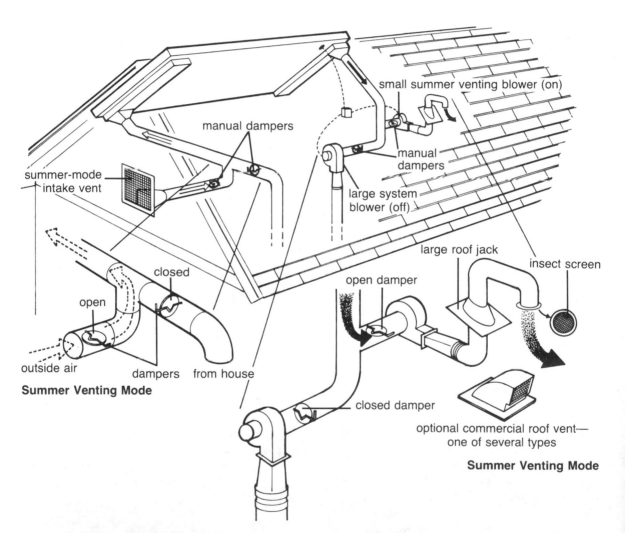

Figure 6-3: Summer venting of a tilted collector is done with a second, smaller blower (¾ to 1 cfm per square foot) that power-ventilates the collector when temperatures reach 180°F. The manual dampers on the ducts that run to the outside must be very tightly sealed or plugged in the winter to eliminate air leaks that would lower the output temperature.

Operating Costs for Active Air Systems

Despite what you may have read in ads for solar heating systems, solar heat isn't really free. It of course costs money to build the system, and with an active system it costs money to operate it. The good news is that blower operating costs don't amount to much—4¢ to 16¢ per year per square foot of collector, depending on your cost per kilowatt hour.

Let's look at the electrical consumption of a blower for a 100-square-foot collector, delivering 300 cfm (at static pressure of 0.5 inch WG). This PSC blower (Grainger's no. 4C666) draws 125 watts in operation, or 1.25 watts per square foot of collector. Assuming a six-hour daily operation for a 180-day heating season, the calculation is:

watts per ft² of collector x hours of use per day x days in year's heating season = watt hrs/yr/ft² of collector

$$1.25 \times 6 \times 180 = 1{,}350 \text{ watt hrs}$$

or

$$1.35 \text{ KWH}$$

At 6¢/KWH the yearly electrical use is approximately 8¢ a year per square foot of collector (1.35 x 6). If the blower is used all year, this figure is doubled. The usage per square foot for larger systems with larger blowers comes out to be about the same figure.

A shaded-pole blower (Grainger's no. 4C444) delivering 300 cfm draws 185 watts, or 1.99 kilowatt hours per square foot per year, so its operating cost is 12¢ per square foot of collector per year for space heating at 6¢ per KWH.

Using Photovoltaics

Many folks have the desire to build a completely independent solar heating system including photovoltaic cells to power blowers and dampers. Since photovoltaics are presently quite expensive, there is no way to justify using them to power an active system if you are currently hooked to the power grid. If you live miles from the power lines and can't install a passive system, give photovoltaics some serious thought and get some expert help.

wiring abilities, get help from someone who knows the basics, or contact a professional electrician.

Materials required to wire simple systems are the following: "bell" wire (18-2 for DTs), 12-gauge cable (Romex 12-2 with ground), conduit (optional), junction boxes with covers, Romex connectors (cable clamps), Wire Nuts and wire staples. Diagonal wire cutters, needle-nose pliers, a slotted screwdriver and wire strippers are the tools needed to do the job. Installation time is three to five hours for a person working alone.

The first and often most difficult step is to locate a live circuit close to your blower location that can handle the extra load of the solar appliances. This line should have a 15-amp (ampere) fuse or circuit breaker and 14-gauge wire, or a 20-amp fuse or breaker with 12-gauge wire. Check the numbers printed on the wire. Avoid tying into circuits that are, or could be, operating major appliances such as the TV, refrigerator or washer. Also avoid tying into a line that is controlled by a wall switch, such as a living-room light. Although a large blower with a ½-hp (horsepower) motor only draws about 8 amps in operation, it can draw three times this much when it first starts up, so use slow-blow fuses. It may be necessary to run a new circuit from the fuse box to the blower, especially if the existing circuit isn't grounded. If you have to do this, *turn off* the entire box before starting work, and use 12-gauge wire and a 20-amp breaker.

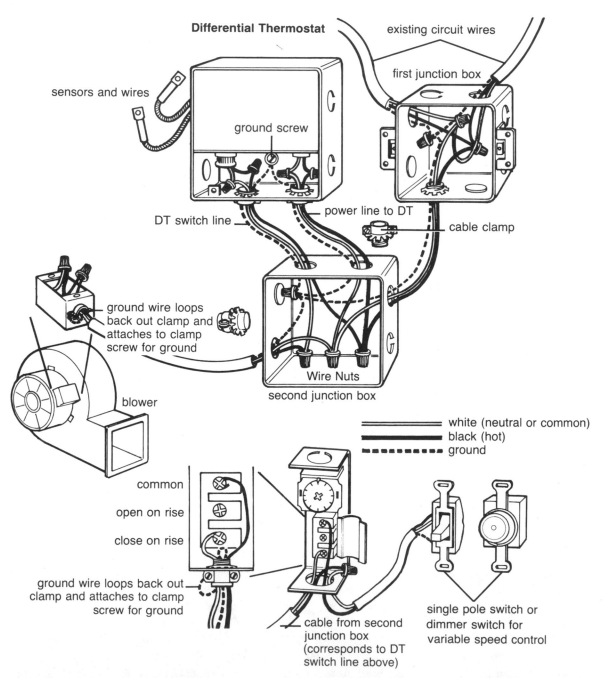

Differential Thermostat

existing circuit wires

first junction box

sensors and wires

ground screw

DT switch line

power line to DT

cable clamp

ground wire loops
back out clamp and
attaches to clamp
screw for ground

Wire Nuts

second junction box

blower

white (neutral or common)
black (hot)
ground

common

open on rise

close on rise

ground wire loops back out
clamp and attaches to clamp
screw for ground

cable from second
junction box
(corresponds to DT
switch line above)

single pole switch or
dimmer switch for
variable speed control

Figure 6-4: When wiring a differential thermostat, two wires must be run from the second junction box to the thermostat. The first junction box can be eliminated if the hot wire you tie into has enough slack in it. Wiring a remote bulb thermostat is easier because only a single line must be run to it—corresponding to the DT switch line above. It is a good idea to include a manual override switch in the line to the remote bulb. Use cable clamp fasteners throughout.

After you locate or create a suitable line, turn off that circuit at the fuse box (plug a test light into the circuit so you will know you have the right one turned off). Cut the cable in half with diagonal wire cutters, and strip back the plastic sheath 9 inches on both wires. Be careful! Don't cut into the covering on the individual wires. Your metal or plastic junction boxes have several knock-out holes punched in them. Locate the most suitable ones for the wires, and knock them in with a screwdriver and twist them out with pliers. Nail or screw the box securely to a wooden framing member and use the cable clamps (or Romex connectors in the knock-outs) if you use metal boxes. Insert the stripped wire through each hole until the outer sheath (which holds all the wires together) is just inside the box. Tighten the cable clamps to secure the wires, and be sure the individual wires extend 6 inches out from the front of the box. Wires of this length are easier to fold into the box than short wires. *All* junctions in your system must be inside a junction box. That's the code. If you have cut in the middle of a line and there isn't enough slack to pull both ends of the wire inside a box, you will need to install another junction box with a short section of wire between the two.

Strip back the covering on the individual wires ½ inch, exposing the copper wire. This is best done with wire strippers but can be done with a pocket knife if you are careful not to score the copper wire with the knife blade. Use pliers to twist together (clockwise) the ends of the wire to be connected, and twist a Wire Nut (clockwise) onto this connection, making sure it covers all exposed copper wire and that it grabs all the wires. Gently tug on each one individually to check. Use green Wire Nuts to connect the ground wires, or fasten them to the box if you are using metal junction boxes.

After connecting all the wires, fold them back into the box (ground wires first) and screw on the cover. When pulling wire from one box or device to another, be sure to leave yourself enough slack wire. Exposed cable must be supported every 4 feet and within 8 inches of all junction boxes with wire staples. Drive these in until they are snug, but don't excessively mash the wire. If your wire is exposed, that is, outside the crawl space, attic or framed walls, the electrical code requires that it be enclosed in metal conduit. Be sure that your wiring practices conform to local codes; they're set up for your safety.

Wiring a Differential Thermostat

Your first step in hooking up a differential thermostat is to read the directions that come with it. Almost all come with detailed and easy-to-understand instructions. Figure 6-4 shows the connection of a DT using two junction boxes. A tie-in to the hot line you have tapped is accomplished in the first box.

The small, low-voltage sensor wires coming from the DT are clearly marked as to which one goes to the collector and which one goes to the point of use and/or the storage component. These sensor wires will have to be extended with 18-2 bell wire (available at hardware stores) to run to the sensors. Make the connections to the sensors with small Wire Nuts. The collector sensor is mounted in the airflow cavity 12 inches from the outlet port. Fasten it with a screw and/or silicone into the collector backing. When heating the north end of a crawl space, mount the point-of-use (storage) sensor on a floor joist in the cool south end of the crawl space. Sensor placement for water tanks and thermal storage bins is discussed in chapters 10 and 11. By the way, you won't be needing a DT for a simple direct-use system, so there are no directions given for placing a "house" sensor.

Wiring a Remote Bulb Thermostat

Hookups to a remote bulb are a bit simpler than for a DT because you need to run only one wire from the second junction box to the thermostat (figure 6-4). This is a switching line that corresponds to the "DT switch line" in the same illustration. The "power line" is eliminated. Other hookups are the same. The wiring shown in figure 6-4 produces a cooling thermostat; that is, it cools the collector and heats the house. The contacts in the thermostat close when the collector temperature rises and complete the circuit to the blower. As the collector cools, the contacts open and the blower turns off. (Ignore the "open on rise" terminal present in most remote bulb thermostats.) Ground the box by pulling the ground wire back through the cable clamp and grounding to one of the clamp screws. It is a good idea to mount an on/off switch near the thermostat to gain a manual override. This junction box is also a good place to mount a line voltage rheostat (light-dimmer switch) if you plan to regulate the blower speed in your system (see chapter 9 for specifics on this).

Blower Hookups

Most blower motors have small junction boxes mounted on them so you can simply remove a knock-out, put in a Romex connector and run the wire in. Small blower motors often have two black wires leading to the motor. Connect your white wire to one, black to the other; it doesn't matter which one goes where. Larger blowers have posts with screws, similar to a wall outlet connection. There will

---

Fishing Wires

If your solar installation requires a wall-mounted, low-voltage thermostat in the living area, you have already asked yourself this question: How am I going to get the thermostat wire to it? It isn't easy, but it can be done.

The easiest way to get a wire into a wall is often from the ceiling. First, drill a ½-inch or 1-inch hole through the drywall in the living area where the thermostat will be mounted. Then, crawl up into the attic and find the top of this wall. (It is very helpful to take measurements from other walls to the stud space you are trying to locate.) After you think you've found the correct stud space (it is very trying to locate), drill a hole through the plates (2 x 4s or 2 x 6s) on the top of the wall, and look for light coming in through the hole you drilled into the drywall. Dangle the thermostat wire down into the wall cavity, and have a helper hook it with a bent coat hanger. Staple your end to a ceiling joist, seal the hole, and you're ready to run the wire back to the control box.

If you have to run your wire into the crawl space, the easiest way is to run it inside the wall and behind the baseboard. Carefully pry an entire piece of baseboard loose. Then drill two small holes, one up into the wall cavity and one down into the crawl space. Be very careful to avoid existing wires. Place the holes so they will be covered by the baseboard. Fish a long, fairly stiff piece of wire up into the wall cavity until you can hook it through the 1-inch hole already drilled in the drywall. Pull one end of the thermostat wire up behind the wall with the stiff wire, and stick the other end through the holes and into the crawl space. Chisel out a groove for the wire in the drywall behind the baseboard, and don't nail through the wire when you put the baseboard back in place. If an electrician is doing your wiring, talk to him in advance and tell him what you have in mind. Then go ahead and fish your wires before he arrives. This will avoid your paying somebody $20 an hour to fish wires.

be a wiring schematic printed somewhere on the motor that tells you where to hook the black and white wires and where to hook up the ground (usually a green screw).

A word of caution: It is possible to wire some types of blowers backwards, so be sure to check the direction of rotation after installation. Running the blower backwards won't hurt the motor, but since some air will move down the ductwork, you may be fooled into thinking the system is operating, but with a poor airflow. Simply switch the appropriate wires in the motor to remedy this situation.

The wiring of more complicated systems is covered in Appendix 2. Two-mode systems are fairly straightforward, but three-and four-mode delivery systems get complicated. Appendix 2 also includes a section on checking out and troubleshooting your wiring, which will be helpful if your system doesn't seem to be operating properly.

Sources for Supplies

Blowers

We use W. W. Grainger's blowers and like them. They are readily available, reliable and easy to replace. (See chapter 9 for part numbers.)

Dampers

Backdraft Dampers

We have successfully used backdraft dampers from Contemporary Systems (which makes a variety of air handlers including a damper with an insulated blade in a steel frame for fan-forced systems) and Hot Stuff, which offers a complete line of reasonably priced dampers, both round and rectangular, which seal very tightly for both active and convective systems.

Contemporary Systems, Inc.
Route 12
Walpole, NH 03608
Phone: (603) 756-4796

Hot Stuff Controls Inc.
P.O. Box 306
406 Walnut St.
La Jara, CO 81140
Phone: (303) 274-4069

Another manufacturer of backdraft dampers is the following:

Arrow United Industries
42-25 21st St.
Long Island City, NY 11101
Phone: (212) 784-7550

Motorized Dampers

Hot Stuff offers round and square motorized dampers and air diverters in a variety of sizes from 6 inches round to 14 inches by 14 inches. All dampers have very tight compression seals, and custom dampers are available.

Heliotrope General offers MD-18S, a tightly sealing 18-by-18-inch, motorized solar damper (see sources for DTs for address).

Other sources for motorized dampers include:

Arrow United Industries
(address listed above)

Delta H Systems, Inc.
16970 Road 36
Sterling, CO 80751
Phone: (303) 522-4300

Differential Thermostats

Differential thermostats are available with a variety of prices and features. The more expensive ones have readouts of the sensor temperatures, and many have features that relate only to liquid-medium collector systems.

Heliotrope carries a large line of reasonably priced, well-built DTs suitable for do-it-yourself installations. The newer models have an adjustable temperature differential, and the few things we didn't like about earlier models have been changed (we can get our 18°F-to-8°F ones by special order). Heliotrope's DTT80 is suitable for simple direct-use systems. To locate a distributor near you (and to order an 18°F-to-8°F differential), contact the following:

Heliotrope General
3733 Kenora Dr.
Spring Valley, CA 92077
Phone: (714) 460-3930

Honeywell offers a nice line of DTs whose part numbers start with R7412. You must locate the necessary precision resistors to set the differentials. Be sure to get the resistors with the thermostats. These well-built DTs are slightly more expensive than Heliotrope General's. Honeywell's no. R7412A is a simple DT that has been used in direct-use systems, but its 3°F cut-off differential is a bit too low, requiring one new resistor.

Other sources for DTs include the following:

Advanced Energy Corp.
14933 Calvert St.
Van Nuys, CA 91411
Phone: (213) 782-2191

Hi Square, Inc.
2611 Old Okeechobee Rd.
West Palm Beach, FL 33405
Phone: (305) 686-8400

Independent Energy, Inc.
P.O. Box 860
42 Ladd St.
East Greenwich, RI 02818
Phone: (401) 884-6990

JBJ Controls, Inc.
1680 Foote Dr.
P.O. Box 1256
Idaho Falls, ID 83402
Phone: (208) 522-2200

Johnson Controls, Inc.
Control Products Division
2221 Camden Ct.
Oak Brook, IL 60521
Phone: (312) 325-7770

Natural Power Inc.
Francestown Turnpike
New Boston, NH 03070
Phone: (603) 487-5512

Now Devices, Inc.
7975 E. Harvard Ave. Unit E
Denver, CO 80231
Phone: (303) 755-9844

Watsco, Inc.
1800 W. 4th Ave.
Hialeah, FL 33010
Phone: (305) 885-1911

Wolfway Products, Inc.
R.D. 1 Box 1135
Tamaqua, PA 18252
Phone: (717) 668-4359

Remote Bulb Thermostats

Our recommendation is Honeywell's line of remote bulb thermostats that begin with the part number L6008C. They have a high range of adjustment. Honeywell's products are readily available, but you can contact the manufacturer for a local distributor:

Honeywell Inc.
10400 Yellow Circle Dr.
Minnetonka, MN 55343
Phone: (612) 931-4266

W. W. Grainger's remote bulb (order no. 2E399), which is available at most hardware stores and sheet-metal shops, is less expensive than Honeywell's thermostats and has been used successfully in hundreds of installations. We no longer use it because of the possibility of failure at stagnation temperatures.

Penn no. A19ABC-4 or White-Rodgers no. 2A38-14 are other remote bulbs that are available through sheet-metal shops and are suitable for low-temperature solar applications. They are more expensive than Grainger's thermostat but have a wider range of setpoints.

For local distributors, contact the following:

Johnson Controls, Inc. (Penn)
Control Products Division
2221 Camden Ct.
Oak Brook, IL 60521
Phone: (312) 325-7770

White-Rodgers Division
Emerson Electric Co.
9797 Reavis Rd.
St. Louis, MO 63123
Phone: (314) 577-1300

Two-Stage Thermostat

A good two-stage thermostat is Honeywell's no. T874F-1015 with subbase no. Q674D-1040.

Snap Disc Thermostat

W. W. Grainger's no. 2E245 is an inexpensive snap disc control. Its contacts close at 110°F and open at 90°F.

7

BUILDING PASSIVE AIR HEATERS: THE WINDOW BOX AND THE THERMOSIPHONING AIR PANEL

In this chapter we get into the nitty gritty of constructing the two basic types of passive, or convective, air-heating collectors: the *window box* and the *TAP (thermosiphoning air panel)*. For each we give step-by-step building instructions and detailed illustrations. Our window box design is made of wood and is glazed with glass. The TAP is typically installed on a wood-frame house and is designed to use two panes of tempered glass. Our example collectors are by no means the only two designs in the world, but they have worked well for us, and they will for you. But if these designs don't fit into your scheme of things, we give other construction options so you can custom-build a system to fit your exact needs.

Building a Window Box Collector

A window box collector is a good project for the first-time solar retrofitter. If you can build a birdhouse or if you know which is the business end of a circular saw, you shouldn't have any trouble putting together this simple solar collector in a weekend. Since convective heaters can't overheat (unless an outlet vent is blocked), using wood for the framework is a reasonable option. Whatever you can't scrounge from the back of your ga-

rage is available at the lumber yard. Using recycled materials helps ensure that your collector is cost-effective. If you buy all new materials, the payback period is much longer. Since it is important that the collector be well sealed and insulated, you will probably need to spend some money for good-quality caulk and high-temperature insulation.

Construction Steps

The Box

The box for a window box collector must be weatherproof and completely airtight. The sides of the box can be built from ½-inch- or ¾-inch-thick exterior-grade plywood or from 1 x 8 or 1 x 10 boards. Tempered, exterior-grade, ¼-inch Masonite or sheets of Masonite siding are good, inexpensive choices for the back of the box. Half- or ⅜-inch exterior plywood can also be used here, but the exposed edges must be sealed (see figure 7-1).

The top of the box that extends into the living area will be supported by two horizontal 1¾-by-¾-inch wooden trim strips and one 3½-by-¾-inch strip. Use your best wood here because it will be visible from inside the house. The strips should be cut so that they just cover the exposed edge of the insulation inside the box but so they don't extend into the airflow cavity, where they would restrict the airflow.

Use 1¼-inch drywall screws and aliphatic resin ("carpenter's") glue to assemble the box. Predrill all the screw holes to avoid splitting the wood. Nails can be used, but they do tend to work out in time. Since the box must be weatherproof and airtight, caulk the inside joints at the sides and bottom of the box with good-quality polyurethane or silicone caulk. Sealing the collector is accomplished at the back, front and sides of the box rather than between the layers of rigid insulation inside the box. Take diagonal measurements across the box to check for squareness, and make sure the inside dimensions will allow a ¼-inch gap between the box and the glass on all sides.

The Insulation

Cut ¾-inch Thermax (foil-faced polyisocyanurate foam) to fit snugly inside the bottom of the box. Use a straight edge and a large, sharp knife to make neat, straight cuts. You can use scrap pieces of insulation for the lower layer in the bottom of the box. A bead of caulk around the edges of the insulation will serve to glue it in place. Insulate the bottom vertical panel of the box as shown in B in figure 7-1. Next cut pieces of insulation for each side of the box, and check them for fit before cutting them into the sections shown in figure 7-1. Install the four wedge-shaped sections (labeled C and D in figure 7-1) on both sides of the box. Use dabs of caulk to hold them to the sides of the box. Place loose insulation into the lower wedge-shaped cavity before installing the turning baffle (made of sheet metal or litho plate) against the wedge-shaped sections. Use several screws to attach the turning vane to the bottom of the box. Push the two insulation pieces (labeled E in figure 7-1) against the ends of the turning vane. The upper end of these pieces will fit snugly against the inlet framework and hold the turning vane securely in place. The vane may bulge out slightly in the center but will be held down by the glazing. Just make sure the sides are flush with the top of the insulation, and that the bottom of the vane lies flush on the insulation in the bottom of the box.

Some Rules of Thumb for Designing a Window Box Collector

- The bottom of the collector should be at least 6 inches above ground level.
- There should be a clearance of at least ⅜ inch between each side of the collector and the window frame.
- The airflow channel behind the glazing should be at least 4 feet long. Shorter lengths won't develop enough "push" to drive the convection loop.
- The collector tilt should be equal to your latitude plus 10 to 15 degrees; in no case should it be less than 45 degrees from horizontal.

- The depth of the airflow cavity should be from one-fifteenth to one-twentieth of the length of the collector to minimize resistance to airflow.
- The bottom of the outlet port at the top of the collector should be above the top of the glazing to help minimize nighttime heat loss.
- The bottom edge of the glazing should be flush to the bottom of the collector, and metal flashing acts as a drip edge.
- A curved baffle should be installed at the low end of the airflow cavity to direct air around this corner.

The Center Divider

This divider (labeled F in figure 7-1) must be insulating, or the cool inlet air will be prematurely warmed and the collector won't operate properly. Use a continuous bead of caulk on the upper edges of both pieces (labeled E). Piece F should fit snugly on top of E, between the sides of the frame and between the vertical (lower) edge of piece E and the upper wooden framework of the box. Bevel the upper edge of piece F, the center divider, to ensure a good fit.

Install the insulation pieces labeled G. They won't want to stay in place so use plenty of caulk to hold them against the sides of the box. The upper section of this piece should be trimmed down slightly (about ⅜ inch) to allow the upper glazing support and the cover panel insulation to be flush with the top of the box. With a pocket knife, cut four small grooves opposite each other in both sides of the pieces labeled G for the ¼-inch absorber support rods.

The Glazing Support

The 1 x 4 upper glazing support, labeled J, must be straight and fit snugly inside the box on top of insulation piece G and flush with the top of the box. Rip a ⅜-by-¾-inch groove in its upper surface for the glass to sit in, as in figure 7-1. Position this piece carefully so that there will be a ¼-inch gap between the top edge of the glass and the support while still allowing the glass to be flush at the bottom of the collector. Secure the glass support to the box with two 1½-inch screws.

The Absorber and Supporting Rods

On a flat surface assemble the absorber from expanded metal lath ("diamond mesh" plaster lath with ¼-by-½-inch holes). The absorber should be five or six layers thick, and the holes in the lath should be positioned so that at least 20 percent of the lath "sandwich" is open space. Cutting the mesh with snips takes perseverance. Take your time and definitely wear a glove. Use the full-sized pieces of mesh for the top and bottom of the absorber, and sandwich the smaller cut-off pieces in between. The absorber should be ½ inch narrower than the inside width of the box and as long as the diagonal cut on insulation piece G. Use thin wire and a pair of pliers to tie all layers of the mesh together. One wire every 8 inches around the perimeter and a dozen in the center should be sufficient. The absorber will show, if you are using clear glazing, so try to be neat and don't mash or distort the mesh. Cut the absorber support rods to a length equal to the inside width of the box minus ¼ inch. Put the rods into the grooves cut in the side insulation and wire them to the absorber. (Wires rather than rods can also be used to support the mesh. See "Building a TAP" later in this chapter.)

Painting

Paint the mesh from both sides with high-temperature, flat black enamel. Make sure all of the mesh has a thin, even coat of paint. Also paint the exposed interior surfaces of the box. Paint the turning vane (front and

Figure 7-1: *On opposite page.* This illustration shows the important dimensions and insulation cuts for a window box collector designed around a 34-by-76-inch pane of tempered glass. Since the most cost-effective window box collectors are built from recycled materials, the dimensions for your collector will likely be different, and you should make drawings like these to plan your installation.

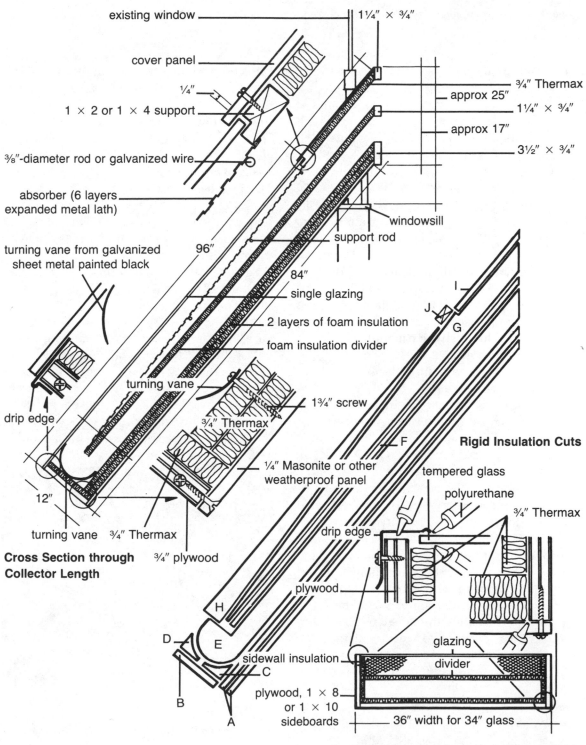

existing window

cover panel

1¼″ × ¾″

¼″

1 × 2 or 1 × 4 support

¾″ Thermax

approx 25″

1¼″ × ¾″

approx 17″

3½″ × ¾″

⅜″-diameter rod or galvanized wire

absorber (6 layers expanded metal lath)

windowsill

support rod

turning vane from galvanized sheet metal painted black

96″

84″

single glazing

2 layers of foam insulation

foam insulation divider

I

J

G

turning vane

1¾″ screw

¾″ Thermax

F

Rigid Insulation Cuts

drip edge

¼″ Masonite or other weatherproof panel

tempered glass

polyurethane

¾″ Thermax

12″

drip edge

turning vane   ¾″ Thermax

plywood

¾″ plywood

glazing

Cross Section through Collector Length

H

divider

D

E

sidewall insulation

B

C

A

plywood, 1 × 8 or 1 × 10 sideboards

36″ width for 34″ glass

Cross Section through Collector Width

125

back on the upper half), the exposed end of insulation piece E, the top of the center divider, the exposed sides of pieces G and H and the upper glazing support. Paint will stick quite well to the aluminum foil facing on the insulation if it is clean and dry.

Installing the Absorber

The top end of the mesh should be butted against the wooden glass support. The lower end of the mesh should stop just short of the end of the divider. Bend this lower end down slightly so that it rests against the center divider. The idea here is to make the air pass through the absorber all along its length and not leave openings at the top or bottom of the absorber where air can short circuit the loop.

Push the insulation pieces labeled H against the absorber and check for fit. Caulk the back sides and put them in place. You can tack in two or three temporary nails at the top of these pieces, if necessary, to hold them down while the caulk cures. Touch up the painting before installing the glazing.

If glass is used for the glazing (our example collector uses 34-by-76-inch tempered glass), it will be installed directly on top of the side insulation and the bottom of the box. The top will be supported by the 1 x 4 glass-support strip. Check the glass for fit and then clean both sides thoroughly.

To install the glass, run a large bead (½-inch round) of polyurethane caulk (or black silicone) on top of the side insulation, the upper glazing support and the bottom of the box. Be sure to use plenty of caulk on top of the wood at the bottom of the collector so the gap here will be completely filled, since it can be difficult to seal this later. Slowly position the glass so there is at least a ⅛-inch gap between the wood edges and the glass edges at the sides and top. Run a bead of caulk around these edges and use your finger to point and seal these edges.

Cover Panel and Trim

After the glass is in place install the unglazed cover panel at the top of the collector. Put a piece of rigid insulation (I in figure 7-1) between the glass support and the top framework of the collector. Bevel the top edge for a snug fit against the wooden frame. Run a bead of caulk on the top edges of insulation pieces G and set the piece in place. The unglazed panel that covers this insulation can be made of exterior ¼-inch Masonite or ½-inch exterior plywood and should extend at least ¼ inch over the top of the glazing. Secure this panel with short cadmium-plated or stainless steel screws rather than nails. Use caulk on the back side of the panel, and reach inside the upper air cavity to push insulation firmly against the panel for a secure attachment here. This panel can be painted to match the collector box or covered with litho plate

Photo 7-1: These window box collectors were built from recycled materials for $20 each by Sam Sampson of Gooding, Idaho. Although they are small in relation to the rooms they heat, Sam estimates they save him 30 percent on his daytime heating bill for each room.

or sheet metal to help reflect more light into the window above. Paint all exposed exterior wooden surfaces on the box with two coats of oil-based exterior paint. Several options are shown in figure 7-2 for installing metal or wooden trim over the sides of the collector. Use screws rather than nails to attach this trim for a permanent attachment and to avoid banging on the glazed collector.

Mounting the Collector

The methods for mounting a collector in a window will vary greatly depending upon window and sill arrangements. The primary goal is to achieve a solid mount and an airtight seal between the collector and the window while at the same time allowing for easy summertime removal of the unit. One good approach is to attach to the outside of the box a permanent wooden or sheet-metal flange that has compressible foam on its back side. This flange can be screwed to the window to make an airtight seal. Filler strips may be needed at the sides of the collector if there is more than a ½-inch gap between the collector and the window. A beveled support strip on top of the windowsill will provide the necessary support and a good seal underneath the collector.

FRP Glazing

Flat FRP (fiberglass) is a good, low-cost glazing material for window box collectors. When you use FRP, the wooden panel at the bottom of the collector is cut flush with the wooden sides. The rigid insulation on the side of the collector is flush with the top of the box. Like the glass glazing, FRP runs flush to the bottom of the collector, and metal flashing acts as a drip edge. Attach the fiberglass starting at one side of the collector. Run a bead of polyurethane caulk on both sides of the FRP perimeter, and sandwich the glaz-

ing between the collector box and the metal flashing cap that lies on top. Drive ¾-inch cadmium-plated or stainless steel screws through the FRP and into the box. Predrilling ⅛-inch holes will help keep the edges of the fiberglass from fracturing. After the first side is secure, lift up the fiberglass and run a bead of caulk along the bottom edge of the box and along the upper glazing support. When installing the screws at the bottom of the collector, drive them through small dabs of clear silicone (chocolate chip size) to insure an airtight and watertight seal. The screws at the top don't need this treatment because they will be removed after the caulk cures but before the cover panel goes on. After the top and bottom of the FRP are secure, attach the second side. Be sure the glazing is completely sealed around its entire perimeter.

Flat Plate Absorbers

Litho plate or thin sheet metal can be used for the absorber plate. Use the double-pass (front and back) design discussed in chapter 4. Center the absorber plate in the airflow cavity, and bend the lower edge at least halfway around the corner at the bottom. Run the plate 2 or 3 inches past the upper edge of the glazing. Overlaps between separate sheets are joined with pop rivets or screws. Overlaps don't have to be airtight but should be snug and attractive. If corrugated metal is used, the corrugation must run across the width of the box in order to turn the corner at the bottom of the collector. This involves a lot of fancy cutting on the side insulation for an attractive installation.

Another option for absorbers is to use a selective surface material, which makes these collectors operate better than those with absorbers that are painted black. Selective absorbers are quite expensive, however, and may be hard to locate. These factors discourage their use in simple window box installations, which are meant to be low-cost anyway. Flat plate convective heaters using nonselective absorbers don't deliver quite as much heat as those with mesh absorbers, but if you have some thin, unpainted metal lying around and are looking for a low-cost installation, use it.

Center Supports

In collectors that are wider than 36 inches, it is a good idea to put in a wooden center support for the insulated divider. This support is a 1 x 6 or a 2 x 6 ripped down to 5 inches wide and inserted in the box before the insulation. It is secured with screws. The insulated divider rests on top of this board, and two turning vanes are used in the bottom of the box.

The cover panel can be supported by a strip of rigid insulation that is secured to the top of the divider panel with a bead of caulk. Reach inside the collector and caulk the up-

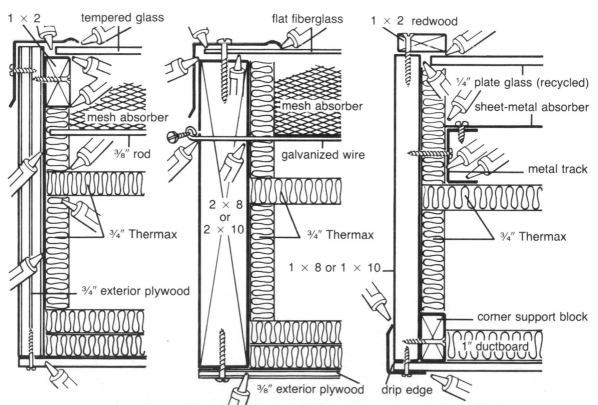

Design 1: Standard Collector   Design 2: Fiberglass Glazing   Design 3: Doublepass Design

Figure 7-2: A wide variety of materials is suitable for window box collectors. When you design your collector, you can mix and match from the three different cross sections shown. Design 1 is our standard collector; design 2 is a collector built from 1 × 8 side and end boards and glazed with flat fiberglass; design 3 is a doublepass design with a sheet-metal absorber built from 1 × 8s.

per joint here as well. In wide window box collectors, use ⅜-inch rods to support the absorber, and use thicker material for the back of the box.

Dimensional Lumber

Dimensional lumber, 1 or 2 inches thick, can be used for the sides of the collector box. Of course, lumber 2 inches thick is sturdier, but lumber 1 inch thick can be used if support blocks are used in the corners. Use two or three 1-by-2-inch blocks, predrilled, glued and screwed in the corners, rather than a continuous block. This will allow for even wood shrinkage. Two-by-6s are suitable for a 4-foot airflow, 2 x 8s for a 5-foot to 6-foot flow and 2 x 10s for a 7- to 8-foot flow.

More Design Options

For a more attractive installation exposed edges of the foil-faced insulation inside the collector can be covered with high-temperature aluminum duct tape. This also makes a better attachment when glass is placed directly on top of the side insulation.

The exposed surface of the turning vane at the bottom of the collector can be covered with flashing. This will allow for insulating behind the vane and make for a cleaner "line" at the bottom of the collector but will reduce

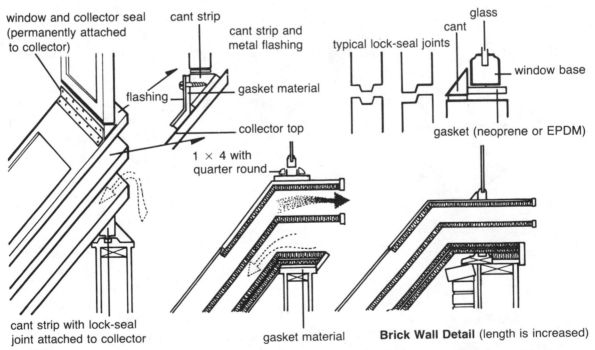

Figure 7-3: It is important to make an airtight seal between the collector and the house window frame. This is best accomplished with flashing, cant (beveled wood) strips and weatherstripping that is permanently attached to the collector. A horizontal window penetration allows a unit to be placed in a window that is lower to the ground. The horizontal extension allows for a proper tilt angle.

the amount of light entering the collector. Another option is to use 20-mil Tedlar or a small piece of thin, clear fiberglass glazing for the top half of the turning vane.

Laying glass directly on top of the rigid insulation is a technique that has been used successfully in many installations, but a better idea is to screw wooden strips (1-by-2-inch or 1-by-1-inch) into the sides of the box above the insulation to support the glass. This will insure a more permanent seal should the insulation shrink slightly after several years of exposure.

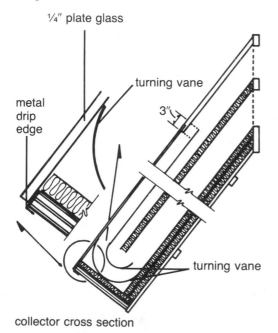

collector cross section

Doublepass Flat Absorber Plate

Figure 7-4: When building a window box collector with a flat (sheet-metal) absorber, bend the absorber to make the curve shown at the bottom of the collector. With this doublepass design the turning vane consists of two pieces of curved metal, one being the tail end of the absorber, the other being the semicircular piece of sheet metal that touches the bottom end of the collector box.

When using small pieces of recycled glass for glazing (old windows, for example), build the collector so that its width allows for their widest dimension. Use H-channel from a glass shop for the horizontal joints (see "Building a TAP," which follows).

Building a TAP

There are basically two ways to build a thermosiphoning air panel (TAP): with a slanted mesh absorber in a single-glazed collector or with a sealed backpass design where the air moves behind an absorber plate in a double-glazed collector. The latter approach is the most-used, but in this section we describe a TAP built with a mesh absorber. Extensive research at the Los Alamos Scientific Laboratory in New Mexico has shown that collectors with mesh absorbers deliver more heat than other absorber arrangements in convective installations.

Mesh collectors are as easy and inexpensive to build as flat plate convective air heaters. However, their one drawback is that dust can collect on the mesh and against the inside of the glazing. Dust on the absorber is primarily an aesthetic drawback and doesn't significantly reduce performance, but dust on the inside of the glazing can reduce the amount of light reaching the absorber. We haven't yet found this to be a big problem, but if you are concerned, build these collectors with vents 6 inches above the interior floor and with full-sized, removable backdraft dampers. Then what little dust that accumulates in the collector can be vacuumed out, and the inside of the glazing can be wiped off every year or two. We feel this simple maintenance is minimal when considering the increased efficiency and the fact that mesh collectors don't require double glazing. Since the construction of a TAP with a mesh absorber shares many details with both the window box collector and a vertical active collector (covered in chapter 8), we will often refer to these sections.

Construction Steps

Positioning a TAP

The 2 x 6 framework and ½-inch CDX plywood backing for the collector are built as a unit and then screwed or nailed to the existing wall studs (see chapter 8 for details on mounting onto masonry). Locate the wall studs (normally 16 or 24 inches on center), and find the center of a stud where one ver-

---

Some Rules of Thumb for Designing a TAP

• Since a TAP is a permanent installation, use good-quality materials for construction. A wooden collector box is suitable, but use glass for glazing and metal trim for flashing.
• The airflow cavity must be large and unrestricted, at least 4 feet long but preferably between 7 and 14 feet long. Collectors connecting two floors need fire dampers. In the event of fire the fusible link in these dampers melts, closing the opening with a metal vane.
• The collector must have full-sized vents at the top and bottom of the collector in every stud space. The bottom vents need tightly sealing backdraft dampers. If your wall is full of pipes and wires, or is of masonry construction and won't allow for large vents, you will be better off building an active collector.
• Locate the lower vents 6 inches above the floor level, and make sure the upper part of the collector is unshaded by the roof overhang during the heating season.
• Install closing doors in the outlet vents to isolate the collector in summer.

---

tical edge of the collector will lie. The outside edge of the collector 2 x 6 is positioned right over this center. Measure 1¼ inches to one side of the center of this stud, and use a level to scribe a vertical line. Measure 70½ inches horizontally from this line to locate the other side of the collector. The collector itself will be 69½ inches wide so it will sit between these two lines with a ½-inch gap on both sides for counter flashing. The second edge of the collector will probably fall between two studs and will require additional support.

Inside the house measure down from a windowsill to the floor level. Transfer this measurement to the outside wall, and use a level to mark the floor level at the collector location. Snap a horizontal chalk line 3¾ inches above this line. This is where the bottom of the collector will be. Snap another chalk line 80 inches above this line to locate the top of the flashing for the collector (see figure 7-5).

Inlets and Outlets

Locate positions for upper and lower vents inside the house, and snap two chalk lines 6 inches apart for the tops and bottoms of both sets of vents. Before cutting the vent openings, turn off the electricity (at the fuse box) to the wall on which you will be working. Cut small test holes where the vents will be located, and fish around inside the wall cavity to locate wires or pipes that could cause problems. Cut out the upper and lower vent openings using a keyhole saw and utility knife (for drywall), and remove the wall insulation where the vents are located. If it is necessary to move electrical wires and outlets at the bottom of the wall, unhook the wires at the outlet box. Reach inside the wall cavity and drill new holes through the studs for the wires.

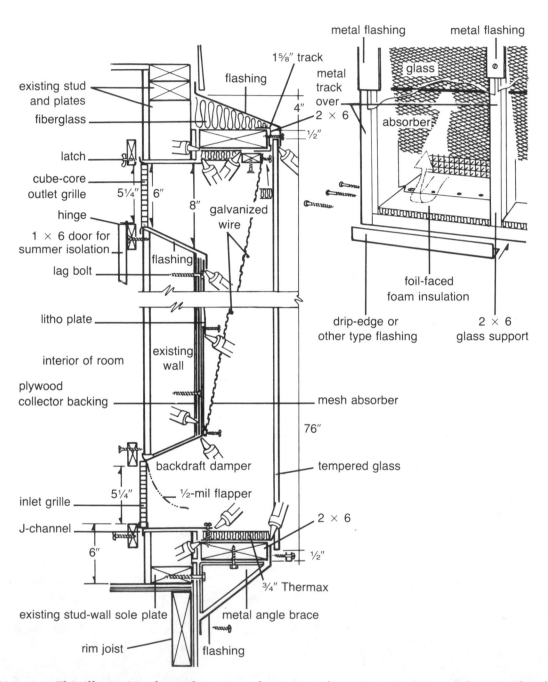

Figure 7-5: This illustration shows the parts and important dimensions in our example TAP. The plywood collector backing is screwed to the 2 × 6 framework and lag-bolted to studs in the wall. Bottom support is required for wood-framed TAPs.

Install a new outlet box that can be mounted through a hole in the drywall.

If you haven't installed jack studs at the edge of the collector that falls between two studs, insert short, 14-inch, 2 x 4s (or whatever the dimension of your wall stud is) into the wall cavity at the top and bottom vents. You can secure these in place with screws through the edge of the drywall and, later on, through the collector's plywood backing once it is in place.

The Collector Box

Build the collector box on a flat, level surface. Its outside dimensions will be 69½ inches wide by 77 inches high. Glue and screw the corners; don't use nails. Insert a 2 x 6 glazing support in the center of the collector and attach it to the top and bottom 2 x 6s with glue and 3-inch drywall screws. Check the box for squareness by measuring the diagonals on each glass opening before attaching the ½-inch CDX plywood backing with 1½-inch drywall screws 8 inches on center. Secure the box to the wall with 3-inch screws driven every 12 inches through the plywood and into the wall studs. Seal the joints between the 2 x 6s and plywood with polyurethane caulk (see "Vertical Wall Mounts" in chapter 8 for more details on these techniques). If the wall behind the TAP happens to be uninsulated, you should install a sheet of rigid insulation between the plywood backing and the house wall.

Outside Vent Openings

To locate the vent openings on the outside of the collector, reach through the openings inside the house and drill small holes into the plywood at each stud. Draw connecting lines on the plywood and cut the openings. These openings will be 8 inches tall (2 inches taller than the 6-inch openings inside the house).

Metal Backing

Cover the back of the collector completely with litho plates. Secure them with screws and seal all edges with black silicone.

The method we prefer for installing tempered glass involves bedding it in silicone against metal and then flashing the collector with metal trim. This is a permanent and attractive attachment and, in our opinion, simpler than attaching glass to wood using Butyl glazing tape. So, at this point cover the top edges of the 2 x 6s and glazing support strip with metal track or two pieces of flat drip edge as shown in figure 7-5. Run a ¼-inch bead of polyurethane caulk on the top of the wooden framework, snap the track in place, and secure the track with ¾-inch drywall screws 16 inches on center on the front of the track. Flat drip edge can be installed in a similar manner with caulk between the two pieces. Don't install track or drip edge on the bottom of the collector since the wood here will be covered with the lower flashing for the collector.

Absorber Support, Insulation and Painting the Collector Backing

Screw a 1 x 2 absorber support at the top of the collector. Its front edge is positioned 1 inch back from the edge of the upper 2 x 6 (see figure 7-5).

Cut ¾-inch foil-faced Thermax (polyisocyanurate insulation) to fit inside the collector on the sides and bottom (5½ inches wide) and at the top behind the absorber support (3 inches wide). Cover the exposed edges of the Thermax with high-temperature aluminum duct tape, and secure the pieces to the box with polyurethane caulk. Use a few 1½-inch screws to hold it in place while the caulk cures. Using a large bead of black silicone, seal the lower joint that the insulation makes with the litho plate on the back of the collector, and point it with your finger.

The mesh in a TAP will butt snugly against the insulation rather than being inserted between two pieces, as in the window box collector.

Paint the litho-plate backing, the exposed foil-faced edges of the insulation and all exposed wood surfaces inside the collector with high-temperature flat black paint. Also paint the front surface of the 1⅝-inch track that covers the front of the box.

Building Backdraft Dampers

A tight-sealing backdraft damper at the bottom of a TAP is vital to proper operation of the system. It is much easier to build this damper as a long single unit than to install individual dampers in each stud space. Since the lower vent in a TAP will always be relatively cool, lower-temperature materials can be used than those required for backdraft dampers for active systems (see chapter 9). The grille that supports the lightweight flapper is standard ⅜-inch-thick, cube-core grille that is typically used for lighting fixtures in suspended ceilings and is available in 2-by-

Materials List for Our Example TAP		
Item	Quantity	Comments
2 x 6	3 each at 8 ft	collector framing
2 x 6	1 each at 12 ft	collector framing
2 x 4	1 each at 10 ft	wall framing
½" CDX plywood	2 4' x 8' sheets	collector backing
1 x 2	1 each at 6 ft	absorber support
¾" and 1½" Phillips screws	100 each	
2½" lag bolts (or 3" wood screws)	20 each (40 each)	collector attachment
6d galvanized nails	100 each	
litho plate	45 ft²	collector backing
litho plate	20 ft²	14 sets of vent liners
polyisoscyanurate rigid insulation	12 ft²	
expanded metal plaster lath	260 ft²	to make 6 layers
heavy-gauge galvanized wire	18 feet	absorber support

4-foot panels at large lumber and home-improvement stores. The metal framework that will line and seal the edges of the grille is ⅜-inch J-channel, which is used to trim out corners on ⅜-inch drywall. The ideal material for the flapper is ½-mil Tedlar film, but thin, uncoated ripstop nylon or 1-mil Teflon film can be used. *Frisket paper*, available at drafting and art supply stores, is another choice. Any film thicker than 1 mil is unsuitable because the flapper will be too heavy to be opened by the convective airflow. Sources for ½-mil Tedlar are listed at the end of this chapter.

With a hacksaw or jigsaw, cut the cube-core grille to the same width (6 inches) as the lower vent opening in the TAP. For best appearances position vertically butted edges of separate grille pieces so they break over a stud. With snips, carefully cut 90-degree angles in the legs of the ⅜-inch J-channel so that it can be bent snugly around the grille. Butt the joints in the channel at wall studs. The side of the channel with the longer, 1³⁄₁₆-inch leg will go inside the wall and the flapper will attach to it, so it must be very flat. Cut litho plates into strips 2½ inches wide and 5¾ inches long to fit under the channel

Materials List—*Continued*		
Item	Quantity	Comments
high-temperature black paint	6 cans	
polyurethane caulk	4 tubes	
silicone (black) caulk	3 tubes	
1⅝″ track (or drip edge)	40 linear feet	
side flashing	14 linear feet	
top flashing	6 linear feet	
bottom flashing	6 linear feet	
bottom angle braces	4 each	
⅜″ tempered glass (34″ x 76″)	2 panes	
cube-core grille	6 ft²	
J-channel	30 linear feet	2 sets of vents
backdraft damper flaps	4 ft²	lower vent
1 x 3 (inside trim)	30 feet	inside trim
1 x 6 (inside trim)	6 feet	inside trim
small hinges	2 sets	
aluminum duct tape	(small roll)	

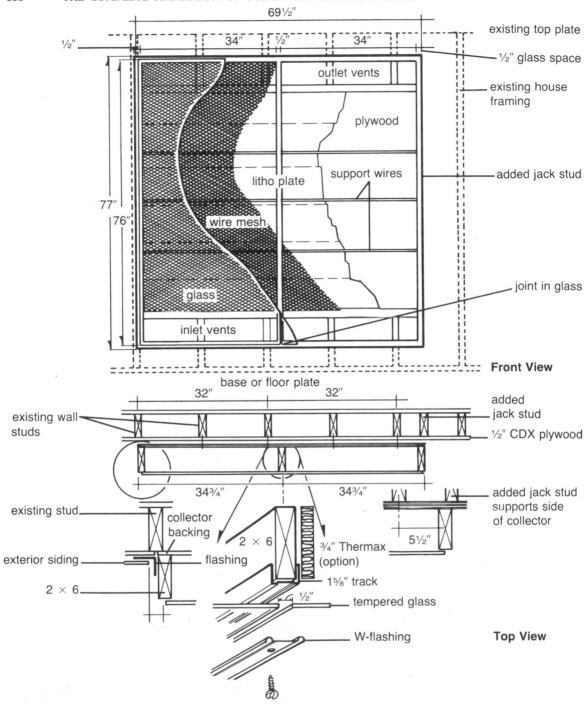

Figure 7-6: This is the typical vent placement for our standard collector when it is mounted on a wall with studs 16 inches on center. In many installations a jack stud must be placed inside the existing wall to support one edge of the collector.

and against the back of the grille at each stud. Hold the grille assembly in place and make sure these 2½-inch strips are positioned so they extend ½ inch on both sides of the stud.

Cut the flapper material to a size that overlaps the 2½-inch strips by ¼ inch, and allow a ½-inch gap at the bottom. Remove the grille and clean the upper inside surface of the J-channel with rubbing alcohol. Run a thin bead of silicone on the cleaned surface ¼ inch from the top edge where the flapper will be attached. Carefully lay the flapper onto this bead. Position it squarely in the opening, and make sure it overlaps the 2½-inch ver-

tical strips by ¼ inch. Allow the silicone to cure overnight before disturbing the damper.

Metal Sheathing for the Vent Openings

Cut and bend litho plates to line the vent openings (with the printing side in) as in figure 7-7. Leave tabs on the inside flanges of the liners to aid in attaching them to wall studs. Paint the exposed surfaces on the liners with high-temperature flat black paint. Use screws to attach the vent liners to the outside of the collector, and use galvanized 6-penny nails and silicone on the inside. Seal joints between the liners and the wall studs

Photo 7-2: The wooden framework for this TAP is screwed and glued over the exterior wall sheathing. Inlet and outlet vents have been cut between the wall studs at the top and bottom of the collector frame. Here the back is being lined with ductboard insulation to minimize heat loss from the collector to the frame wall behind it. Although it is not an insulator, litho plate is another, less expensive choice for a collector backing.

with black silicone. The exposed wall studs can be covered with metal and/or litho plates, but this isn't necessary.

Install 6-inch-wide cube-core grille at the top of the collector as you line the upper vents. Cut the grille so that both horizontal surfaces are flat, eliminating the need to wrap this vent with J-channel. Bond it to the flashing with several dabs of silicone and allow the silicone to cure before trimming out the vent.

Test and Install the Backdraft Damper

Once the silicone attaching the flappers on the backdraft damper cures, hold the backdraft damper in place and gently blow into each opening. Make sure it opens easily and doesn't rub on the vertical studs. If necessary, carefully trim the flapper with scissors, but be sure that, when sealing, it overlaps all edges as much as possible.

Cut 1-inch strips of aluminum duct tape and tape the exposed edge of the J-channel

Interior Vent Details

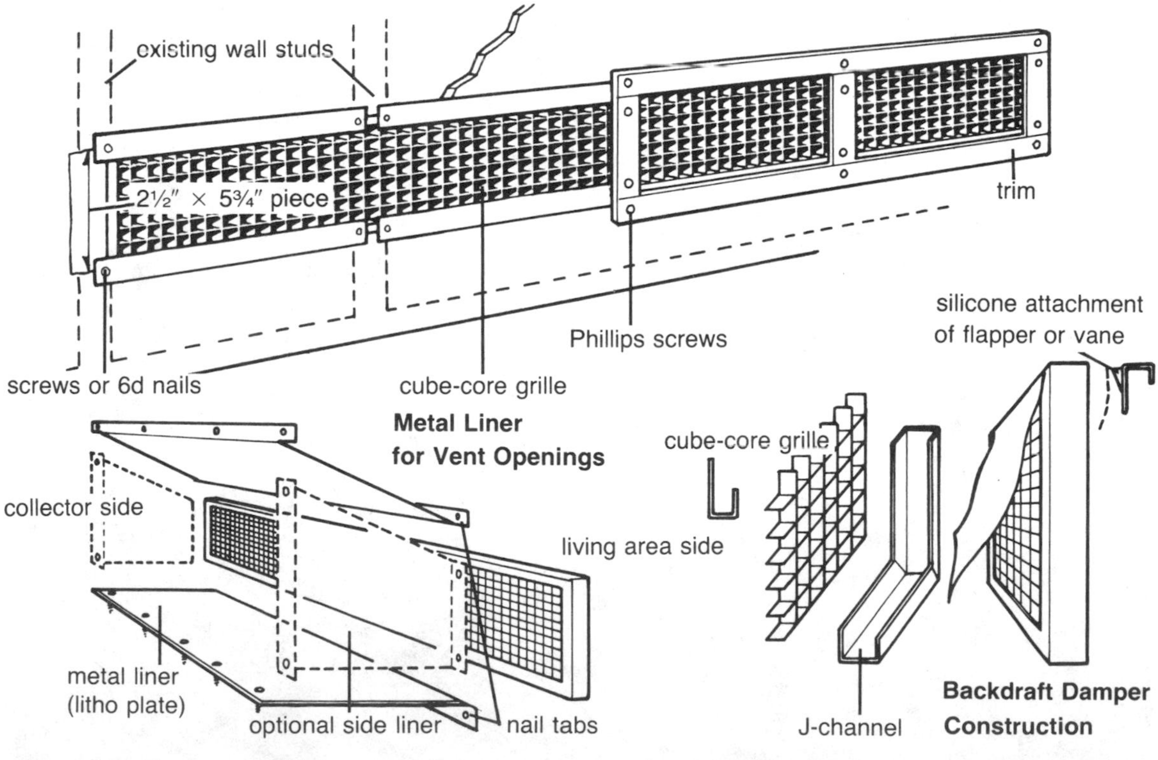

Figure 7-7: The upper and lower vent openings are lined with litho plate. The exposed wall studs can also be covered with litho, but this is optional. Continuous removable backdraft dampers are required in TAPs that use mesh absorbers. The lightweight flaps for the dampers fit between studs and are attached to the metal J-channel with silicone. Aluminum duct tape seals the damper in place while allowing for easy removal.

on the backdraft damper to the exposed edge of the litho-plate vent liner. This insures an airtight seal and allows for easy removal of the damper when cleaning the bottom of the collector or replacing flappers, if necessary.

The Absorber

The mesh absorber is made of six layers of expanded metal lath. It attaches to the front of the upper absorber support and to the litho-plate backing (½ inch above the vent opening) and is supported by wires.

Snap chalk lines diagonally on the insulation at the sides of the collector and along the center glazing support. These lines are where the absorber will be located. Drill three sets of ⅛-inch-diameter holes through the sides of the collector and the glass support for the absorber support wires. Run 16-gauge galvanized wire through the holes, pull it tight, and secure it to the sides of the collector with countersunk screws. In some installations it is desirable to install these wires after the absorber is in place.

Assemble and paint the absorber as described in the previous window box section. It must be just wide enough to fit inside the collector box (¼-inch gap minimum). Screw the top of the absorber to the upper absorber support with ¾-inch screws 6 inches on center. Pull the lower edge of the absorber down and screw it through the litho plate and into the plywood ½ inch above the lower vent opening with ¾-inch screws 6 inches on center. Attach the absorber to the support wires with small wires about every 8 inches to prevent sagging of the vertical absorber. Touch up the paint inside the collector before installing the glazing.

Glazing and Trimming the Collector

Attaching the glass will be very straight-forward if you have covered the outside of the wood frame of your collector with metal. Follow the directions given for building an active collector in chapter 8. Paint the drip edge black and attach it at the bottom of the collector with 6-penny galvanized nails or 1¼-inch drywall screws, 4 inches on center, to cover the wood before installing the glass. Install temporary support screws to hold the bottom of the glass, and bed it against the metal with a large bead of black silicone. We prefer metal rather than wood for trimming out and flashing collectors.

Photo 7-3: Interior vents for a TAP can and should be unobtrusive, yet airflow to them must not be restricted. The couch in this room should sit a few inches away from the wall to provide for a free flow of air.

Vent Trimming inside the House

Trim out the upper and lower vents in the collector with your choice of wood trim. At the lower vents the trim should be installed so that it just covers the exposed edge of the J-channel on the backdraft damper. Screw the trim into place to allow for easy removal if the backdraft damper requires service or the collector needs vacuuming.

Make the vertical opening between the trim strips at the top of the collector exactly 5¼ inches. This will cover the exposed edges of the cube-core grille and allow a 1 x 6 to be used for the upper vent door. This door can be made from a single straight board hinged to the lower trim strip and held closed by a latch. It should fit snugly to prevent unwanted heat from entering during the summer, but it doesn't have to be completely airtight.

Operation

You can check out the operation of the collector by burning some incense. Hold a stick of incense near the closed backdraft dampers at night to indicate leaks and improperly sealed flappers. If you burn several sticks at the same time, you can observe the convective airflow through the TAP during the day. Make sure the flappers are completely open during sunny periods.

Some Construction Options

All-Metal Construction

The mesh absorber TAP we described can be built easily using an all-metal framework. Six-inch structural, exterior track, available in 18- or 20-gauge, with short legs is used for the perimeter. The glazing support is stiffened by inserting three 6-inch pieces of track inside the long piece of track used

as the support. Heavy-gauge, 8-inch track is also available for collectors over 8 feet high.

The construction steps for assembling the collector are similar to those of the all-metal collector described in chapter 8. The litho-plate backing is screwed onto the plywood backing, and the legs of the track are attached over the litho plates in a bed of silcone. The mesh absorber is installed by bending and screwing its top edge to the 6-inch perimeter track at the top of the collector. Screw penetrations here are sealed with silicone and

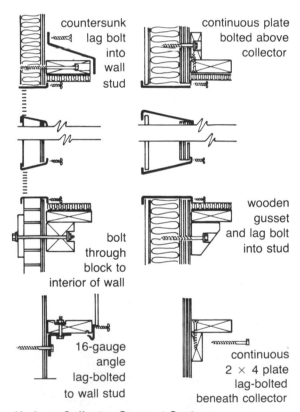

Various Collector Support Systems

Figure 7-8: These cross sections show several options for supporting and flashing the top and bottom of a vertically mounted collector.

the mesh is supported with wires. The collector is insulated outside the track with rigid insulation, and it is trimmed with metal flashing.

Other Glazing Materials

This collector uses 76-inch-long tempered glass because this size fits easily on 8-foot walls, but 92-inch glass (another standard size) can be used as space permits. If you use longer glass, you may need to place the lower vent closer to the floor level or the upper vent 3 or 4 inches below the top of the collector in order to fit the upper vent under the plate at the top of the stud wall. Never cut through this structural plate. Collectors can also be built using 46-inch-wide tempered glass rather than the 34-inch glass used in our example.

Corrugated fiberglass is another choice for glazing low-cost TAPs (see chapter 8 for installation details).

Horizontal Mullions

Because of limitations in standard glass lengths, TAPs that are taller than 8 feet will need a second piece of glass and consequently a horizontal support piece. When using mesh absorbers, it is important that this support doesn't extend into the air cavity and interfere with the airflow. The best way to have this support is with an aluminum "H" extrusion, available from a local glass shop at about 45¢ per foot. Notch out a groove for the channel in the wood-frame side piece, and cut the H-channel as in figure 7-9. Install the lower piece of glass in the collector. Put small pieces of EPDM weatherstripping into one side of the channel to act as a cushioning expansion joint, run a bead of silicone into the channel, and gently tap it over the edge of glass. Secure the channel at each edge

of the collector with screws. Run a small bead of silicone in the upper channel, and lower the next pane into place. Silicone seal the upper and lower exposed legs of the channel where it joins the glass. H-channel can be attached to metal frame collectors using screws or pop rivets.

Tempered glass installed in this way will have some give at the center of the collector but will withstand strong winds nonetheless.

Backpass Designs

A convective air heater with a sealed absorber and a backpass airflow path can be

Horizontal Mullions

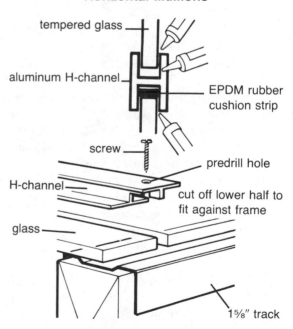

Figure 7-9: Larger thermosiphoning collectors sometimes require horizontal supports (mullions) between panes of glass. To minimize interference with the airflow, continuous sections of aluminum H-channel are a good choice for these mullions. The ends are notched (as shown) to attach to the edge of the collector, and clear silicone seals the exposed edges.

Photo 7-4: These TAPs provide daytime heat to the most frequently used rooms of this mobile home. This simple application is within the range of any do-it-yourselfer.

built much like the wood-frame collector described in chapter 8. The airflow cavity needs to be larger than for an active collector but can be smaller than for a mesh collector. About 2 to 2½ inches deep is a good figure to shoot for in a 6- to 8-foot flow. The absorber is installed in two sections so that it extends from both sides of the collector to the center glass support. It is further attached to a small, diamond-shaped, wooden absorber support or piece of track in the center of the collector.

A selective surface absorber can be used in a backpass design. B. J. Rodgers of Dixon, New Mexico, has tested convective heaters side by side and has found that single-glazed units using a selective surface in a sealed backpass design outperform mesh collectors. We haven't built convective collectors using selective surface, but a half-and-half design with a selective surface in the top half of the collector and single glazing should work well (see chapter 5). If you use a selective surface, line the airflow cavity with metal. Otherwise, wooden absorber supports are suitable for withstanding the normal operating temper-

ature. If you choose to use a doublepass design, leave a 4-inch gap at the top and bottom of the absorber to allow for an airflow in front of the absorber. The absorber will be unsupported at the top and bottom.

Double Glazing

Convective backpass collectors using nonselective absorbers require double glazing to prevent heat loss out the front of the collector. Mesh collectors in very cold climates can realize a performance benefit with double glazing.

The easiest, but more expensive, way to double glaze a TAP is with standard tempered glass units. They are installed in the same manner as single pane glass, using a silicone seal onto metal trim.

A more inexpensive way to achieve a double glazing is by using a solar film such as 1-mil Teflon for the inner glazing. This film is attached to the wooden framework over a two-sided transfer tape adhesive such as 3M Transfer Tape no. 465. This should be done on a sunny day. Tape is applied to the wooden framework, the adhesive backing strip is removed, and the film is then laid over the tape. As the collector heats up, the film will expand and can be lifted off the tape and pulled tight over the collector. Prepainted 1 x 2 furring strips are then screwed over the framework through predrilled holes with 1¼-inch drywall screws, 8 inches on center, creating a very tight sandwich. A bead of polyurethane caulk is then run around the perimeter and at joints in the 1 x 2 to further seal the dead air space. The wooden strips are then covered with prepainted metal flashing, and the glass glazing is attached with a silicone seal.

Most solar films (especially Teflon) are prone to expansion and sagging when the collector gets hot. In slanted mounts that are double glazed with film, small galvanized wires can be run across the collector behind the film, 2 feet on center, to support it and keep it from touching the absorber. This step is unnecessary in vertical mounts if the film is pulled tight across the warm collector during installation.

Glazing with Butyl Tape

We prefer the silicone-to-metal method of attaching glazing, but many professional glaziers and some collector builders use Butyl glazing tape to attach glass to wood. Butyl tape should only be used for the outside glazing on a double-glazed collector because it can't withstand the temperatures present in single-glazed units, and it will tend to degrade and leak. Since the glass won't be self-supporting when glazing tape is used, it must be supported at the bottom of the collector, and the bottom must be braced to the wall as in figure 7-11. Two ¼-inch neoprene setting blocks are placed under each pane of glass, Butyl tape is placed on the wooden framework, and the glass is set in place. Another

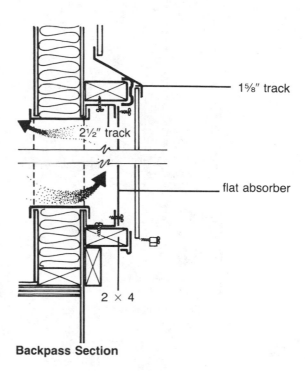

1⅝" track

2½" track

flat absorber

2 × 4

Backpass Section

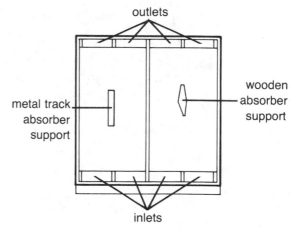

outlets

metal track absorber support

wooden absorber support

inlets

Figure 7-10: TAPs with a backpass airflow are built much like active collectors except that they have a larger airflow cavity. Making the perimeter and glass supports with metal track is a good option, though wood is also commonly used. Pieces of metal track placed vertically in the center of the collector are needed to support and stiffen the sheet-metal absorber.

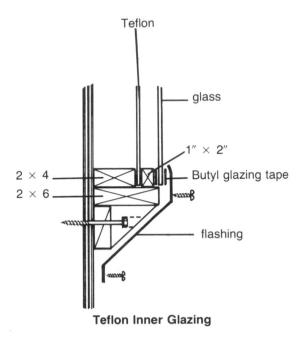

Teflon Inner Glazing

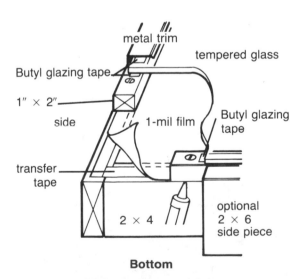

Bottom

Figure 7-11: Thin solar films are an inexpensive choice for double-glazed backpass collectors. This film is pulled tight across transfer tape and held in place with 1 × 2s and urethane caulk. This illustration also shows how to use two layers of Butyl glazing tape for attaching tempered glass glazing.

layer of glazing tape is run around the edge of the glass, and everything is secured with heavy (24-gauge) flashing bent to an 85-degree angle. Seal the edge of the flashing to the glass with clear silicone, using special care at the bottom of the collector to prevent water penetration here.

Butyl glazing tape (¼-by-³⁄₃₂-inch tape) and neoprene setting blocks (¼-by-⁵⁄₈-by-2-inch) are available at most major glass dealers.

Trimming Out Vents with Louvered Grilles

Louvered grilles can be used to trim out TAP vents inside the house. Registers with closing louvers can be installed at the top of the wall, allowing the TAP to be closed off in the summer. At the bottom of the collector, backdraft flappers can be attached to the back side of return air grilles. Since the horizontal face bars in most grilles extend behind the grille, it is necessary to seal the edges here with duct tape (or weatherstripping and silicone). This creates a flat, airtight surface around the perimeter on the back side before attaching the flapper vane. Be sure the lower vents are easily removable to allow for servicing of the backdraft damper. When using louvered grilles, oversize the total vent area by about 40 to 50 percent because the louvers will cut down somewhat on the "free area" available for thermosiphoning airflow. The duct size should also be increased accordingly to minimize resistance to airflow. The local sheet-metal shop or large hardware store can help select and order grilles for your particular application. Grilles aren't available in all sizes, so be sure to buy them before cutting holes.

Thermal Storage

Designing a convective rock-bed storage system is tricky. The distribution and return ducts need to be very large, and properly sized

Photo 7-5: Large, tilted thermosiphoning systems that incorporate rock storage are becoming very popular in new construction in the southwest. The thermosiphon engineering for this system was by W. Scott Morris. Architecture and construction was by Mark Jones, AIA, of The Mark Jones Corporation. It can be difficult to retrofit a system like this because the collector has to be located below the house.

storage must be used. If your site is suitable for a system like this, we recommend getting professional help in designing it, since the construction is a considerable investment in time and materials, and you won't have a blower to bail you out of trouble. The Mark Jones Corporation (which designs only passive systems) and W. Scott Morris, both of Santa Fe, New Mexico, have been designing and building convective rock-bed systems with great success and are good resources.

Mounting on Masonry Walls

If your existing south-facing wall is uninsulated masonry, a Trombe wall is your best choice if you plan to glaze all or most of it. But if you're thinking "small collector," you don't want to glaze just a small portion of the masonry. Heat would migrate from such a collector through the masonry and back to the outside, and the solar benefit would be too small to justify the effort. So for a limited-area collector on a large masonry wall, a TAP is a better choice because it can be thermally isolated from the masonry, which is practically a bottomless pit for heat absorption. If you want more collector area than a TAP, but still less than the wall area, you can consider building an active collector as described in the next chapter. One of the pitfalls in making a TAP work right is in not cutting large enough vent openings. Yes, it's harder to cut through the masonry, but you can't "go light" on vent area without compromising overall performance.

Construction of U-Tube Collectors

U-tube collectors have the advantages that only one set of vents must be cut into the building and that no backdraft dampers are needed since these collectors are self-damping at night. They are a hybrid between a TAP and a window box collector and thus are built accordingly.

U-tubes need to be about twice as thick as TAPs (2 x 10s are suitable for the framework for an 8-foot-long-airflow), so bottom supports and proper wall attachment become important construction details. Lag bolt a horizontal 2 x 6 support to wall studs below the collector, and use heavy metal angle braces here. Attach the top framework to the wall with metal angle brackets at each stud. Line

---

Room Air Circulators

If a large TAP delivers more heat than can be used in the room immediately behind it, a room air circulator can be used to move this excess heat to another room in the house. The wall cavity inside an interior stud wall is used as the air race to move air from room to room. An inexpensive register grille and a house thermostat (such as Grainger's no. 2E158) are mounted at the top of a stud space in the overheated room. A small computer fan (such as Grainger's no. 4C549 with a cfm of 100) is installed near the floor in the cold room. Wiring (Romex 14-2) between the thermostat and the fan is run through the wall behind the fan mounting. The fan is mounted in a large bed of silicone (to reduce vibration and to enable it to run more quietly), on a small square of painted plywood. A wire, with a plug attached, extends from the fan mount to the nearest electrical outlet.

Whenever the ceiling of the overheated room reaches a certain temperature, for example 80°F, the warmest air will be blown out of this room and across the floor of the adjacent room (which is the coldest part of that room). These small blowers are inexpensive to install and very effective in equalizing the temperatures between two rooms. Fans delivering 55 cfm are quieter than those delivering 100 cfm and are adequate for most applications. For safety reasons order a finger guard (Grainger's no. 4C551) along with the fan.

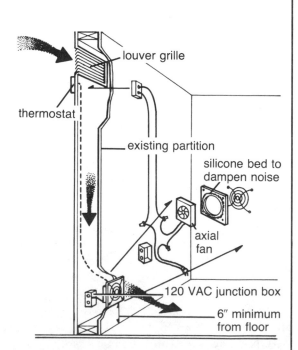

Figure 7-12: When a large TAP produces more heat than can be used in the room behind it, an easily installed room air circulator can blow excess heat into adjacent rooms.

the back of the U-tube with litho plates, and use ¾-inch Thermax to insulate the sides of the collector and for the insulated divider strip. Install turning vanes at the top and bottom of the collector. A turning vane for the cold return air at the top of the collector is made by mitering Thermax and attaching it with aluminum tape and silicone. Install the absorber as for a TAP. Use a single cube-core grille to trim out the vent opening.

Sources for Information

Convective Rock-Bed Storage Systems

The Mark Jones Corp.
826 Camino Del Monte Rey
Santa Fe, NM 87501
Phone: (505) 983-7037

W. Scott Morris
P.O. Box 4815
Sante Fe, NM 87502
Phone: (505) 982-8205

Sources for Supplies

Frisket Paper

This is available at large art or drafting supply stores.

Room Air Circulators

Room air circulators are easy to build, but if you are interested in a commercially available item, consider Tjernlund's model no. CA-7, Convect-Aire.

Tjernlund Products, Inc.
1620 Terrace Dr.
St. Paul, MN 55113
Phone: (612) 636-7500

W.W. Grainger, Inc. is another source for room air circulators.

Tedlar

One-half-mil Tedlar for convective back-draft dampers is available from the following:

Contemporary Systems, Inc.
Route 12
Walpole, NH 03608
Phone: (603) 756-4796

Hot Stuff
P.O. Box 306
406 Walnut St.
La Jara, CO 81140
Phone: (303) 274-4069

Teflon

One-mil Teflon for damper flaps or inner glazing is a Du Pont product. Write to the following for a local dealer:

E. I. du Pont de Nemours & Co., Inc.
Plastic Products and Resins Dept.
Room D-11096-1
Wilmington, DE 19898
Phone: (800) 441-7515

Transfer Tape

Industrial Specialties Division
3M Center
Building 220-7E
St. Paul, MN 55144
Phone: (612) 733-8302

8

BUILDING ACTIVE AIR COLLECTORS

There are many different designs for active air-heating systems, but there are some basic steps that are the same for all. They include: 1) making a preliminary design and a materials list, 2) evaluating your site and removing any obstructions that will shade the collector, 3) making detailed drawings of the collector and the entire ductwork system, 4) securing all the materials, 5) comparing the materials with the design plans to be sure they are compatible, 6) installing the ductwork, 7) hooking up the blower and controls, and 8) building the collector. It is best to install the ductwork and hook up the blower and controls before building the collector because it is important to begin moving air through the collector as soon as the absorber plate is installed, to keep it cool during construction.

Andy Zaugg has assisted in the design and construction of over 45 site-built systems, and no two of them have had the same design. It would be misleading to present the perfect do-it-yourself design since the best plan depends on such factors as the site, the heating needs of a house, the time and money an owner can invest and the availability of certain materials.

In this chapter we give step-by-step instructions for building an active air-heating collector that we think has a good design that is widely applicable. After detailing the construction steps for this system, we give several design variations, such as using different

materials and installation methods, to help you to design and install a system that best fits your needs and resources.

Perhaps the most important design consideration in our example collector (see photo 8-1) is the ease of construction. All of the special "solar" materials used are readily available in any major city or by mail from solar suppliers. The materials are simple to use, and there are no complex construction techniques required. This system is also easy to seal, and if care is taken in construction, it will be an attractive installation capable of lasting the lifetime of the house. In the design stage you should also be very certain the added load of the collectors can be carried by your existing roof structure. An experienced carpenter or a structural engineer can advise you if you're not sure, and the advice will be well worth any cost if it saves you from the possible consequences of an overload problem.

This collector is small enough to be built as a first solar project but large enough to save substantially on your fuel bill. It is 120 square feet, mounted on a roof pitch, and it is designed to carry seven panes of 34-by-76-inch tempered glass. The collector has a black-painted, aluminum absorber plate over the first two-thirds of the airflow channel and a selective-surface copper plate in the last third. Air flows 20 feet horizontally behind the absorber through a 1-inch-deep cavity. Three cavities (separated by baffles) are supplied by floor register boots (6 inches in diameter to

Photo 8-1: The design of our example collector is widely applicable, and the collector is easy to build.

2 inches by 12 inches) extending from manifolds on both ends of the collector.

A tilted mount was chosen for our example system in order to show how the tilt-up framing goes together. If your roof is pitched steeply enough (at least to a 40-degree pitch), you can build the collector right onto the roof plane. This same collector design can also be mounted vertically to a south wall or tilted back to the wall, with the lower end supported just above grade level by a footing. As for end uses, this collector can be used for crawl-space heating, direct-use heating and, in summer, domestic water heating.

Let's look at some of the basic design characteristics of this collector. First of all, the air makes one straight pass through the air channel behind the absorber plate so there are very few places where stagnation will occur in the airflow cavity. Long, rectangular inlet and outlet ports provide comprehensive air distribution through the collector. The single, straight pass of air avoids complex baffling and manifolding and ensures optimum heat transfer from the absorber. Each baffle space receives the same airflow, so there is no need for balancing dampers in the manifold ports. The selective surface installed in the hot end increases delivery temperatures by reducing heat loss back through the glazing.

The collector is insulated on the back and sides with ductboard ducting (R-6). The spaces between the framing members behind the plywood sheeting are filled with 3½-inch fiberglass batts. This is especially helpful in the hot half of the collector, where heat losses are highest, but may not be required in more moderate climates.

Building a Roof-Mounted Collector

Building your own collector takes time, especially for the first-timer. A contractor adept at building collectors can complete a fairly complex system in a week, but it can take the do-it-yourselfer several weeks, even months, to finish a similar project. The ductwork installation and collector framing can

be done at a leisurely pace, but when it comes time to finish out the collector and install the glazing, get some help and try to complete these steps fairly rapidly. For each construction step we've indicated the number of people required and approximately how long it will take to complete that step. Of course, these hourly figures will vary somewhat with experience and enthusiasm.

The standard tools required for building a roof-mounted collector are the following: aviation snips, carpenter's level, cat's-paw, 2 caulking guns, chalk box, electric drill/screwdriver, extension cords, felt-tipped markers, hammer, large slotted screwdriver, measuring tape (15 feet or longer), pencils and paper, skill saw (or hand saw), wrecking bar, try square or bevel square and utility knife. Make detailed drawings using a graduated ruler and paper and pencil. This step will take about three hours for a person working alone.

Photo 8-2: When using trusses for a roof-mounted collector, build the first one and make sure it fits everywhere along the collector mount before using it as a pattern. Note that the shingles have been removed before the base plates have been nailed down. Metal flashing will close up any potential leaky spots.

Detailed drawings are an absolute necessity when building a custom collector. In this section we include exact dimensions of this specific collector to make you aware of how critical the measurements are when you design your own unit. You will need a detailed cross-sectional drawing of the collector, along with several different front views—one for each layer in the collector. The glazing material you choose will probably be the

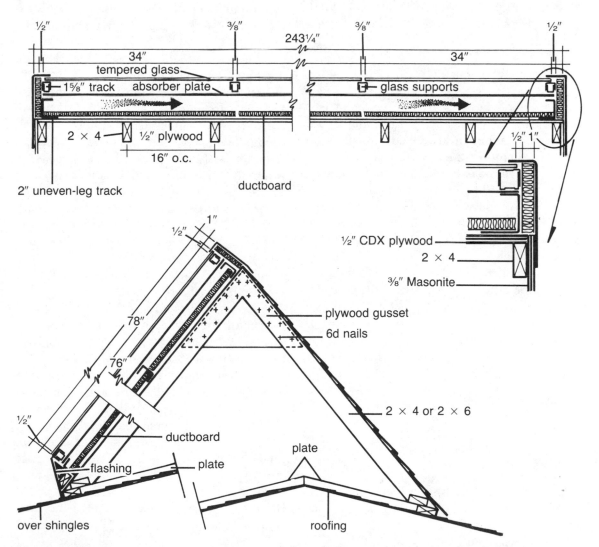

Figure 8-1: It's always a good idea to make detailed dimensional drawings for your collector to avoid surprises. Be sure to allow for ⅜-inch gaps between glass panes and for insulation on the sides of the collector. Our standard collector is 120 square feet in size and built around seven panes of 34-by-76-inch tempered glass. The outside dimensions are 78 inches by 243¼ inches. Note that any gaps between components are for visual clarity. All parts should be tight except where expansion is a factor.

limiting factor in determining size. Tempered glass, the recommended choice, is very unforgiving. It can't be cut, and any collector designed to use it must be built within close tolerances ($\frac{1}{8}$ to $\frac{1}{4}$-inch). The collector is $\frac{1}{2}$ inch longer, top to bottom, than the 76-inch glass. It is as wide as seven sheets of 34-inch glass plus an additional $\frac{3}{8}$-inch gap between the panes, or

$$(34 \text{ inches x } 7) + (\tfrac{3}{8} \text{ inch x } 6) = 240\tfrac{1}{4} \text{ inches}$$

The plywood backing for the collector extends 1 inch past the edge of the collector on the top and sides to allow for edge insulation. The lower edge of the collector is built up 2 inches above the existing roof where there is likely to be little snow at the site. (At sites with deeper snow cover, the collector may need to be raised as much as a foot.) This makes the dimensions for the collector backing $79\frac{1}{2}$ inches by $242\frac{1}{4}$ inches (see figure 8-1). The collector shown runs over the roof peak to the other side because of the site, but whenever possible the collector should be designed to run just to the peak and not over it.

Construction Steps

Attach Base Plates

Materials
2 x 4 or 2 x 6 boards (long)
16d nails
shim shingles
$1\frac{1}{2}''$ x 6'' counter flashing (sheet metal)
Tools
utility knife
skill saw with old blade
cat's-paw
hammer
wrecking bar

Labor
1 person @ 3 hours

Whenever possible the base plates of the collector mount are fastened directly to the roof framing, not just the sheathing. They must be straight and should form a truly square rectangle.

Measure out the collector perimeter on the roof, adding an extra inch to both dimensions. Snap chalk lines on the shingles, and measure diagonals to check for squareness. Make sure the finished collector will be adequately drained and that there is enough room for flashing. Remove the shingles only where the base plates will be attached. Don't nail these plates to the roof through the shingles. Use a utility knife or a circular saw with a dull blade to cut through the shingles, and set it to avoid cutting into the roof sheathing. Tear off the shingles with a wrecking bar or straight-claw hammer, and remove roofing nails with a cat's-paw.

Sweep off the exposed sheathing and snap new chalk lines where the collector base will actually be. (This time you don't add an inch to the perimeter length or width.) Locate the roof rafters and mark their locations. Cut 2 x 4 or 2 x 6 plates to fit around the perimeter of the collector mount so that joints between them rest over rafters. Tack the plates with 16-penny nails. Hold a string line along the lower base plate to check for sags in the roof, and use shim shingles under this plate to make it flat, if necessary (tolerance: $\frac{1}{4}$ inch). Drive two 16-penny nails through the plate into each rafter.

Before attaching the side base plates, carefully lift up the shingles, pulling nails if necessary, and insert $1\frac{1}{2}$-by-6-inch flashing under them. The 6-inch leg goes under the

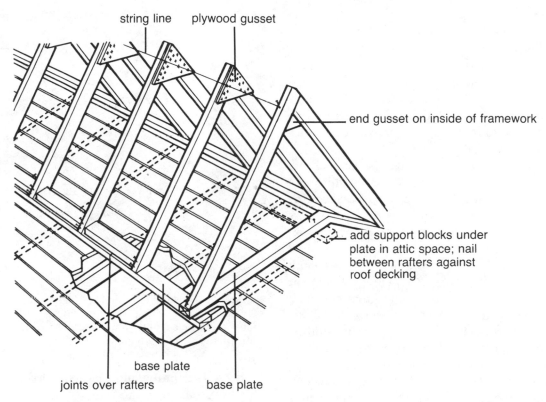

string line   plywood gusset

end gusset on inside of framework

add support blocks under plate in attic space; nail between rafters against roof decking

base plate

joints over rafters   base plate

Figure 8-2: If the base plate on the side of the collector breaks between roof rafters, add blocking between the rafters to support it. Joints in the base plate must be placed over underlying rafters.

shingles. This flashing doesn't have to be pretty so you can make it yourself. Don't nail this flashing in place yet. If a side base plate doesn't lie over a rafter, use 2½-inch screws to secure it to the sheathing. If the rafters are accessible, it is a good idea to nail 2 x 4s between the rafters under these side plates for added support. After the side plates are secure, nail the 1½-inch leg of the flashing to the sides of the plates. Carefully lift up the shingles and smear the undersides with generous gobs of roofing tar.

Lay out the truss placement on both horizontal plates, either 16 (with 2 x 4 trusses) or 24 inches (with 2 x 6 trusses) on centers.

The first truss will be located 15¼ or 23¼ inches from one edge, so that 4-foot plywood will run to the edge and still break over the middle of a truss edge. Plan your truss placement to allow for the manifolds at both edges of the collector, and place closely spaced trusses in the center.

Build and Install Collector Trusses

Materials
2 x 4s or 2 x 6s
6d and 16d nails
½" or ¾" plywood scraps

Tools
bevel square
Labor
2 people @ 4 hours each

Trusses are easier to install than frame support walls, but they must be installed carefully so that the collector mounting surface is very flat.

Make detailed drawings of a sample truss and build the first one. (Trusses are much easier to build on the ground.) Depending upon your design, both boards for the truss could have angles other than 90 degrees on both ends. The upper member should extend over the end of the lower member. Make both ends of the first truss ½ inch longer than your drawing calls for. Cut scrap ½-inch plywood for a triangular gusset, and tack it (don't secure it permanently) into place. After hauling it up onto the roof, hold the truss in place on top of the horizontal base plates, and make sure this truss fits everywhere along the collector mount perimeter. Cut one or both ends to get the right fit and tilt angle for the collector. Disassemble this truss and use it as a pattern to make the other trusses.

The plywood gussets should be quite large (14-inch triangles) and nailed to both truss members with six or seven 6-penny nails. Build the first truss from your pattern. Make sure the butt joint between the two legs is tight (tolerence: ⅛ inch) before nailing the gusset. Check the first finished truss for correct fit along the length of the collector mount perimeter before building the rest. On one truss nail the gusset on the opposite side so it can be placed on one end of the collector and not interfere with the siding.

Toe nail the two end trusses to the base plates with 16-penny nails (gussets inside). Plumb them with a level, and brace them

Photo 8-3: Blocking is nailed to the base plate between trusses to help support them. The base plates must be securely attached to roof rafters with plenty of long nails.

with 2 x 4s running onto the roof. Drive one 16-penny nail into the roof if necessary (it won't cause a leak). String a horizontal string line 3 inches below the peak of the trusses and pull it *tight* (see figure 8-2). Toe nail the other trusses onto the plates. Hold them just behind the string line and right on the edge of the base plate. Sight down the truss line to make sure that the collector plane is absolutely flat, all trusses within ⅛ inch of the line. Nail 2 x 4 blocking into the upper and lower plates between the trusses. Nail the trusses to the blocks as you put them in.

Cut Duct Holes and Sheath the Collector

Materials
½" CDX plywood
8d nails
Tools
standard tools

Labor
1 person @ 2 hours

Before covering the collector framework with plywood, cut holes through the roof for the ductwork. Consult your drawings and determine the approximate locations where inlet and outlet ducts will run through the roof. Tear off the shingles in this area; there's no need to be neat about it since the holes are inside the "room" formed by the trusses. Locate the roof framing and cut away the sheathing between rafters, but don't cut the rafters, of course. Make holes large enough to accommodate both the ductwork and the insulation that will be wrapped around it. An extra hole may be necessary for you to crawl in and out of the attic space. If rafters are in the way of your proposed ductwork installation, you will need to use ductwork elbows to go around them.

Sheath the collector surface with ½-inch CDX plywood, but don't sheath the north roof at this point. Use 8-penny nails, 8 inches on center, nailing into every truss. The face grain of the plywood *must* be perpendicular to the truss framing. Install it so that the best (C) side of the plywood faces out (south). Seal all plywood seams with beads of silicone.

Frame in the ends of the collector mount. Place the studs 24 inches on center, and toe nail them to the base plates and the end trusses. Allow for a 24-inch-wide access door in one end of the collector mount.

Build the Collector Manifold

Materials
button-lock ducting
4 end caps
6 floor register boots; each 6″ diameter to
   12″ x 2¼″

½″ or ¾″ hex-head sheet-metal screws
silicone caulk (any color)
Tools
standard tools
Labor
1 person @ 2 hours

The inlet and outlet manifolds should be made up in advance to allow the silicone seals in them to cure before installation. They must be completely airtight.

Cut the two L-shaped halves of 8-by-10-inch standard *button-lock* rectangular ducting to length. Lock the two halves together.

Photo 8-4: After the plywood collector backing is in place, you can position the manifold and accurately mark the holes for the inlet and outlet ports. Note that the north roof of the collector mount hasn't been completely sheathed, to allow for easier access when installing the manifolds behind the collector.

It may be necessary to slightly open the female side of the joint with a screwdriver and tap the joint together with a hammer. Haul the manifold under the collector to make sure it fits before putting on the end caps (see additional information on building manifolds in chapter 9). Three floor register boots will be installed on each manifold. They should be placed at one side of the manifold, to be close to the outer edge of the collector. All the boots are positioned to be centered in their respective baffle spaces (see figure 8-3). Locate and mark the placement of these ports on the manifold with a marker. Cut out the round holes for them with snips. Also cut a 10-inch-diameter hole for the distribution ductwork in the back of the manifold. The 10-inch hole for the inlet manifold is below center, and the hole for the outlet manifold is above center to help create the most even distribution through the airflow cavity. Insert the round, 6-inch ends of the floor register boots (tabs in and out), secure them with screws, and seal them with silicone. Let the silicone cure three or four hours before moving the assemblies.

Cut Holes through Plywood and Install Manifolds

Materials
manifolds
½" sheet-metal screws
silicone caulk
plumber's strap
Tools
jigsaw (or keyhole saw or reciprocating saw)
Labor
2 people @ 1 hour each

The manifolds are mounted onto the plywood back of the collector (truss sheathing) at this point. The ports extending into the collector will later be fastened permanently.

After the manifolds are built, hold them in place on the collector, and roll them over 180 degrees so that the boots rest on the plywood near the collector perimeter but at least 3 inches inside the perimeter. Carefully mark the holes for the ports on the plywood, and cut them out with a jigsaw. Do a neat job so the registers will just fit through the holes (tolerance: ¼ inch or less). When in place the ports should extend at least 2 to 3 inches beyond the face of the plywood. Make a temporary 1-inch-wide tab on each port. Bend these tabs over onto the plywood and screw or nail into place. These tabs will later be bent back up when the insulation is put on the collector. Secure the manifold to the trusses behind the collector with a plumber's strap. Fill any gaps between the metal and the plywood with silicone. Avoid the areas around the tabs for now.

Sheath and Shingle the North Roof

Materials
15-lb felt
asphalt shingles
1½" roofing nails
flashing
drip edge
½" CDX plywood
Tools
utility knife
staple gun
Labor
1 person @ 4 or 5 hours

The north roof is sheathed and shingled before the next step in constructing the collector. Sheath the north roof as you did the collector surface with ½-inch CDX plywood. There is no need to caulk any seams, though. Staple 15-pound felt onto the new roof. Vertical and horizontal overlaps on the felt should be about 3 inches. Nail metal drip edge along the sides of the roof over the felt. Allow a space on the underside of the drip edge so that siding can be slipped up under it. Nail on the shingles so they overhang the drip edge ½ inch. If you've never dealt with shingles before, read the directions on the packaging and get advice from someone who knows. A double row of shingles or flashing is a good idea where the new roof joins the existing roof.

Install the Solar Ductwork, Blower and Thermostat

Materials
duct tape
round ducting
elbows
blower assembly
½" or ¾" hex-head sheet-metal screws
Romex (electrical wire)
thermostat
plumber's strap
silicone caulk
Tools
snips
crimpers or needle-nose pliers
electric screwdriver or variable-speed reversible drill w/hex bit
slotted screwdriver
diagonal cutters and wire strippers
Labor
2 people @ 8 hours each (varies greatly with system)

Assemble all the ductwork and the blower mount before continuing work on the collector. This is a big job and should be done early on so the blower can be turned on as soon as the absorber plate is installed (for specifics see chapter 9, which deals with ductwork).

To tie into the manifold, cut slits into the male end of one 10-inch elbow and into the female end of another elbow. These slits should be about 1 inch apart and 1 inch deep. Bend four or five of the tabs out to make an in-and-out attachment (detailed in chapter 9).

Insert the elbow with the modified male end into the inlet (cold) manifold and the one with the modified female end into the outlet (hot) manifold (male ends always point in the direction of airflow). Position both elbows so they will attach to the ducts that come up through the roof. Reach up inside these elbows and bend over the tabs inside the manifolds so they hold the elbows in place.

Crimp one end of a 2- or 3-foot section of ducting and push it through the "inlet" hole into the attic, and repeat this with a section of duct for the "outlet" hole, connecting both sections with their respective manifold elbows. When everything is connected, secure the ductwork. Start at the manifold and screw the four or five exposed tabs to the manifold. Caulk all cracks at this junction. Tape all joints in the elbows and the overlaps, and put ¾-inch sheet-metal screws through the tape at all joints. Run all of the hot and cold ducting in the system (see chapter 9 for details). Also, at this point install the blower and do all of the electrical and thermostat hookups (see information on installing blowers in chapter 9 and information on installing thermostats in chapter 6). You can wait to install the distribution ductwork if you choose, but the ductwork must be finished to the blower, and it must be operating properly before you continue work on the collector.

Install the Perimeter
Framework and Collector Baffles

Materials
2″ uneven-leg track
1″ Phillips screws or ¾″ hex-head screws
silicone caulk
⅛″-diameter-by-¼″ pop rivets
Tools
awl punch
pop-riveter
Labor
1 person @ 1½ hours

Both the perimeter framework and airflow baffles will be made from 2-inch uneven-leg drywall track. The longer leg will attach to the plywood. Before working with it, clean the outside of the 2-inch track with rubbing alcohol.

Install the sides of the collector first. Carefully measure out the collector on the plywood truss sheathing. The top and sides will be 1 inch in from the edges, and the bottom will be 2 inches above the lower edge of the plywood. To repeat, the dimensions for this collector are 76½ inches by 240¼ inches. Snap chalk lines at the perimeter, and measure diagonals to check for squareness. Cut a small V-notch (10 degrees) in both legs of one 10-foot section of 2-inch track about 2 feet from one end. Bend to form a 90-degree angle (see the information on working with sheet metal in chapter 5). Measure down the track 76½ inches (the height of the collector), and cut and bend another 90-degree angle at that point. Snip both ends to allow the next sections of track to telescope into the joints. With an awl punch or hammer and nail, start

pilot holes through the long leg of the track every 8 inches to help start the screws if they won't start themselves.

Run a bead of silicone on the down side of the track, and screw it into place. Work rapidly. After you run the bead of silicone, it will skin over in about five minutes (see the information on silicone caulking in chapter 5). Cut the next section of track and install it in the same manner.

Cut and install the baffles inside the collector. Snap chalk lines to keep them straight, and attach them to the perimeter track by inserting the baffle track inside the perimeter track. The joint is pop-riveted together, and the baffle track is secured to the plywood with screws spaced about every 8 inches. The baffle track doesn't need silicone between it and the plywood backing since it is inside the collector.

All joints, overlaps and angles in the track should be secured with pop rivets on the upper (sun side) surface and screwed together and to the plywood through their lower legs. After all the track is installed, run a bead of silicone along the inside joint, where the leg of the track rests on the plywood, and inside all joints in the track. Smooth and fill joints and gaps with silicone using your finger. Also, run a bead of silicone along the outside edge of the perimeter track where it joins the plywood. Smooth and fill. Use plenty of silicone. The perimeter of the airflow cavity must be completely airtight.

Check the overall measurements of the perimeter track after installation to make sure they conform to your plans (tolerance: ¼ inch). Sight down the long sections of track and make sure they're flat and straight. The top surface of the track will probably look a bit wavy, especially where it has been joined. Make it as flat as possible, but don't be overly

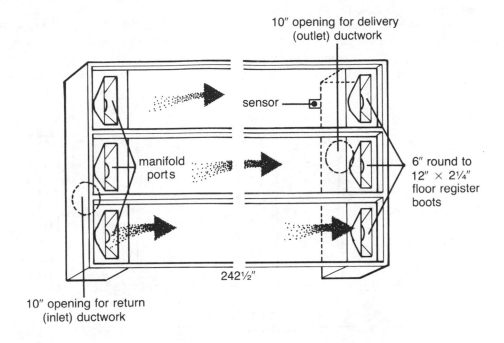

10″ opening for delivery (outlet) ductwork

sensor

manifold ports

6″ round to 12″ × 2¼″ floor register boots

242½″

10″ opening for return (inlet) ductwork

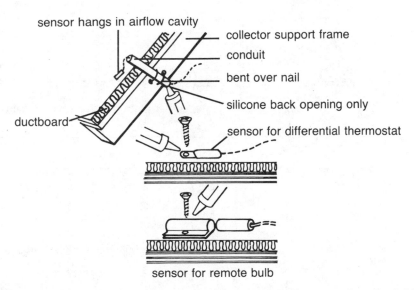

sensor hangs in airflow cavity

collector support frame

conduit

bent over nail

silicone back opening only

ductboard

sensor for differential thermostat

sensor for remote bulb

Figure 8-3: The return duct from the house ties into the lower end of the inlet manifold while the delivery ductwork ties into the top of the outlet manifold. This helps to ensure an even airflow through all three baffle spaces. A thermostat sensor is placed in the airflow cavity near a hot outlet port. Running this sensor through a short piece of metal conduit, sealed with silicone, allows for easy removal if the sensor fails. Another possibility is to simply fasten the sensor to the collector backing.

concerned as long as it is within ¼ inch for its entire length. Once again, measure diagonals to check for squareness (tolerance: ¼ inch). If the absorber plate for your collector is something other than galvanized metal, smear a film of silicone with your finger to barely cover the upper (sun side) surface of all the track. This will separate the two dis-

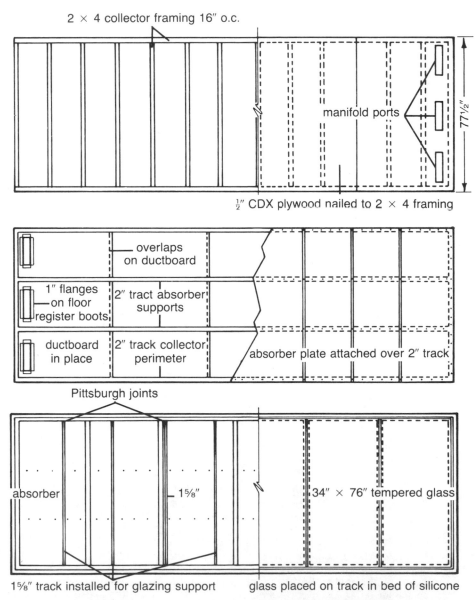

Figure 8-4: This illustration shows the different layers involved in building our example collector. It is best to completely install and seal each layer before starting on the next one.

similar metals and inhibit galvanic corrosion. Silicone paint, available in quart cans, works best for making this thin film, but the silicone in your caulk gun is perfectly adequate (see information on galvanic corrosion in chapter 5). This step can be done before assembling the track or as the last task of the day. It can also wait until you have installed the insulative backing in the collector. You will end up with a mess if you try to install the absorber over fresh silicone.

Install the Insulative Backing

Materials
ductboard
2″ Phillips-head screws
1⅝″ screws
sheet-metal scraps
silicone
Tools
long, serrated knife

Labor
2 people @ 1½ hours each

This collector design uses ductboard ducting material for the insulative backing. It is installed with the foil facing the airflow cavity, and all edges and joints are sealed with silicone. This step can be done before the manifolds are installed if the ductwork has been completed and the manifolds are ready for connecting with the distribution lines. Cover the ductboard in case of rainy weather, or it will become waterlogged.

Cut off the laps on both edges of a sheet of 1-inch ductboard ducting, leaving the foil on one edge to act as an overlap. Use a sharp serrated knife with a long blade and a sawing motion for best results when cutting ductboard. Your cuts must be neat and not leave the edges ragged. Cut the ductboard into 47¼-inch pieces that will fit snugly inside the perimeter and the baffles (tolerance: ⅛ inch).

Photo 8-5: A 2 × 4 temporarily nailed across the collector aids in installing the perimeter and baffle track. Insulating the ductwork for this collector was a real challenge because it was right in the way of the access door. Better planning would have eliminated the problem before it became a problem.

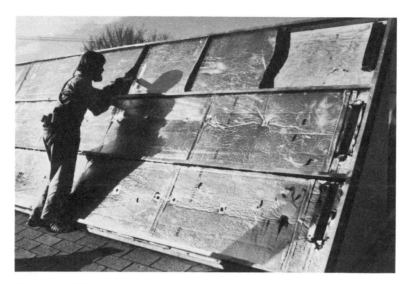

Photo 8-6: The foil overlaps on the ductboard backing must be sealed with silicone. The ends of the manifold boots have been bent over and sealed to the ductboard, and ductboard support washers have been installed. Note that the boot in the lowest baffle space is above center to make the manifold shorter and easier to install behind the collector.

Start at the hot end of the collector and work towards the cold end. Overlaps in the foil facing should point towards the hot end, the direction of airflow. It will be necessary to bend the ductboard slightly in the center to get it to pop into some of the baffle spaces. Before installing the ductboard over the manifold ports, remove the screws on the tabs that are holding the manifold in place. The manifold should be self-supporting at this time, and the temporary tabs can be straightened. Finish sealing the joints between the ports and the plywood around the tabs with silicone. Cut holes in the ductboard to fit snugly around the manifold ports, and put these sections in place. Snip the edges of the ports back to where they contact the ductboard, to form flanges. Bend the flanges over onto the ductboard, barely compressing it, and screw each one through the ductboard and into the plywood with five or six 1⅝-inch screws (see the information in chapter 9 on attaching ductwork to the collector). Silicone all seams and holes to completely seal the ports to the ductboard. Use plenty of silicone, and point it into all gaps with your finger.

Another method of covering joints in the ductboard involves stripping the foil facing off 2-inch-wide ductboard scraps and thoroughly scraping the insulation off the back sides. These "Band-Aids" can then be siliconed over joints. High-temperature aluminum duct tape can also be used. Don't use regular duct tape or the iron-on tape that is typically used to attach ductboard. Overlaps work better, so use them whenever possible. Whether you use overlaps, ductboard "Band-Aids" or aluminum duct tape to seal joints, run a generous bead of silicone along all exposed edges and flatten it to firmly seal the entire joint. Run a bead of silicone along all edges of the ductboard where it contacts the track and baffles. Smooth and point this bead with your finger, pushing it into the crack to get a good seal. When sealing the ductboard, take your time, and crane your neck and carefully inspect each joint to make sure there are no gaps between the edge of the ductboard and the track. This is important.

There should now be two completely airtight seals between the airflow cavity and the air behind and outside the collector: one under and outside the perimeter track and the other one where the ductboard butts against the track.

Cut sheet metal or litho scraps into 2- or 3-inch squares. Punch holes in the centers of them with a hammer and nail. Put silicone on the back sides, and screw them through the ductboard and into the plywood backing with 1⅝-inch screws. These screws and their support washers should be placed about every 16 inches and over ductboard joints in between the baffles. They should just barely compress the ductboard. Put a glob of silicone on top of each screw head to insure an airtight seal. Reseal the overlap edges on the ductboard joints if necessary. At this point, spend 15 minutes carefully going over the entire surface of the ductboard, and make doubly sure all joints and overlaps are completely sealed with silicone.

Install the Thermostat Sensor

Materials
thermostat sensor
silicone
Tools
drill with ½" bit
Labor
1 person @ 15 minutes

The thermostat sensor should be mounted in the upper outlet end of the collector, 1 foot from the top outlet port. Drill a ½-inch hole through the ductboard and plywood backing and insert the probe into the airstream from behind the collector. Secure the probe as shown in figure 8-3. Seal both sides of the hole with silicone. Mount the thermostat, and make the connections with the blower if this hasn't already been done. Cycle the blower to make sure the wiring is complete and correct and that the thermostat and sensor are not defective.

Lay Out the Absorber Plates and Glazing Supports

Materials
none
Tools
felt-tipped pens (2 colors)
Labor
2 people @ ½ hour each

If your drawings are complete, this step will go quickly and offer no surprises. You know what your glass size is, so you can determine where the glass supports will fit on top of the absorber plate after it is installed. Make very careful measurements to locate the center of the ⅜-inch gap between each of the glass pieces, and mark the perimeter track, top and bottom, with one color. Next determine where the vertical joints in the absorber plate will be located. Since the glass supports will rest directly against the absorber plate, the bulging Pittsburgh joint in the absorber surface must not be located under a glass support. Mark the locations of the joints in the absorber on the perimeter framework (top and bottom) with a different color.

Prepare the Absorber Plate

Materials
absorber plate materials

baking soda and water
high-temperature paint
Tools
mop
Labor
1 person @ 2 hours

Cut the absorber plates to length, ½ inch shorter than the width of the collector's perimeter track. Take them to a sheet-metal shop and run Pittsburgh joints on one edge. If you are using a selective surface absorber, run it through the machine with the selective side up. With litho plates, run them through with the printing side up (see the information on Pittsburgh joints in chapter 5). On the way home, handle them carefully and keep them flat in order to avoid kinking the Pittsburgh joints.

Lay the aluminum absorber on a flat surface (such as a piece of plywood), and clean the exposed surface (the side that will be painted) with a baking-soda-and-water solution. When installed, the rolled-over part

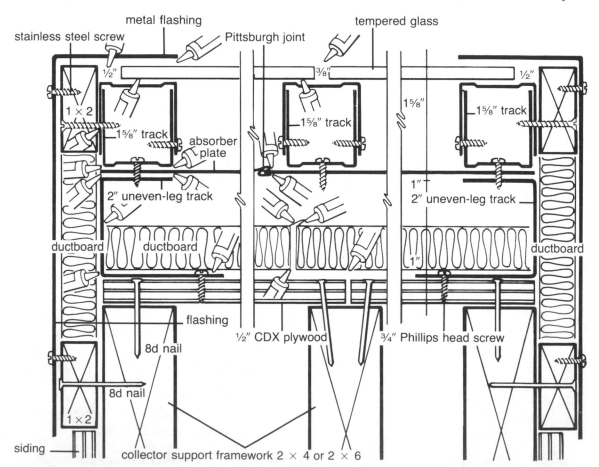

Figure 8-5: *Above and on opposite page.* All points requiring a silicone seal are indicated in these cross sections and cutaway views of our standard collector. For clarity, this illustration shows gaps between some materials that don't exist in the finished collector.

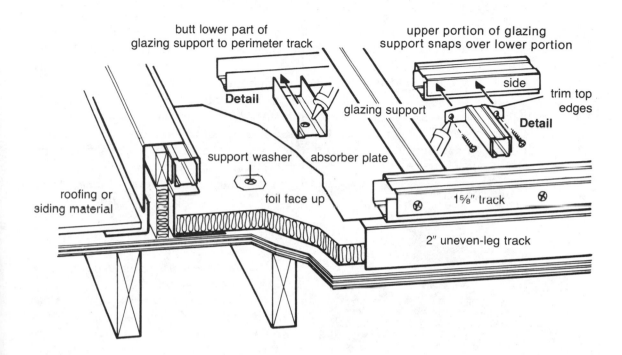

of the Pittsburgh joint will be facing the sun. Make sure you know which side is up. Rinse the absorber well and allow it to dry completely, especially in the joint (see the information on cleaning the absorber in chapter 5). Paint the exposed side of the aluminum absorber plates with high-temperature paint. Use a thin even coat and allow it to dry 10 or 15 minutes in the sun (see the information on painting in chapter 5). Run a slotted screwdriver down the Pittsburgh joint, and twist it very slightly to open up the joint just a bit. This will allow the next plate to fit easily into the slot. All of this can be done in advance.

Install the Absorber Plate

Materials
absorber plate with Pittsburgh joint
black silicone
7/16" Phillips screws

Tools
electric screwdriver (screw gun)
snips
Labor
3 people @ 2 hours each (installation time)

Before you install the absorber plate, all of the ductwork should be completed, and both the blower and thermostat should be mounted and operating. All debris should also be cleaned from the airflow channels.

We have two different methods for installing the absorber. Whichever you use, you will need at least one helper, and you will have to work fairly rapidly. It is important to place the glazing on the collector as soon as possible after the absorber is installed, especially if you use a selective surface.

Method 1: Start installing the absorber plates at one end of the collector. Hold the first plate in place, with the non-Pittsburgh edge flush with the outside edge of the perimeter track. Once you are satisfied with the placement, screw the plate to one of the horizontal baffle tracks with two $7/16$-inch saber-pointed screws. These screws, and all screws holding the absorber to the underlying track, should be placed near the fold edge in the track, not near the cut edge, which can bend under the pressure of drilling and screwing.

Photo 8-7: The absorber in the inlet end of this collector is aluminum mobile-home siding with a wood-grain texture. A selective surface absorber was used in the hotter, outlet end. The plate is being sealed and screwed to the underlying baffles and perimeter track, which has been covered with a film of gray silicone to act as a barrier against the possibility of galvanic corrosion. The Pittsburgh joint on its left edge has been opened slightly and is ready to receive the final absorber plate.

If a screw happens to miss the track below, fill the holes in the plate with black silicone.

Open a tube of black silicone with a large $1/4$-inch hole. Make sure you use a screw gun, and have plenty of $7/16$-inch screws handy. Lift the edges of the absorber, and run a flat bead of caulk onto the perimeter track. Pull the edges of the absorber to make sure it's tight and smooth, and then screw it to the perimeter track about every 4 inches. Work rapidly. The silicone will begin to skin over in five minutes, and the plate needs to be well attached within this time limit. A small amount of silicone should ooze out around the edges of the absorber plate if you have used the correct amount.

After fastening the top, bottom and sides of the first piece of absorber, continue fastening it to the baffle tracks. The screws here must be imbedded in silicone. Put very small dabs (chocolate chip size) on the absorber every 6 inches on top of the horizontal baffles. Drive $7/16$-inch screws through the dabs and into the baffles. The screws around the perimeter don't need to be imbedded in silicone because they will be removed after the silicone cures and before the glass supports go in place.

Install the next piece of absorber in a similar manner. Move your ladders into place, and open the Pittsburgh joint slightly with a slotted screwdriver. Carefully run a small bead of silicone inside the Pittsburgh joint. Move the absorber plate into position, and wiggle the edge of the plate being installed all the way into the Pittsburgh joint. Start at the top and work your way down the joint. Have all hands available to steady the plate. When the flat edge is completely inserted, drive two screws through the plate into the baffle track. Run a bead of silicone on the top and bottom perimeter track and screw down these edges. Fasten the plate to the baffle tracks using "chocolate chips" of silicone. Remove the two initial support screws, and put silicone under them before reinserting these screws. If it is necessary to cut the last absorber plate to size, do this before installation. After all the absorber plates are installed, run a very small bead of silicone along each Pittsburgh joint to ensure a total seal and to make a neat appearance. This is best done after the silicone in the joint cures a bit. Avoid smearing a lot of silicone on the plate, but make an airtight seal.

Method 2: Putting in and then removing screws along the perimeter of the absorber plate is somewhat time-consuming, and there is another approach for fastening the perimeter. Don't run a bead of silicone under the plate at the perimeter track at this point—simply support it about every 2 feet with a $7/16$-inch screw. Securely attach the absorber to the baffles as outlined above. When the glass supports (the next layer) are cut and ready to be installed, remove the perimeter support screws, lift up the edge of the absorber, and run a large bead of silicone onto the perimeter track. Lay the plate down into this bead, and run another bead on top of its perimeter. Screw the glass-support track into place through the absorber and into the underlying 2-inch track. This will require rapidly inserting screws every 4 inches. Cover the screw heads in the glass-support track with silicone (see the following section on installing glass supports).

It is almost impossible to install all of the absorber plates so that they lie completely flat, and it is inevitable that there will be a few bulges and waves in the absorber surface that can be seen through clear glass. These can be kept to a minimum by installing the absorber plates while they are warm, by pulling and holding them tight, by making sure they are square in the collector and by

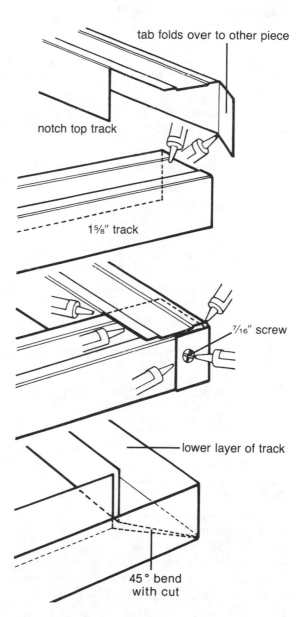

Figure 8-6: Corners for the $1\frac{5}{8}$-inch support track are cut and bent in the top layer of track as shown. Forty-five-degree angles are cut in the corners of the lower leg of the track. These cuts at each corner are the most difficult ones you will have to make.

having plenty of help during this part of the installation. Using nonreflective, low-iron glass helps to hide the waves.

Once all the plates are installed, do a smoke-bomb test (described later in this chapter) to check for air leaks. Turn on the blower as soon as possible to prevent undesirable stagnation. Before installing the glazing, adjust internal balancing dampers, if your system has them.

Install the Glass Supports

Materials
$1\frac{5}{8}''$ track
black silicone
1″ Phillips screws
$\frac{7}{16}''$ Phillips saber-pointed screws
$\frac{3}{4}''$ hex-head cadmium-plated screws
Tools
snips
electric screwdriver
Labor
2 people @ 1½ hours each

The glass supports are a rigid framework made from two pieces of $1\frac{5}{8}$-inch track, one inserted into the other. They are installed on top of the absorber plate and screwed to the perimeter track and the baffle tracks.

Before proceeding, clean the track with rubbing alcohol. Paint the three outside surfaces of half of the track with high-temperature, flat black paint. This is for appearance' sake only and, therefore, it isn't necessary to paint the inside surfaces because they won't show. Paint the bottom edge only on the other half of the track. This edge will rest against the absorber plate, and the coating of paint will help prevent any galvanic corrosion that could occur.

First cut the pieces of track that will be used for vertical glass supports. Cut two pieces for each support, 76 inches long, and set them aside. These pieces must be continuous, while joined pieces can be used for the long perimeter supports.

Install around the perimeter the sections that were painted on one side. Cut 45-degree angles on the track, as shown in figure 8-6, to form the outside corners. Remove the perimeter screws from the absorber if you used method 1 to install the absorber. Run a large bead of black silicone around the perimeter of the absorber ½ inch in from the edge. Put an extra glob on each screw hole that penetrates the absorber. Screw the precut pieces of track in place with ¾-inch, hex-head, cadmium-plated screws, either 8 inches on center (the absorber has silicone under it in method 1) or 4 inches on center (the lower surface of the absorber perimeter is being sealed at this point in method 2). Butt the joints in this layer of track rather than overlapping them. Cover all screw heads with silicone.

Once the first layer of perimeter track is installed, fit the prepainted sections of track over the installed pieces. The upper track should be snapped in place over the lower track and then pushed down into position. Don't try to bend in the legs on the lower track because this will make the job more difficult. Join the corners as shown in figure 8-6. The joints in this second layer will overlap each other and be sealed with black silicone. They shouldn't fall over the butt joints in the lower layer. After the second layer of track has been run around the perimeter, secure it to the first layer with ⁷⁄₁₆-inch screws (2 feet on center) through the sides of the track on the inside and outside of the collector. Cover these screw heads with silicone, and run a large bead of black silicone around the inside of the track where it contacts the absorber plate. Smooth the silicone with your finger to completely fill this joint. Also run a bead around the joint on the outside of the perimeter track to insure that an airtight space will be created between the glass and absorber.

The vertical glass supports are each made as a unit and inserted inside the perimeter track if you are using small panes of glass, but it is better to attach the lower track to underlying baffles. Take one of the precut pieces you set aside earlier that is painted on one side, and cut it to just fit inside the perimeter track. Locate and mark the locations of the baffle tracks. Remove the track from the collector and drill two ⅛-inch holes through it where the baffles are located. Put large globs of silicone over the holes on the back side of the track, and carefully lower the track onto the absorber. Drive two ¾-inch hex-headed, self-tapping screws through the track into each underlying baffle, and cover the heads with silicone.

Cut the upper track 2 inches longer than the lower track, and cut and bend a tab on each end as in figure 8-6. Predrill a ⅛-inch hole in each tab and snap the upper section over the lower section. Make sure the upper track is flush with the perimeter track before screwing the tabs to it. Cover the screw heads and the edges of the tabs with black silicone. There will be spaces between these vertical supports and the absorber (since the absorber won't be perfectly flat), but these spaces present no problems, and the perimeter of the dead air space is the only place a seal is required. The lower track of the glass support may bow slightly (in or out) since it is attached to the baffles. Make sure that the upper piece of track is flush with the perimeter (tolerance: ⅛ inch) before screwing the two pieces together on the sides.

Before installing the glass, attach flashing to the glass support at the lower perimeter. The glass will lie on top of this flashing, allowing water to run off the bottom of the collector. To make a nice appearance, paint the edge of the flashing that will lie under the glass flat black. Insulate under the bottom of the collector before attaching the flashing. Run a large bead of black silicone on the bottom glass support, and pop-rivet (don't screw) the flashing into place. Flatten the silicone that oozes out of the joint with your fingers. Other flashing options and some ideas for supporting the bottom of the collector are discussed later in this chapter.

Glaze the Collector

Materials
³⁄₁₆″ tempered glass
2″ drywall screws
litho-plate scraps
black silicone
Tools
standard tools
Labor
2 people @ 3 hours each

The glazing for this collector is ³⁄₁₆-inch-thick tempered glass in 34-by-76-inch panes. The glass will be set into a bed of silicone with a ⅜-inch gap between each pair of panes. Check the measurements for all of the glass supports to make sure they are in the proper place. Clean both sides of the glass. A squeegee (rubber blade) is very helpful. This is your only chance to clean the inside of the glass. Streaks will show, so do a thorough job. Always install nonreflective solar glass with the rough surface inside the collector.

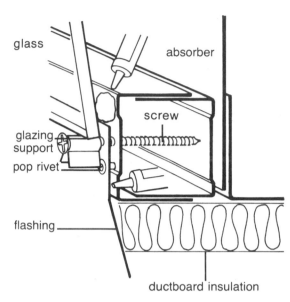

Figure 8-7: Two padded support screws hold each pane of glass at the bottom of the collector during installation. Carefully drive in the 2-inch screws until they just make contact with, but don't penetrate the back of, the 1⅝-inch track. Cover the exposed threads on the screws with small scraps of light-gauge metal. Flashing at the bottom of the collector is installed under the glass with pop rivets and black silicone.

Determine the lower edge of the glass at both ends of the collector and snap a chalk line the entire length. Insert 2-inch screws into the track just below this line and near each glass corner to act as supports. Cut and bend small pieces of litho plate to fit over the ½ inch of screw threads that should be exposed. This will allow the glass to slide smoothly over them. Read the rest of this section, and do a dry run with a pane of glass before bedding it in silicone.

Run large beads of black silicone onto the glass supports ¼ inch out from the inside edges. It is important to work quickly so that this step is completed before the silicone skins

over. After applying a large bead of silicone (½-inch round) on the entire perimeter, rest the lower edge of the first piece of glass against the support screws. Avoid touching the inside surface of this glass with fingers. Slowly lower the upper end into place. Two or three people will be required to do this. By pushing and pulling slightly on the glass, you can insert the nozzle of the caulking tube into the crack between the glass and track to work

Photo 8-8: Although one person can set glass by himself, it is very handy to have help with the siliconing. Take care not to smudge the inside of the glass during installation, and remember to put the textured side in if the glass has a textured side.

Photo 8-9: This is a sure sign of a well-sealed collector.

more silicone into the seam to fill it completely. It is, of course, important to completely fill any gaps between the glass supports and the glass and make an airtight seal. There should be a continuous silicone seal at least ½ inch wide all around the glass. Install the next pane in the same manner, leaving a ⅜-inch gap between the panes. You will need about one tube of silicone for each pane. If the panes are slightly out of square when placed on the support screws, put small wooden shims between the support screw and the glass on the side that is low to compensate. There must remain at least a ¼-inch gap between the panes. If necessary, move the glass slightly off center on the glazing supports to obtain this gap.

Don't try to clean silicone smears on the outside surface of the glass at this point. After the silicone has cured, smears can be scraped off with a single-edged razor blade. After two or three days, remove the support screws and caulk the screw holes. Fill the gap between the panes with black silicone, and seal well around the perimeter of the glass with silicone. You need a continuous, watertight seal around all edges. This type of silicone seal allows for expansion and contraction of the glass and for easy replacement if a pane breaks.

Trim Out the Collector

Materials
1 x 2 furring strips
1⅝" Phillips screws
silicone (black and clear)
ductboard or duct insulation
flashing
siding
½" Masonite
8d nails
¾" stainless steel screws
Tools
standard tools
Labor
1 person @ 6 hours

It is important to weatherproof the collector as soon as possible after it's assembled.

Cut 1 x 2 furring strips equal to the length of the sides of the collector. These will be attached to the glass-support track. Drive 1⅝-inch Phillips-head screws into this furring strip until the points just come through the wood. They should be spaced about every 16 inches. Put a glob of silicone on each screw point, and fasten the strip to the track so it

Photo 8-10: The frame for this collector was built on a driveway and hauled into place. Although there is an adequate footing, the bottom of the collector would be better located 8 inches above grade, especially in areas with a lot of snow. The brick cornice at the top of the collector was a limiting factor in this case.

is flush with the glass surface. The silicone will seal the screw's penetration into the track. This furring strip is needed for fastening the flashing and provides some edge insulation. This space will seldom reach temperatures above 130°F so it is all right to put wood here. On the sides of the collector, screw 1 x 2 furring strips parallel to those on the glass supports and slightly below the ½-inch CDX plywood (see figure 8-5). These strips will be used for attaching the bottom of the flashing and will act as trimmers for siding. Cut ductboard or duct insulation to fit snugly inside this space and install it, foil face inward. The ductboard insulates the sides of the airflow cavity. Insulate the top edge of the collector in a similar manner.

Cut ½-inch Masonite siding to enclose the ends of the collector frame. Nail this in place with 8-penny galvanized nails.

Install the flashing (have it folded or "broken" at a sheet-metal shop) in 4-foot sections, working from the bottom of the collector upward. These relatively short sections allow for expansion and contraction and usually make for a neater appearance. Seal the 1-inch overlaps in the flashing with clear silicone, and use clear silicone to seal the edge of the flashing to the glass. It may be necessary to tape the flashing to the glass for an hour or two to hold it down while the silicone cures. Screw the flashing to the 1 x 2 furring strips using ¾-inch stainless steel screws at each overlap.

Completing the Job

In most construction work, whether it involves building a house or a simple do-it-yourself backyard project, it usually seems like the last 10 percent of the work takes 50 percent of the time. Collector building is no exception. Some construction steps, such as the distribution ductwork or installing an air-to-water heat exchanger, can wait, but always be sure to finish your system up to the point where the collector is operating properly and the collector and its mount are completely weatherproof.

Smoke-Bomb Testing

Air leaks, even very small ones, can hurt collector performance. A good, inexpensive way to detect leaks is to drop a smoke bomb

---

Materials Checklist for a Simple Active Collector System

This is a general list of materials. Use it to determine the materials needed for your system.

Collector

Framing materials
dimensional lumber for collector mount
½" CDX plywood for collector backing
8d and 16d nails
Silicone caulk
1 10-oz tube per 4 ft² collector (mostly black, a few tubes of clear)
Track
2" uneven-leg track for airflow cavity and baffles
1⅝" track for glass supports
Ductboard ducting
area of collector plus 6" strips for entire collector perimeter
Absorber plate (with Pittsburgh joints)
area of collector plus scraps for ductboard support washers (½ ft² per 10 ft² collector)
Paint, high-temperature flat black
1 can per 20 ft² collector
Tempered glass
Screws
2" drywall screws (2 for each pane of glass plus 1 per 2 ft² of collector for support washers)
¾" hex-head screws with saber points (8 per ft² collector)
⁷⁄₁₆" Phillips screws with saber points (8 per ft² collector)
1¼" drywall screws for flashing (1 per ft² collector)

¾" stainless steel screws (2 per linear foot of flashing)
Pop rivets
¼" long (1 per ft² collector)
Flashing

Ductwork

round ducting
duct tape (1 roll per 100 linear ft of ductwork)
½" or ¾" hex-head screws (3 per ft of duct run)
adjustable elbows
transitions to and from blower
vibration collars
silicone (2 or 3 tubes)
plumber's strap
insulation (area of duct plus 20%)
manifolds:
    button-lock ducting with end caps
    floor register boots
    silicone (1 tube)

Devices and Wiring

thermostat (remote bulb or DT) plus thermostat wire
blower
backdraft damper
14-gauge, 3 wire, Romex—14-2 w/gnd or 12-2 w/gnd
junction boxes
Romex connectors for ½" knock outs
Wire Nuts
electrical wire staples

into your collector after the absorber is installed but before it is glazed. Smoke tests are a sure cure for overconfidence. No matter how much silicone you have used for sealing, you will probably find some spots that need more. For the test use a 30-second bomb that is nontoxic and which leaves no residue.

Before doing the test, notify the fire department and your neighbors. A small, 30-second bomb produces lots of smoke, and for several minutes it will look like your house is on fire.

To do the test, plug the outlet duct with a plastic garbage bag and duct tape. Pressurize the collector at the return duct by using a hand-held hair dryer. Place the smoke bomb in a tin can, ignite it and set it inside the inlet duct. Quickly tape plastic over the inlet, and begin looking for leaks. Check the manifolds as well as the collector perimeter and the joints in the absorber. A 30-second bomb should produce enough smoke to enable you to seal all the leaks before it is exhausted. Larger collectors may require a bomb at both ends, and if your collector is very leaky, a second test may be needed.

Once you are satisfied that all the leaks have been fixed, reconnect all of the ductwork, and cycle the collector thermostat to turn on the blower and to pull the smoky air out of the collector. Although the smoke is nontoxic, try to direct it somewhere other than the living area.

Mounting Collectors

We have covered the construction sequences for a roof-mounted collector in our example, but a few more details are in order. Many liquid collector installations have panels mounted on rigid metal racks which are raised above the roof. Insulated pipes are run through roof jacks to the house. Installing panels in a *stand-off* mounting of this type can be simpler than integrating a collector into a roof because the whole roof doesn't have to be torn up. The supporting framework for the panels attaches to the roof in only a couple of places, and these are sealed quite easily. Rack mounting of air collectors, however, isn't recommended. The air ducts as well as the manifolds in an air collector are quite large, much larger than water pipes, and need to be well insulated and weatherproofed. A 12-inch insulated duct is hard to fit through a roof jack. The best way to mount an air collector is to enclose the entire back side of the collector in a wooden framework that is securely attached to the roof. *Attached mounts* offer much better wind resistance than stand-off mounts, since they are very rigid and well attached to the roof. Stand-off mounts, however, are sometimes torn off roofs by windstorms. Attached mounts are only slightly more expensive to install than open metal stand-off mounts.

When installing an attached mount, design it to run up to the peak of the roof whenever possible. In this way the north-facing shingles can be run continuously up the back side of the collector, insuring a watertight attachment, and collectors mounted in this way are better looking. They look like part of the house rather than a "solar thumb" that has been tacked on. If this isn't possible, or if the collector is mounted across the surface of the roof for better orientation, it may be necessary to build a *cricket* (a small drainage roof) between the back of the collector and the roof. A minimum roof pitch of 3-and-12 (14 degrees) will be required to insure proper drainage. If your roof is shingled with wooden shakes or tiles (anything other than asphalt roofing), it is especially important to run an attached collector mount to the peak of the roof, or flashing around the collector can become a real nightmare.

Scab Roof Mounts

When the roof added for a collector mount runs to the peak of a continuous roof line and the north side of the solar roof can be the

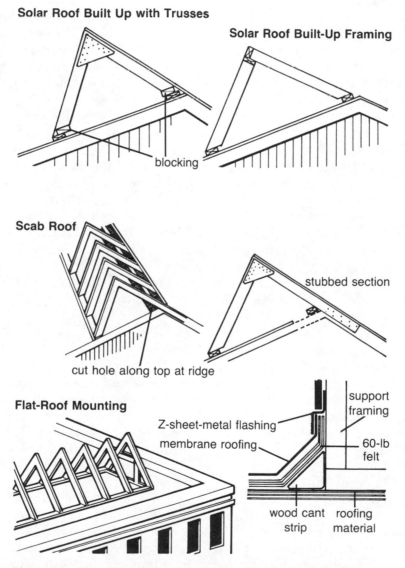

Solar Roof Built Up with Trusses

Solar Roof Built-Up Framing

blocking

Scab Roof

stubbed section

cut hole along top at ridge

Flat-Roof Mounting

Z-sheet-metal flashing

membrane roofing

support framing

60-lb felt

wood cant strip   roofing material

Figure 8-8: This illustration shows several framing options for roof-mounted collectors. A collector retrofitted onto a pitched roof should run to the peak, if possible.

same pitch as the existing north roof, a scabbed-on solar roof may be suitable (see figure 8-8). This technique usually works best if the existing north roof has a fairly steep pitch.

Remove the shingles and plywood from the top 12 inches of the south-facing roof, and cut holes for the ductwork in the south roof and to provide easy access to the attic space. Make detailed drawings and consult them to determine the angle needed on the overhanging ends of the 2 x 6s to be scabbed on. Find the "crown" on the boards so that any bow in the board will be faced upward before cutting this angle. These boards must be long enough so that 4 or 5 feet of board can be nailed (scabbed) to the north rafters. Working with a partner inside the attic space, insert the two boards on opposite ends of the collector so they stick out the correct distance, and have your partner nail the 2 x 6s alongside the existing roof rafter with four 16-penny nails through each side. String a line between the ends of these two overhanging boards, and hold the rest of the boards to this line while nailing them in place (tolerance: ¼ inch). Nail a 2 x 4 plate onto the south roof rafters snug against the bottom of the overhanging boards. Attach the lower plate and side plates to the south roof as in our example collector. Insert precut boards under the ends of the overhanging boards, and secure with plywood gussets.

Tilted Collectors on Flat Roof

Enclosed, attached mounts are suitable for flat roofs (see figure 8-8). The base plate should be securely attached to the roof framing. Hot tar roofs that have been penetrated can be difficult to reseal properly. After the framework has been built, angled cant strips are nailed to the base plate, and Z-flashing is used under the exterior siding. Sixty-pound felt is run under the flashing, over the cant strip and onto the roof. Fiberglass flashing membrane (available at large lumber stores) goes over the felt, and plastic roof mastic covers this membrane. The membrane gives more strength and durability to the mastic and helps to keep it from cracking or leaking. Two coats of mastic should be used over this membrane. Your local roofer can seal the perimeter of a flat roof mount with hot tar for about $1.50 to $2.00 per linear foot, which may be money well spent if the installation is large.

If you are installing more than one collector array on a flat roof, be sure to check your winter sun angles to make sure that one won't shade the other. Make detailed drawings of your installation using a protractor and the sun path charts in Appendix 1.

A final note of caution about working on the roof: Be very careful to avoid electrical wires. Don't work near them and never touch them. If wires must be removed to allow for the collector mount, call an electrician or the local utility.

Tilted Ground Mounts

The construction sequence for a tilted ground mount is much the same as for a roof mount. Because it's on the ground, a footing must be poured to support the base plate, and the collector should be designed and built so the bottom is at least 8 inches above grade to keep it dry and to keep snow off of it in the winter.

The footing to support the collector should be continuous around the collector perimeter, about 10 inches wide and at least 10 inches deep, or deeper depending on the frost line at your locale. The entire foundation should be reinforced with rebar. To build this footing, lay out collector dimensions with stakes and a string line. The footing should extend 3 or 4 inches longer and wider than the collector. Measure the diagonals to check for squareness, and then dig a trench 10 inches wide and 10 inches deep. Cut and bend ½-inch rebar to size, support it in the middle of the trench, and pour the concrete. Put 8-inch foundation bolts into the pour every 3 feet.

Installing the Foundation Plate

The foundation plate makes for a secure attachment between the collector and the foundation. It should rest at least 8 inches above ground level, and redwood or pressure-treated wood should be used for this plate to prevent rotting. If the footing is poured level with the grade, you can add a course of concrete block to raise the sill 8 inches.

After the footing has cured a day or two, strip off the forms and lay out the collector on the footing. Check diagonal measurements to make certain it's square. Snap chalk lines on the footing where the outside edge of the plate will be. Cut 2 x 6s to the proper length for the plate. Make sure that each 2 x 6 piece will be attached to at least two

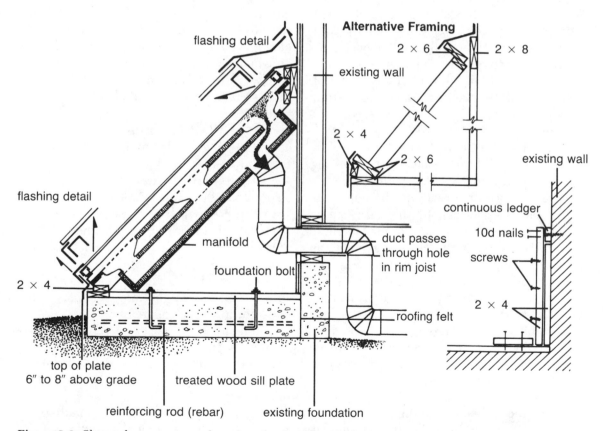

Figure 8-9: Shown here are several options for framing tilted, ground-mounted collectors. The collector framework must be securely attached to the existing wall with lag bolts. When the sides of the collector mount don't break over wall studs, attach them as shown in the detail at right.

foundation bolts, and drill the sills at the appropriate points for the bolts. Put ½-inch fiberglass sill-sealing insulation on the foundation, ¼ inch in from the chalk lines, and put the 2 x 6s in place and bolt them down. It may be necessary to chisel out a depression with a wood chisel to allow for the bolts. Tighten the bolts down very snugly with a socket or crescent wrench. This plate should be within ¼ inch of the chalk line.

The collector framework is attached to the existing wall with lag bolts. Locate the top of the collector along the wall, and snap a chalk line where the top of the wall mounting plate will be located. Make sure this line is level. Lag bolt the 2 x 6 plate into the existing studs. Attach vertical 2 x 4s to the wall where the edge of the collector will be. It may be necessary to remove some types of siding to get a secure fit to the wall. If these vertical sections can't be attached to an existing stud, attach them as shown in figure 8-9.

Making Holes to Crawl Space

At this point you want to cut holes for the inlet and outlet airflow paths, either to the crawl space or directly to the house. In both cases the rim joist is a logical place to run the ductwork in from a ground mount. Locate the rim joist and cut through it with a reciprocating or circular saw. Check the inside as well as the outside of the rim joist for electrical wires or other possible obstructions. Make the hole through the joist as neat as possible and slightly larger than the duct or boot that will fit through it. Be sure to cover the holes to keep stray cats out of your crawl space during construction.

Framing the Collector Mount

The framework for a ground mount can be assembled as a rectangular box on a flat driveway and carried over and set into place. Or, the tilted members can be individually cut to length and nailed directly to the wall and foundation plates.

When the framework is built on a flat surface, the individual 2 x 4 studs are all cut to the same length and nailed to plates top and bottom, 24 inches on center. If the top and bottom plates are not long enough to span the entire length of the collector (the recommended procedure), joints in them must be centered on the 2 x 4 tilted members. If the tilted 2 x 4s are attached directly to the wall and to the plate, they are all cut to the same length with the appropriate angle cuts at each end. A bevel square is helpful in determining this angle. Cut one 2 x 4 and make sure it fits everywhere along the length of the collector mount. Then use it as a pattern for the rest. Insert blocking between the members, as in the roof-mounted collector.

Vertical Wall Mounts

Mounting a collector directly on a south-facing wall is easier than building a slanted mount, especially if the wall is of frame construction. When the frame wall is covered with shingles, lap siding or other siding that isn't flat, all the siding inside the collector area must be removed in order to get a flat mounting surface. Lay out the collector on the wall, and be sure to include the edge insulation on the top, bottom and sides of the collector in your measurements. Try to locate the collector so at least one vertical edge is over an existing wall stud. It is even better if both ends match up with studs, especially if the underlying sheathing is insulation fiberboard.

If your wall covering is lap siding, cut through it with a circular saw, avoiding nails in the siding. Also try to avoid cutting deeper than the thickness of the siding. Remove the siding with a wrecking bar, cat's-paw and framing hammer. If the sheathing is made of horizontal 1-inch dimensional lumber (1 x 8s, for example), the plywood collector backing can be nailed or screwed directly to these boards. If the backing is plywood, the collector can be built directly on this layer. If insulation board underlies the lap siding, as in most new construction, it is a good idea

to remove it and to nail ½-inch CDX plywood in its place. This will insure a tighter fit to the wall, though the plywood backing can be screwed (not nailed) to the studs over the insulation board. If one of the sides of the collector must fall between studs, it will be necessary to insert a vertical stud (or two) into the wall behind this joint to make it sturdier. It will probably be difficult to install this stud, and it will probably be necessary

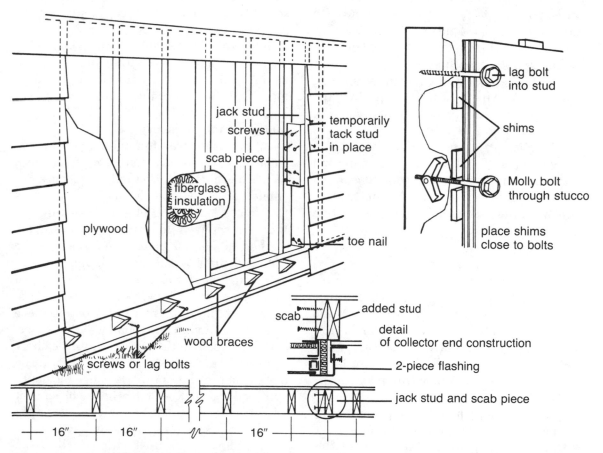

Figure 8-10: Design a vertically mounted collector so that at least one side falls over an existing wall stud. Insert an additional stud in the wall at the other edge of the collector, using the scab-on technique shown, if necessary. When attaching the plywood collector backing over stucco, use lag bolts into studs and large Mollys through the stucco at one edge of the collector. Wooden shims can be used to plumb the plywood backing and take out dips and humps in the wall.

to move the existing wall insulation out of the way before inserting it. To get this stud in, it may be necessary to cut it in half, insert the halves separately, tack them in place through the siding, and then screw a "scab" in place over the joint in the stud (see figure 8-10). Don't nail the siding securely to this stud yet, but do attach the collector plywood to its center.

Counter flashing can now be slipped in vertically under the lap siding and over the underlying layer. This flashing should be about 2 inches by 4 inches with the 2-inch leg sticking out from the wall. The collector flashing will go over this counter flashing. At the top of the collector, the counter flashing will go over the collector flashing. Once the counter flashing has been installed, nail the siding back into place.

A collector can be mounted directly on flat wooden siding if it is in good shape and the wall is flat. It is always a good idea to put a few extra screws through flat siding to make sure it is well secured before construction begins.

Stucco Walls

When mounting a collector to a stucco wall, install the plywood to extend 1 inch past the sides and bottom of the collector and 1½ inches past the top. The collector flashing can then run over the insulated sides and down over the plywood. Once again, locate the collector so that one edge breaks over a wall stud.

Hold the first piece of precut plywood in place, and drill two holes through it and the stucco below with a ¼-inch masonry bit. These holes will be located over wall studs. Use 3-inch lag bolts with washers to fasten the plywood to the wall studs. Plan ahead and avoid putting lag bolts in places where

their heads will interfere with the 2-inch track that will go on next.

If one edge of the collector falls between two studs, insert large Molly bolts into the cavity behind the stucco. This, of course, will have to be done before putting lag bolts into the studs. It will also involve careful measurements for predrilling holes in the plywood so that the Mollys will line up with the holes drilled in the stucco. Before tightening the lag bolts all the way down, make sure the plywood is flat, especially around the perimeter. It may be necessary to drive wooden shims between the plywood and the stucco to achieve this. A string line held along the plywood edge of any side shouldn't show more than ¼ inch of unevenness.

Mounting into Masonry

The easiest way to mount a collector to a masonry wall is to attach a 1 x 4 framework to the masonry with lag bolts and masonry shields or other types of masonry fasteners (see figure 8-11). The plywood backing can then be screwed to this framework. This method is simpler than trying to attach the plywood directly because it is much easier to line up holes through the 1 x 4 than through large sheets of plywood. It also makes it easier to shim the 1 x 4s and plywood to make a flat surface for accommodating the collector.

Drill holes in the masonry joints exactly as deep as the shields are long. Put a piece of tape on the bit to properly gauge the depth you need to drill. Keep the drill and bit straight while drilling so that the shields fit snugly in the hole. Clean out the holes with a lag bolt, or blow out the dust with a small plastic tube. This is important. Insert a shield and make sure it fits properly. You will need shields and bolts about every 2 feet around the perimeter of small collectors. Larger col-

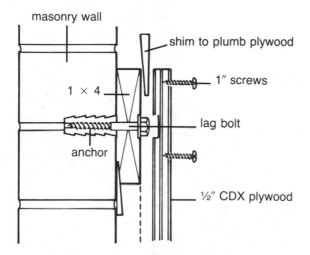

masonry wall

shim to plumb plywood

1" screws

1 × 4

lag bolt

anchor

½" CDX plywood

Figure 8-11: When mounting a collector on a masonry wall, avoid placing lag bolt shields near corners of the wall. This can cause the masonry to crack as the bolts are tightened in the shields. Attach 1 × 4s to the wall with lag bolts, and screw the plywood backing to these boards. Size anchor holes and lags correctly. Through-bolting is time-consuming but guarantees a secure placement. Beware of soft bricks, which don't hold anchors well and fracture easily.

lectors will need additional support in the center.

Cut the ½-inch plywood to size, and use a wood chisel or drill to make holes in the back of the plywood so it fits over the heads of the lag bolts. Screw the plywood to the 1 x 4s at several points with 1¼-inch screws, but don't snug them up. Check the surface of the plywood with string lines to make sure it is flat. Loosen the screws where necessary, and place shims behind the plywood to bring out the low spots. Put in additional screws every 8 inches and through the shims.

What if your existing wall isn't flat? Now that we have stressed that the mounting surface for your collector must be absolutely flat, we will tell you that it doesn't have to be.

We have retrofitted collectors on existing vertical walls that had as much as a 1-inch bulge in them in just 10 feet. It is a bit tricky, but if the 2-inch perimeter track is cut part of the way through so that it "breaks" where the joints in the glass (and flashing) are located, it will work. These collectors definitely have a site-built look to them.

Bottom Supports

Installing strong bottom supports for vertically mounted collectors is always a good idea, especially if the collector is tall. In our design the collector is securely fastened to the wall, but most of its weight, which comes from the glass glazing, is pushing straight down. Triangular gussets that support the bottom of the collector can be cut from 2 x 6 lumber. Size them to fit to the edge of the glass-support track, and drill holes in them. Secure them with screws to wall studs after the glass supports are in place but before the bottom flashing or glass is installed. Place the gussets at every stud along the bottom.

The local welding shop can also make some braces for you out of 1-inch angle iron. Holes are drilled in the back, and they are lag-bolted into the studs.

Ladder Arrangements

It is easy to build vertically mounted collectors that are less than 8 feet above the ground by working from two stepladders. If the collector is over 8 feet high, it is well worth it to rent or borrow one or more sets of scaffolds. The scaffold arrangement should be wide enough to extend to within a foot of both sides of the collector, or it should be mobile.

A slanted ground mount or a collector added to a roof with a gentle pitch (less than

4-in-12 or 18 degrees) can be built by leaning a 2 x 10 upright against the collector and placing a ladder over it. It is very handy to have two such ladder setups, especially when installing the absorber plate and glass. If you are working on a roof, it is a good idea to tack a 2 x 4 to the roof at the bottom of the ladder to prevent it from kicking out.

If you plan to build a collector on a steeply pitched roof, it would be nice to have had some training in acrobatics, but you can devise safe, convenient access. Scaffolding should extend to the bottom of the roof to make it easier to pass parts up (especially the glass) and to act as a safety net. Two overhanging ladders can be made from 2 x 4s so

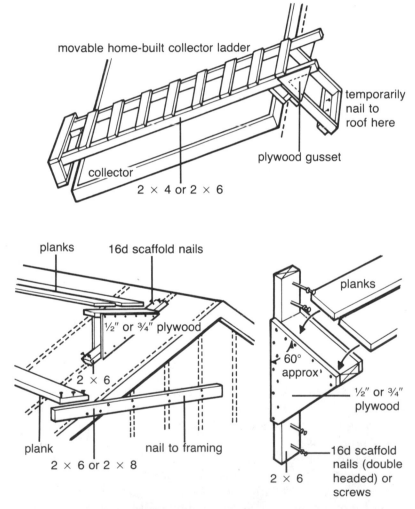

Figure 8-12: Movable ladders or angle braces for building large roof-mounted collectors are easy and inexpensive to build. On relatively narrow collectors, planks can be supported by 2 × 6s tacked to both sides of the collector frame.

Photo 8-11: This is a good ladder arrangement to use for building tilted ground mounts. The clear tempered glass has been installed on this collector, and it is ready for trim. Notice how flashing runs over the redwood sill at the bottom of the collector.

that they stand out over the profile of the finished collector. They rest against the north roof and are moved back and forth over the collector during construction. Three people will be needed to install glass with a setup of this type. If a tilted collector is long and narrow, angle braces can be tacked to the plywood on the sides of the collector with a plank run between them. These angle braces can be rented or made up from 2 x 6s and ½- or ¾-inch plywood.

When building on a steep roof pitch it is very handy to have at least a 1-foot space

on both sides of the collector. Ergo: Don't extend the collector all the way to the side edge of the roof. It is also nice to be able to stand on the north roof to do work on the top of the collector. Extra care must always be taken when working on roofs or scaffolds. Look twice and step once. Avoid dangling tools from your nail belt, and don't leave tools on planks where they can be kicked off onto somebody's head. Wear hard hats and carry insurance.

Some Options to Our Design

The previous pages detailed the construction steps for a collector design that we feel is good and one that can easily be adapted by others. The following are some options that we have not included in our collector, but your installation might require them.

Installing Turning Vanes

Turning vanes should be placed in all square corners and where the airflow changes direction inside the collector in order to discourage eddies of air and stagnant spots in the air cavity and reduce excessive static pressure (our example collector has a straight airflow path). Turning vanes are put into place when the ductboard backing is installed on the collector.

Turning vanes are made from the same track that is used for the perimeter and baffles. Cut tabs 1 inch wide in both legs of the track, and bend the track in a curve so that the tabs spread apart from each other (see figure 8-13). Cut the ductboard backing to size, but don't secure it in the baffle space. Position the vane on the ductboard in the collector, and mark its locations with a felt-tipped pen. Cut a curved slit in the ductboard, and work the vane into this slit. Insert and secure the ductboard and the turning vane into the baffle space. Cover all seams with a foil-and-silicone "Band-Aid," and seal the joint be-

Photo 8-12: "I'm not goin' up there!" Working on a steeply pitched roof can be a little nerve-wracking, to say the least. At this job site, working from the gently sloping north roof, using movable ladders, and a scaffold "safety net" all make the job go smoother. Always plan for job-site safety first, with convenience running a close second.

tween the turning vane and the ductboard with silicone. It usually isn't necessary to attach the vane to the plywood backing, but use a few ductboard support washers to hold the ductboard down near the turning vanes.

Airflow Disruptors

If available materials dictate building a collector with an air cavity deeper than 1 inch, *airflow disruptors* should be attached to the back of the airflow cavity to increase the static pressure, air turbulence and heat transfer in the airflow cavity (our example collector has a 1-inch airflow cavity). They are inexpensive and easy to install and will definitely help performance.

These disruptors are simply 30-gauge galvanized angles made at the local sheet-metal shop and installed in the collector perpendicular to the airflow every 12 inches. If litho plates are used for the collector backing, the disruptors can be screwed directly to the back of the collector. If ductboard is used for backing, screw the angle through it to the plywood backing with 1⅝-inch drywall screws. Cover the screw heads with silicone.

Some collector builders use pop rivets to attach *airflow disrupting fins* to the absorber plate to improve heat transfer. This technique increases the surface area of the absorber slightly but not much because the fins make poor contact with it. It also complicates construction, so we prefer instead to install the disruptors.

Absorber Options

Pittsburgh joints, used in our example collector, are a very good method for joining absorber plates, but other methods are also suitable. *Hook joints* can be folded on the

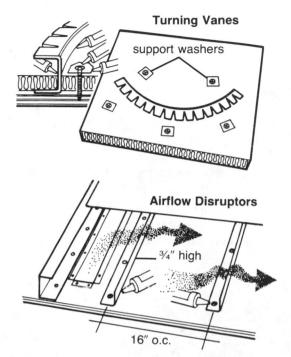

Figure 8-13: Turning vanes help direct airflow around corners and are made from the same metal track used for the baffles and perimeter. Seal them with silicone and use plenty of duct-board support washers. Airflow disruptors are installed 16 inches on center to create a 1-inch airflow cavity and to increase air turbulence in collectors with oversized cavities. Seal all fastening penetrations in the collector backing with silicone.

edges of absorbers with a bending tool (see information on useful tools in chapter 5). These ½-inch hooks are useful when installing lightweight, flat absorber plates that are textured or dimpled as some selective surface materials are. The inside of one hook is filled with silicone, and the next absorber plate is simply hooked into place. With lightweight absorbers it is often helpful to *carefully* drive ⁷⁄₁₆-inch self-tapping screws through the attachment to help make the joint tight. After the silicone cures, run another small bead along the seam to ensure a tight seal.

When using litho plates for an absorber, design the baffles in the collector to lie under joints in the litho plates. In this way, half of the litho-plate edges will be overlapped and secured on top of baffles (use silicone and screws every 3 to 4 inches), and the unsupported joints can be attached with Pittsburgh joints. It may be desirable to put in "dummy" baffles when installing the 2-inch track. These won't direct airflow but will serve to support joints in the litho plate.

Photo 8-13: Galvanized sheet metal is a permanent and inexpensive flashing material. It does look a bit "tinny" but is suitable for less-visible roof or back-yard installations, or it can also be painted. Short sections allow for expansion and contraction and for ease in replacing glass, should a pane be broken. However, the sections used in this collector are too short.

Glazing

The only other glazing method we can recommend is the CYRO Universal Glazing System that is widely used in greenhouse applications. This technique uses preshimmed glazing tape under the glass and a clamping bar with EPDM gaskets over the glass. It makes a very attractive installation and allows for the expansion and contraction of the glass, but it is much more expensive than using a silicone seal. Since the CYRO system was primarily designed for installations with wooden frameworks, it is also sometimes difficult to attach it to metal-frame collectors.

We have seen several collectors that were glazed by setting the glass onto the sticky glazing tape used for auto windshields. The glass was then held in place with small glazing clamps. All of these installations leaked and had to be resealed with silicone, so avoid this technique. Also don't use a caulking material other than high-temperature silicone to glue the glass in place in active installations. Other caulks are liable to deteriorate with time.

Double glazing an air heater can be very costly and time-consuming and is only justified at very cold or windy sites. If you are

Photo 8-14: Solar glass and coated metal flashing make this a very attractive site-built collector. Both the outlet and inlet ducts for the two separate collectors tee together in the attic in this direct-use system.

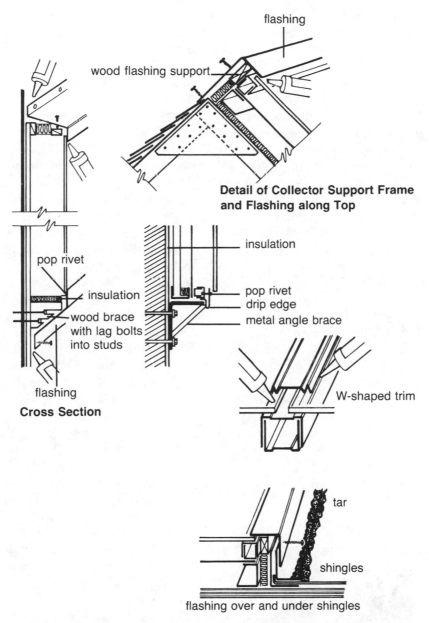

flashing

wood flashing support

Detail of Collector Support Frame and Flashing along Top

pop rivet

insulation

wood brace with lag bolts into studs

flashing

Cross Section

insulation

pop rivet
drip edge
metal angle brace

W-shaped trim

tar

shingles

flashing over and under shingles

Figure 8-14: This illustration shows several flashing options for site-built collectors. The flashing should run over the glass at the top and sides of the collector and under the glass at the bottom. W-shaped trim can be installed over glass joints to improve the appearance, but it isn't necessary. Double flashing over shingles is optional but can be good insurance on retrofits.

worried about heat loss from the front of your collector, using a selective surface absorber is almost always a better idea than double glazing. Many commercial manufacturers of air panels have had problems with double-glazed units, and most now use collectors with a single glazing and a selective surface. Double glazing a collector on site can present even more problems.

If you feel your active collector absolutely needs double glazing, we recommend using manufactured, double-pane tempered glass units. Outside of their weight, double-glazed units are as easy to install as single panes of tempered glass, following the steps in this chapter. Check with the supplier of your glazing to see if the lower edge requires a support. Don't try to build your own double-glazed tempered glass units since they are sure to be a disappointment.

Solar films such as 1-mil Teflon are another reasonable choice for inner glazing when glass is the outer glazing. Films are less expensive than sealed glass units but are more difficult to install in metal-frame collectors. It is important to keep films from being in direct contact with metal, and high-temperature glazing tapes are required (see the section on double glazing a TAP in chapter 7 for more details).

Flashing

The flashing for your collector is best made at a sheet-metal shop (see chapter 5 for information on flashing materials). Make detailed drawings of your flashing arrangement before starting construction so you don't end up with surprises after the collector is completed. In most cases it is a good idea to have the flashing made after the collector is finished so the dimensions you order will be exact ones. Do this immediately. You shouldn't leave the collector without flashing for more than a week. Remember that the flashing at the bottom of the collector must be installed before the glazing.

It is best to insulate the sides of the collector on the outside (as we did in our example collector) to prevent the metal track from carrying heat out of the collector, but in some installations it may be necessary to run flashing directly over the outside of the collector and insulate inside the collector. To do this, cut ductboard strips to fit against the sides of the airflow cavity, foil face towards the air stream, and seal in place with silicone. The glass supports can be insulated in their centers, but this won't help much in preventing heat loss.

Joints between the panes of glass are filled with black silicone. This seal is permanent, watertight and not too unsightly. We usually don't put flashing over these joints in the collectors we build, especially on roof mounts, because the joints aren't visible from the ground. It also makes the collector look like it was built from panels, which we don't like. If you wish to cover your joints, have a W-

Figure 8-15: *On opposite page.* The local availability of metal track can affect your collector design. This illustration shows several configurations using different-sized track. Design 1 is built from 3⅝-inch and 1⅝-inch track. Airflow disruptors provide desirable static pressure in the airflow cavity, and galvanized angles support the glass. Design 2 uses 2½-inch track for the perimeter and baffles. A layer of polyisocyanurate is placed under the ductboard to create a 1-inch airflow cavity. The collector in design 3 uses 1⅝-inch track throughout the collector. Since this collector is lined with polyisocyanurate, it is unsuitable for collectors with selective absorbers.

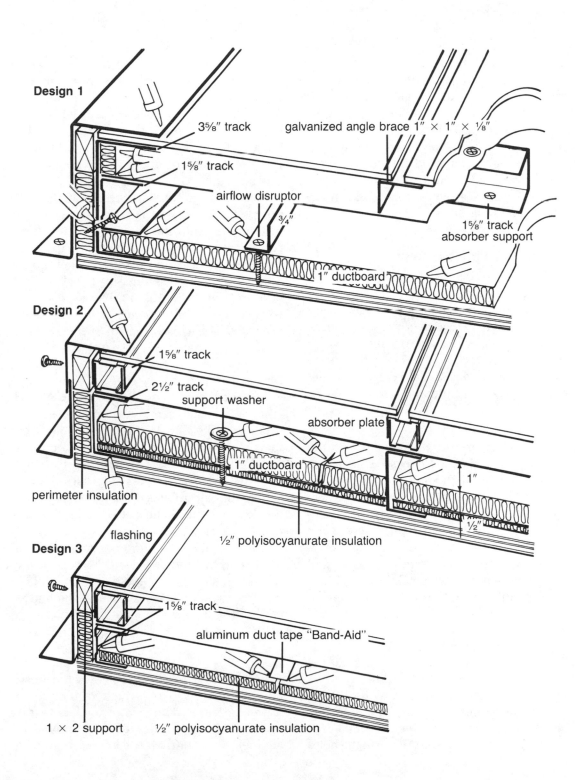

Design 1

3⅝" track

galvanized angle brace 1" × 1" × ⅛"

1⅝" track

airflow disruptor

¾"

1⅝" track
absorber support

1" ductboard

Design 2

1⅝" track

2½" track
support washer

absorber plate

1" ductboard

1"

½"

perimeter insulation

½" polyisocyanurate insulation

flashing

Design 3

1⅝" track

aluminum duct tape "Band-Aid"

1 × 2 support   ½" polyisocyanurate insulation

shaped piece of flashing made up at the sheet-metal shop, and attach it to the glass with clear silicone. Try to avoid driving screws between installed panes of glass because one slip can yield two exploded panes. Use duct tape to hold the flashing temporarily in place and put silicone under both edges. Point with your finger and scrape off the cured excess with a razor blade. Flashing over horizontal joints will require an especially good silicone seal on the upper edge.

All-Metal Collectors: Some Options

There are many different ways to build a permanent, efficient, all-metal active air collector. Figure 8-15 shows cross sections of three good designs (designs 1, 2 and 3), which use slightly different materials from those used in our collector design, although the construction details are much the same.

Design 1

This collector design, used successfully in many installations, is a reasonable choice if 2-inch uneven-leg metal track is unavailable. The 1⅝-inch airflow cavity is larger than what's normally used but can work in collectors with a higher airflow rate (about 3 cfm per square foot at sea level). For collectors with a standard airflow rate (2 cfm), install ¾-by-1-inch galvanized metal *airflow disruptors* perpendicular to the airflow cavity every 12 inches to increase turbulence and improve heat transfer.

The 1⅝-inch track is screwed to the 3⅝-inch perimeter track and snugly against the ductboard. Screw penetrations are sealed on both sides of the 3⅝-inch track with silicone. Track for the baffles and angles for the airflow disruptors are attached over the ductboard and screwed to the plywood. Predrill the holes

for the screws, don't compress the ductboard excessively, and put a glob of silicone under the track at each screw penetration and on top of each screw head.

The absorber plate is screwed to each baffle. At the perimeter the plate is held against the 1⅝-inch track by inserting a tight-fitting piece of ductboard under the 3⅝-inch track, foil face out. This eliminates the need for screws at the perimeter. Strip ½ inch of the fiberglass off each strip of ductboard so it sits further back into the cavity. After insertion, completely cover the ductboard strip with a large bead of silicone. Point with your finger. Sixteen-gauge or heavier 1-by-1-inch angle is used to support joints in the glass. It is pop-riveted to the 3⅝-inch perimeter track. Don't use panes of glass wider than 34 inches when using angle irons for glass supports.

Design 2

This collector uses a 2½-inch track, which is easier to locate than 2-inch uneven-leg track. It is built like the standard collector except that ½-inch Thermax (polyisocyanurate) is installed *under* the ductboard to create a 1-inch airflow cavity. This increases the cost slightly but eliminates the need for both extra insulation behind the collector and airflow disruptors in the air cavity. If you use this design, install plenty of large support washers to hold the ductboard and Thermax firmly in place. Don't worry about sealing joints in the Thermax, but be sure to seal the bottom edge of the 2½-inch track and all ductboard seams.

Design 3

This collector design shows the possibility of using 1⅝-inch track for both layers in the collector. The size of track is readily available, but since ductboard is only available in 1- and 1½-inch thicknesses, ½- or ⅝-

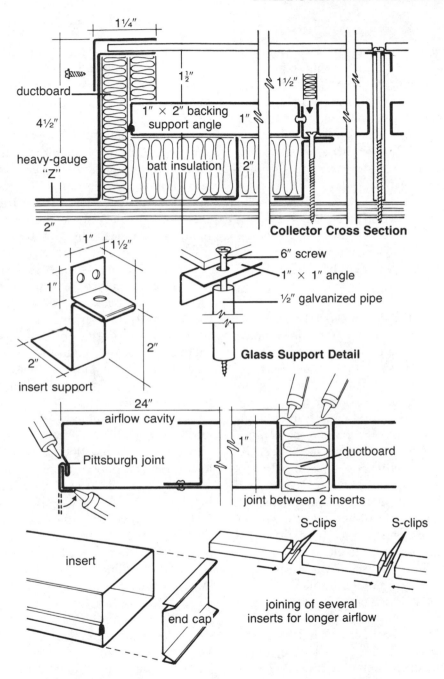

Figure 8-16: In this collector design, the airflow cavity is a long, narrow metal box that is prebuilt as a unit and inserted into the collector framework. The ends of each section are connected with S-clips and silicone to make a box 20 feet long. Since they float inside the collector and aren't under any stress, inserts require very little structural support.

inch foil-faced Thermax (polyisocyanurate) is used for the collector backing to give a 1-inch airflow cavity. In our opinion, however, Thermax is *not* a good material for use in active collectors although many commercial manufacturers use it. It is rated to 250°F and should never be used with a selective surface. Be sure to cover all seams with foil-and-silicone "Band-Aids."

Another drawback to this design is the difficulty in screwing the lower leg to the plywood backing. The space available between the legs in 1⅝-inch track is small, and driving screws here can be difficult and frustrating.

Design 4

The idea of using metal inserts for air heaters was developed by our solar mentor, J. K. Ramstetter. He is currently designing all of his systems to use them, and they look like a very good option. With the insert system, the absorber surface, collector backing and entire airflow cavity are made from a single sheet of aluminum litho plate or galvanized sheet metal. The 1-inch-thick units are easily joined together with S-clips and silicone, eliminating 60 percent of the screws and silicone normally needed in collector construction. This greatly simplifies both the sealing and installation work. Since the lightweight insert "floats" behind the glazing and is under almost no load, its structural support can be minimal. The perimeter framework is the only thing that must be securely attached. This design also allows for insulation on the sides and back of the air cavity, reducing heat losses from the collector and allowing for expansion and contraction of the inserts inside the collector. Collectors using inserts must be externally manifolded, and since turning the airflow is difficult, inserts work best in collectors that have a 16-to-20-foot-long, straight airflow path.

The inserts are best made at a local sheet-metal shop from 30-gauge galvanized sheet metal, aluminum litho plate or flat aluminum mobile-home siding. A Pittsburgh joint is located near one edge (see figure 8-16), and individual units are put together using silicone in a bent-over Pittsburgh joint. Separate units are then joined together with S-clips and silicone on the long sides (see chapter 9). Three ⅛-inch holes are predrilled through each S-clip to allow for ½-inch screws to hold the clips in place. The 1-inch sides are simply overlapped, sealed with silicone and screwed together. A 1-by-1-inch metal angle (of the same material as the insert) is pop-riveted to the back side of the insert inside the air cavity to prevent sagging. Floor register inlet and outlet boots are installed on each insert with "in-and-out" flanges (see chapter 9) before putting on the end caps. These W-shaped caps are also made at the local sheet-metal shop. They snap onto the ends in a bed of silicone and are held in place with two screws on each long side.

Inserts are installed on the wall or collector framework before the perimeter framework is attached. First, 6-inch-long, 1-by-2-inch metal angles (backing supports in figure 8-16) are screwed to studs in a row behind the center of each insert. Two-inch fiberglass insulation is then placed over the entire back side of the collector. Two-inch-wide angle brackets (figure 8-16) are pop-riveted to the sides of the inserts, 2 feet on center, and the predrilled tabs on these brackets are attached to the collector backing with 3-inch drywall screws. The perimeter framework is made of 24- or 22-gauge sheet metal bent to a Z-shape. One-inch ductboard insulation is placed inside the Z (foil face out) before securing it to wall studs or the collector backing. The Z at the bottom of the collector is supported under the glass joints.

Next 1⅛-inch-wide strips of ductboard are cut and pressed into place between the inserts so they rest ¼ inch below the top surface of the insert. All edges are sealed with black silicone. These strips are painted black when the absorber is painted and help to seal the dead air space in the collector.

Glazing supports are made from 1-by-1-by-⅛-inch galvanized angle that is pop-riveted to the collector's framework. These angles are supported in the center of the collector with 4⅜-inch sections of ⅜-inch galvanized pipe held in place with 5-inch, flat-headed wood screws countersunk in the angles. They are painted black, and glass is attached with a silicone seal as outlined earlier in this chapter. Correct positioning of the angles is very important.

Turns in the airflow are difficult to make with inserts, and if your collector requires them, you may be better off using the construction techniques for our standard collector. Turns can be made, however, by joining two inserts together by bending tabs and adding narrow strips of metal. The connecting section should be 1½ times as wide as the individual inserts. A 1-inch angle is snipped, bent and placed inside the insert to act as a turning vane (secure with screws or pop rivets) before putting on the end caps. Take care to seal all joints and overlaps with black silicone.

Self-adhesive, selective-surface foil can easily be attached to the absorber side of the-hot end of the inserts to improve collector performance.

Building Collectors from Panels

When first considering a solar installation, many do-it-yourselfers like the idea of building small, modular air heaters in their workshops and then installing these panels in a bank on their houses. They think that since commercial systems use them they must be all right. This sounds very handy, but we don't like panels. They are not only more expensive than site-built collectors, but they are also difficult to install and seal properly.

Some air panel manufacturers have developed methods of internal manifolding in which air moves from the side of one panel into the next. This requires elaborate sealing mechanisms that are out of reach of the do-it-yourselfer, who must rely on external manifolding, and this gets complicated and expensive. A bank of seven panels (34 inches by 76 inches each) requires two large manifolds, 20 feet long with seven ports in each one. The same size site-built collector can be built using two manifolds 5 feet long with three ports each. The commercial panels will have 130 feet of collector perimeter to seal, while the site-built system has 55 feet.

System Maintenance

A well-built, all-metal collector requires very little maintenance, another reason why we like it so much. Regular cleaning of the glass is required, however, because dirty collector glazing will block out much of the available sunlight. The glass should be cleaned when the collector is not in direct sunlight. Vinegar can remove troublesome spots.

Should a pane of glass break, treat it as you would a broken window in your house and replace it immediately or your collector will soon become filled with dust. Remove the flashing, scrape off the remaining pieces of glass and silicone with a chisel, vacuum out the inside of the collector, and put in the new pane in the same way as the original.

The only other maintenance most systems need is oiling the blower twice a year, once when it first comes on in the fall and again at the New Year. If you have a water-heating subsystem, celebrate the summer solstice by oiling the blower. Use SAE20 or 10W40 motor oil and put two drops (no more) in each oil port. Oiling the blower is very important and will greatly increase its life, but overoiling is as bad as not oiling it at all.

Building a Wood-Frame Collector

Although we have helped design a few, we haven't built a wood-frame, FRP-glazed collector in over four years. We *have* built many in the past but haven't been completely satisfied with them. Wood isn't a high-tem-

perature material and is more difficult to seal than metal. FRP, even the highest quality, isn't permanent, and it requires replacement after several years of use on a collector. These collectors do, however, have applications where they may be preferred over those built of metal and glass. FRP-glazed collectors can have more flex in them and are a good choice for portable collectors that are moved around throughout the year for different heating functions. The glazing is easier to repair if someone backs a car into it, and FRP glazing should always be used if vandalism could be a problem. We generally feel you are better off with a metal-and-glass collector, but in

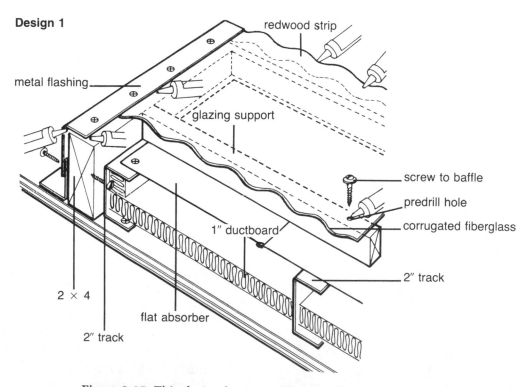

Figure 8-17: This design for a wood-frame collector uses a flat absorber and corrugated FRP (fiberglass) glazing. The perimeter is lined with metal track, and wooden glazing supports are installed over the absorber.

this section we will give some tips for building a wooden collector.

First of all, never use a selective surface absorber or selective paint on a wooden collector. If the collector ever experiences prolonged stagnation (such as when the power fails), the wood in the collector can become hot enough to smolder or ignite. Never allow an unvented, slanted wood-and-fiberglass collector to stagnate in summer. The fiberglass will be banana yellow in three years, and most of the seals will deteriorate after the first summer. Power venting is a good idea with these collectors to keep them below 180°F at all times.

Don't use nails anywhere in the collector. Use screws. Collectors experience tremendous heat fluctuation and nails will work loose. We have seen 16-penny cement-coated nails used in collector frameworks that worked out over an inch in three years. If you use ring-shank nails rather than screws to attach the FRP, you are in for a big chore when it needs replacing in six or seven years.

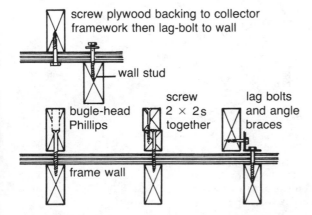

Figure 8-18: Shown here are several methods for making a secure attachment of wooden frame collectors to an existing wall.

Silicone caulk doesn't adhere as well to wood as it does to metal unless the wood is very dry and/or well primed. High-temperature urethane caulk (see chapter 5) is a better choice when sealing wood. Never use Butyl or latex caulks.

We feel that it is always a good idea to use metal track rather than wood inside the airflow cavity. Wood is all right for the glazing supports that lie inside the dead air space because this space will seldom reach temperatures over 160°F even when the collector is stagnating. The airflow cavity, however, can reach temperatures near 300°F. Lining the airflow cavity with metal is not difficult and makes for a much more durable installation.

Many wooden collectors are built with flat FRP and corrugated absorber plates. We have had better luck with flat absorbers and corrugated fiberglass glazing. Flat FRP isn't actually flat and looks ripply and quite ugly unless it is installed in short sections. Corrugated fiberglass has a better appearance.

Retrofitting a Black-Roof Collector

If you have ever been in an uninsulated attic on a cold, sunny, winter afternoon, you were probably amazed at how warm it was up there. Wouldn't it be nice if there was some way to use this available heat? There is—with a *black-roof collector*. This system pulls the available heat from the attic and blows it into an insulated crawl space. According to builder Roy Moore of Las Cruces, New Mexico, there are over 300 installations in southern New Mexico that operate this way. They use an unglazed, dark-colored roof as the collector and are used for both winter heating and summer cooling. As it turns out, these systems work even better for cooling than they do for heating, and many black-roof houses in southern New Mexico meet nearly 100 percent of their cooling needs with them. There have also been several units installed in the San Luis Valley that are for heating only (see chapter 12). Most of these systems are retrofits and deliver enough heat to be good sources of supplemental heat.

If your house has a large south-facing roof area and the wind at your site doesn't blow too hard on cold, sunny winter days, a black-roof system may be a good retrofit for you. Since the "collector" is already in place, you can complete the installation during a weekend. The major components required are some ductwork, a fairly large blower and a

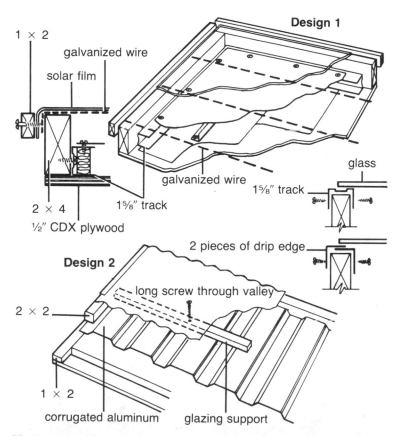

Figure 8-19: This illustration shows two design options for wood-frame collectors. In design 1 the collector has been temporarily glazed with a solar film supported with galvanized wires. As the detail shows, this collector is designed to accommodate tempered glass at a later date. The corrugated absorber in design 2 is installed directly on top of the insulative backing to create a ¾-inch deep air-flow cavity. The wood glazing support in this collector is installed on top of the absorber's ribs with long screws driven into the valleys. Penetrations through the absorber are sealed with black silicone.

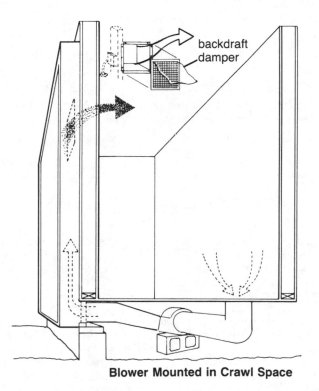

Blower Mounted in Crawl Space

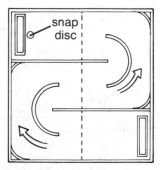

Figure 8-20: Blowing air into a small collector and delivering solar-heated air directly into an adjacent room is a good choice for low-cost systems. The blower can be mounted in a crawl space (shown here) or in a closet and is controlled by a snap disc thermostat placed near the outlet grille. If mounted in a closet, the intake grille should be in the closet door. A lightweight flap mounted over the grille acts as a backdraft damper (see the box "The $300 Collector").

differential thermostat. The cost of the installation should be under $200.

There are several factors that affect the overall performance of a black-roof system. The south-facing roof must be large. These collectors operate very inefficiently (5 to 10 percent versus a collector operating at 35 to 45 percent) so what they lack in efficiency must be made up in area. Roofs smaller than 300 square feet won't deliver much heat. The roof rafters must be uninsulated so that heat is transferred into the attic space. This usually isn't a problem since most homes have insulation in between the ceiling joists rather than on the rafters. Roofs built from rafters rather than from trusses are easier to retrofit.

Black-painted tin roofs are the best roofing for this type of collector, but asphalt shingles are also suitable. The color of the roof will make little difference as long as it isn't

The $300 Collector

In this book we stress fairly large systems that provide a good percentage of a house's heating needs. In most retrofits, we encourage the homeowner to build as large a system as will fit conveniently on his house, but this certainly isn't the only approach. If you are on a tight budget or looking for a simple installation to get started in solar heating, there is a lot you can do with $300.

We've designed a $300 collector for direct-use heating of the living area. It is a small, vertically mounted active system that is built around two panes of 34-by-76-inch tempered glass door replacements. All-metal construction makes it a durable collector. Recycled litho plates are used for both the collector backing and the absorber plate, and 1⅝-inch metal track is used for the framework, baffles, turning vanes and glass supports. Since absolute sealing is necessary, as in all collectors, silicone caulk is one of the major expenses.

The small blower is controlled by an inexpensive snap disc thermostat and is placed so that it pushes air into the bottom of the collector. The coolest air in the room will be taken off the floor level, and heated air will be blown directly into the same room. The collector is baffled as in figure 8-20 and requires airflow disrupters in the air cavity. Ductwork is almost entirely eliminated (unless you want to run a duct from the top of the collector through the attic to heat other rooms). A lightweight curtain mounted over hardware cloth (or a cube-core grille) at the top of the collector acts as a backdraft damper to prevent reverse thermosiphoning at night.

The blower can be mounted in the living area or in the crawl space. Blower noise is the only drawback to mounting the blower in the living area. To avoid this, purchase a blower that operates at a low rpm, and place it in a closet with a grilled vent cut through the wall or door (or cut off the bottom of the door), and mount it on top of cut-up inner tubes. Buy a vibration collar, or use six wraps of duct tape to act as a collar, to dampen the vibration between the blower and floor register boot that feeds the collector.

If the blower is mounted in the crawl space, it will be quieter, but you will need to install a floor register boot, an elbow and a short section of ductwork to feed the collector. Plan for the 90-degree turn the air will make as it enters and leaves the blower.

Whichever blower arrangement you use, an air filter should be placed in the inlet. If the collector is fed from a floor-mounted register, cut furnace filter material to fit in the register below the grille, and support it with hardware cloth. If the blower is mounted in the living area, fabricate a sheet-metal funnel and attach the filter material to it with a large rubber band.

In this simple active system, wiring will probably be the biggest job. You will need wires running from an outlet box to the snap disc thermostat and to the blower. One of the easiest white or a light color. Light-colored shingles can be painted with a dark roof paint.

Since these systems deliver very low-grade heat (60 to 80°F), the airflow from a black-roof system must be isolated from the living area. The heat can't be blown directly into the house, and the delivery temperatures are seldom high enough, even in summer, to use in a water-heating system. These systems have their best application in heating an in-sulated crawl space to about 55 to 65°F. These temperatures may seem low, but many crawl spaces are 40°F or colder during much of the winter. If you can heat your floor to even 60°F on every sunny day, it can make a sizeable dent in your fuel bill.

You can collect a good deal of heat from your attic by simply running two ducts from the attic to the crawl space at opposite ends of your house and moving the air in a loop.

---

ways to get wires to these points is to cut grooves in the exterior wall studs and run wires through metal conduit behind the plywood collector backing.

Snap disc thermostats, like all thermostats, must be mounted (and wired) where they can be replaced. The best place to locate them in this system is near the outlet port in the upper end of the collector. Cut a 2-inch-diameter hole through the collector backing next to the outlet port and line it with litho plate. Attach 12-inch-long asbestos wires (replacement cords for irons) to the snap disc with crimp-on connectors (use a special tool) or solder. (Teflon-covered wire is the best material to use here, but it is hard to locate.) Cover the outside of the asbestos wire that is in the airstream with silicone. Completely cover the back (not the front) of the snap disc and the wire connections with a large glob of silicone to insulate the connections. Allow the silicone to cure for two days before dangling the thermostat in the airflow cavity through the 2-inch hole. (The snap disc itself isn't fastened to the absorber, to facilitate its removal in the event of a malfunction.) Knock out the back hole in the junction box, pull the asbestos wires through, and screw-mount the box over the hole. Seal its perimeter and the hole in its back with silicone. Hook the wires together inside the junction box with Wire Nuts, and place a cover over the box. An oversized grille will fit over the hot port and the junction box for the thermostat, allowing for easy servicing of the thermostat. This is a safe way to install a snap disc, but it may not be code-legal in all locations, so be sure to check local codes.

Materials and Costs for $300 Collector

glazing: 34″ x 76″ tempered glass patio door replacements (seconds), 2 panes @ $20 each	$40
plywood backing: ½″ CDX, 2 sheets @ $12 each	24
backing and absorber: litho plate, 100 ft² @ $0.25/ft²	25
metal track: 1⅝″, 100 linear ft @ $0.26/ft	26
snap disc thermostat: 110°F to 90°F, Grainger no. 2E245 @ $9 each	9
blower: 130 cfm @ 0.3″ WG, Grainger no. 4C564 @ $32 each	32
silicone: 12 tubes @ $5 each	60
screws: ¾″ hex heads, ⁷⁄₁₆″ Phillips, pan head zip-ins, 2″ bugle heads	15
flashing: 28 ft. @ $1/ft	28
ductwork and transitions	15
paint: Grainger no. 2X717, 2 cans @ $4.50 each	9
misc materials, trim and wiring	35

A few simple procedures, however, will help increase the efficiency of your system without greatly increasing its cost or complexity.

Baffling and Manifolding

On a sunny winter day the north-facing roof will often be 25°F cooler than the south-facing roof, so it is important to isolate the rafters of the south roof. Do this by stapling a sheet of 6-mil plastic at the top and bottom of the rafters to create a large bag. Nail three or four horizontal rows of 1 x 2 furring strips to the rafters over this bag to create smaller, 12-inch horizontal bags for the air to flow through. This will hold the airflow close to the south roof and cause more air turbulence for better heat transfer. At both ends of this arrangement staple the plastic along the length of one rafter to form a large manifold bag to feed the baffle spaces. Sheet-metal ductwork can be attached to this plastic manifold with duct tape. It will be impossible to make this bag completely airtight, but do the best you can.

Airflow Requirements

We have done tests on black-roof retrofits, and it appears that a 40- to 50-foot-long flow path is optimal. Most roofs can use horizontal baffling to achieve this flow length. If your flow will be less than 30 feet, consider using up-and-down serpentine baffles for the plastic bag.

Most black-roof systems will present a static pressure close to 0.4 inch WG. Size the blower to move about one-third the amount of air required for a regular active air-heating collector mounted in place on the roof, or ⅔ cfm per square foot of roof at sea level (see the static pressure section in chapter 9). This rate of airflow will produce about a 10°F temperature rise from the inlet to the outlet. The blower for most installations will need to be large, but remember it is much more economical to move heated air than it is to heat the air. Lowering the airflow rate won't increase the temperature differential that much. Using a very large blower will increase efficiency but will also increase operating costs.

Controls and Insulation

A differential thermostat with a 15°F-to-5°F differential will be suitable for controlling the blower, but an ideal thermostat will have adjustable high and low set-points. This will allow fine tuning of the blower control and keep the system from running too late into the afternoon.

Dampers will be required on both duct lines to prevent heat loss from the crawl space at night. The return line to the roof can often be framed in a race in the back of a closet that has a home-built backdraft damper mounted horizontally inside it. Reverse airflow down the hot duct can be prevented by blowing the solar-heated air into a short plastic tube which deflates at night. Using a motorized damper in these low-cost installations can nearly double their cost, so their use isn't justified.

It may be a good idea to place 4-foot-wide pieces of 1-inch duct insulation over the plastic bag at the hot end of the "collector." This would add only $40 to $50 to the cost of the installation and should help raise the delivery temperatures a few degrees. We haven't done enough tests to know if this is really cost-effective. Insulation can always be added later if you feel your system could benefit from it.

Summer Cooling

Black-roof collectors work very well for summer cooling, perhaps better than they do for wintertime space heating. Since there is no glazing to slow radiant heat loss from the roof on a clear summer night, a dark-colored roof will be considerably cooler than the outside air temperature and cooler than a collector would be that was mounted in its place. The blower and thermostat can be arranged so that cool, nighttime air is pulled from the attic and blown under the house. Summertime cooling works best if there is a slab-on-grade or other type of storage under the house to store the *coolth* (see the super crawl-space

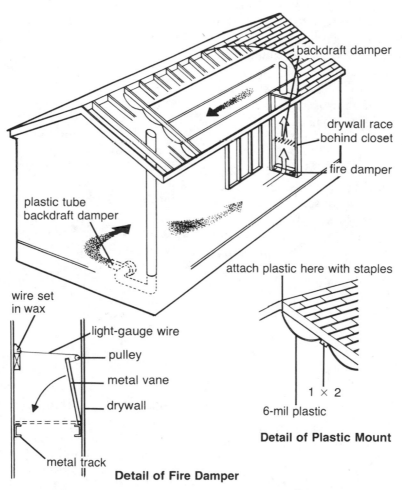

Figure 8-21: Using a south-facing, dark-colored roof for a collector is an easy retrofit project. Simply enclose the south rafters in a plastic bag, and move air through this bag into an insulated crawl space. Black-roof retrofits are simple to install but quite inefficient, so keep costs to a minimum for a good return on your solar investment. The detail shows an easy-to-build, inexpensive fire damper, which is a good idea in black-roof retrofits.

design described in chapter 11). Evaporative coolers or ceiling fans operating at low speed during the day can move air within the house and bring it in contact with the cool floor.

A Final Word

Don't expect miracles from a retrofitted black-roof collector. It is a quick, inexpensive project that can get you started in solar heating and provide you with 10 to 15 percent of your heating needs, depending upon its size. It won't provide more than this amount, so keep the installation inexpensive. When you get ready to build a more elaborate system to provide for more of your needs, you can use the blower, the differential thermostat and the ductwork and races in the new system.

Sources for Information

If you are considering installing a black-roof system on a new house, you would do well to consult Roy Moore, who has been building these systems for over eight years and has helped to design systems throughout the country. For a reasonable cost Roy can revise your blueprints to accommodate a black-roof system. His systems have qualified for tax credits.

Roy Moore
5400 Jornada Rd. South
Las Cruces, NM 88001
Phone: (505) 522-0055

If you are interested in building a collector system from panels, Earth Sun Resources offers a complete set of plans for the panels we mention in this chapter.

Earth Sun Resources
P.O. Box 1407
Alamosa, CO 81101
Phone: (303) 589-3605

Sources for Supplies

Smoke Bombs

Superior Signal Co., Inc.
P.O. Box 96
Spotswood, NJ 08884
Phone: (201) 251-0800

For Further Reference

Conrad, George R., and Pytlinski, G. T. *The Design and Prototype Testing of a Solar Heating and Cooling Roof/Collector/Radiator Module.* Albuquerque: New Mexico Energy Research and Development Institute, 1981. Order no. EMD2-68-2228. This publication, which includes the results of extensive testing on the feasibility of using different roofing materials for heating and cooling in unglazed collector systems, is available for $7.25 from the following: New Mexico Energy Research and Development Institute, Information Center, University of New Mexico, 117 Richmond Drive, NE, Albuquerque, NM 87106. Phone: (505) 277-3661.

Small Farm Energy Project. *Portable Solar Collector Plans.* Hartington, Nebr.: Small Farm Energy Project, 1982. This publication is available for $3 from the Small Farm Energy Project, P.O. Box 736, Hartington, NE 68739. Phone: (402) 254-6893.

Temple, Peter L., and Adams, Jennifer. *Solar Heating: A Construction Manual.* Radnor, Pa., Chilton Book Co., 1981.

9

DESIGNING AND INSTALLING DUCT SYSTEMS

Installing ductwork for an active system may not be as exciting as building the collector, but doing it correctly is very important to the overall success of the system. In some installations with elaborate air distribution schemes, installing the distribution system can actually be as much or more work than building the collector. Difficulties arise for do-it-yourselfers because they know little about ductwork and may be hesitant to get specially made items and/or help from a local sheet-metal shop. But installing ductwork is essentially a series of simple tasks. After basic how-to knowledge, what installers need most is patience since retrofitting ductwork usually involves working in awkward places. Ductwork must be properly sized and sealed airtight, or the performance of even a very efficient collector will be impaired. It is usually a good idea to install all of the ductwork and the blower before building the collector so that air can be pulled out of the collector as soon as the absorber plate is installed, to prevent the absorber and backing from overheating.

Ductwork Design

Proper design of the distribution system for your collector is vital. Solar air systems require large ductwork in order to move large volumes of air. It is important to realize that more Btu's are delivered in a strong blast of 110°F air than in a weak flow of 180°F air, and a lot of air must be moved through your system to get good overall efficiency.

Air encounters resistance to its flow when it moves through a collector or ductwork, and minimizing this resistance (static pressure) is the first consideration when planning your duct system. Static pressure is dependent on the length, size and configuration of the ducting and on the volume and velocity of air moving through this ducting. If a system has large, short, straight ducting with gradual changes in direction and a small airflow, its static pressure will be low. If the ducting is extensive, small in cross-sectional area and has many abrupt changes of direction, it will have high static pressure. It is important to make gradual transitions, to use rounded corners and not to choke the airflow at any point.

Preventing heat loss is another important consideration. To avoid excessive losses, design your ductwork runs to be as short as possible, especially the hot (outlet) ducts. Moderately insulated return (inlet) ductwork can be run through an unheated attic space, but outlet runs located here must be very well insulated (R-11).

Locating Ductwork

Finding a suitable place in your house to accommodate large air ducts can present problems. The following are some ideas for where to run your ductwork.

Photo 9-1: Shown here is a variety of 8-inch duct sections and fittings. A square backdraft damper attaches to a square-to-round transition with S-clips. Round ductwork is cut to length before snapping it together. An adjustable elbow attaches to an 8-inch tee, which feeds two 6-inch connections. Note that the male ends always point in the direction of airflow. Ductworking tools include a duct crimper, felt-tipped pen, aviation snips, measuring tape and a slotted screwdriver.

Rules of Thumb for Duct Sizing for a Simple System with 120 Feet of Ductwork

Size of Collector	Size of Ductwork
70 ft² with 3 right-angle, curved elbows	5″ round or equivalent cross-sectional area
70 ft² with 6 elbows	6″ round
150 ft² with 3 elbows	10″ round
150 ft² with 6 elbows	12″ round
250 ft² with 3 elbows	14″ round
250 ft² with 6 elbows	16″ round

Rim joists. Rim joists are the band of 2 x 10s or 2 x 8s that runs around the perimeter of your floor joists. They are the best access to the crawl space.

Closets. These are a natural choice. An insulated 12-inch duct will fit easily into a back corner.

Corners of a room. A corner can be framed-in to create an uninsulated race for return air or to enclose insulated outlet ductwork.

Attic. The attic is a good choice for ducting for roof-mounted and vertically mounted collectors.

Through a room. Uninsulated spiral ductwork can be painted a decorator color and run directly through a living area if your home and furnishings work with the high-tech look.

Here are places not to run ductwork:

Outside the house. Eliminate plans for outside ductwork runs or keep outside runs as short as possible and very well insulated. Weatherizing exposed ductwork can be a problem.

Underground. Design your system so that it isn't necessary to move air a long distance between the collector and point of use. Heat losses from underground ductwork can be great, and the ductwork must be enclosed in a concrete shell or similar arrangement to keep it from collapsing. It must also be completely waterproof, which makes the installation all the more expensive.

In short, give plenty of thought to where your ductwork will go.

Duct Materials

Round sheet-metal ductwork is probably the best choice for the home-built system. It is economical, durable, easy to work with and readily available in many sizes. A wide variety of adapters, transitions and fittings

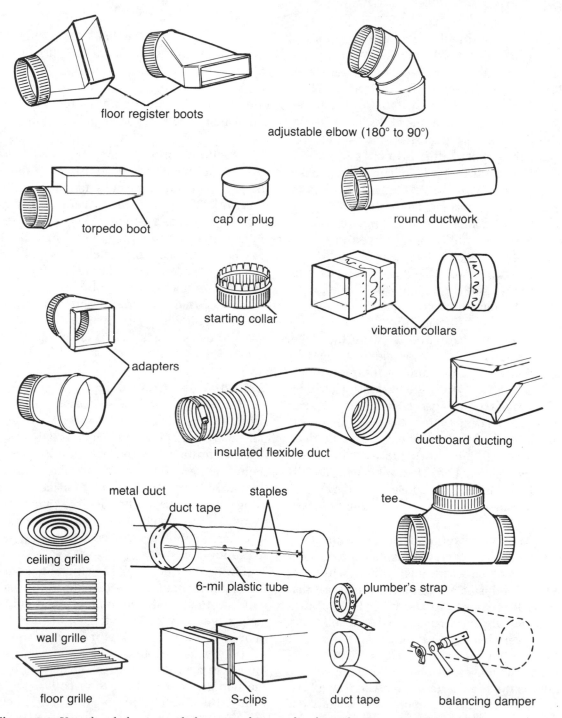

floor register boots

adjustable elbow (180° to 90°)

torpedo boot

cap or plug

round ductwork

starting collar

vibration collars

adapters

insulated flexible duct

ductboard ducting

ceiling grille

metal duct

duct tape

staples

tee

6-mil plastic tube

plumber's strap

wall grille

floor grille

S-clips

duct tape

balancing damper

Figure 9-1: Your local sheet-metal shop or solar supplier has a large stock of inexpensive duct system fittings and devices that will greatly simplify your solar installation.

are all standard and can be ordered quickly if the local sheet-metal shop doesn't have them in stock. Twelve-inch-round duct costs about $1.90 per foot, and 6-inch duct is about $1.00 per foot (1983). Some installations may require rectangular sheet-metal ductwork. It can be ordered from a local sheet-metal shop, but it is a little more difficult to install.

It may be necessary to have a shop make up one or two special transitions or plenum boxes for your system. These should be made of galvanized sheet metal and not ductboard. Ductboard is a foil-faced, pressed fiberglass material that is easy to use and is self-insulating, but typically the fibers are exposed to the airflow and it tends to disintegrate with time, allowing fiberglass fibers to enter the airstream (see "Air Quality" in chapter 5).

Sheet-metal ducting is easily insulated on the outside. We recommend using 3½-inch batts for outlet ducts running through unheated spaces and 1½-inch batts when they run through heated spaces. Return (inlet) ducts for space-heating systems usually don't need insulation unless they're exposed to outdoor temperatures, but return ductwork for water-heating systems should be well insulated. Foil-faced insulation is more insulative and doesn't "shed" fiberglass fibers so it isn't as messy to install.

Another material that can be used for ducting is *flexible duct* (also called flex-duct). This is a plastic- (or metal-) lined, wire-reinforced, preinsulated, round duct material that is, as the name implies, very flexible. It is considerably more expensive than regular duct but is very handy for getting around the tight spots in an installation. Other than its expense the biggest drawback to using flexible duct is the higher static pressure it presents. Its inside surface isn't smooth so it offers about three times more static pressure than does the same length of conventional round duct-

work. If you use flexible duct, try to keep the runs short and the turns gradual. If you use it for your entire installation, it is important to over-size it so that static pressure problems don't arise. Flexible duct may require additional insulation if it runs outside a heated area.

We know of a few simple, crawl-space heaters that have been built using a 16-inch-diameter polyethylene tube for distribution ductwork. The tube is easily made by folding over the edges of a 4-foot-wide piece of plastic and stapling them together. This idea is very handy for low-cost installations or where the crawl space is very small. The 160°F air from the collector won't hurt the bag and, unless exposed to light, it should last forever. When the blower is off, the bag deflates and also serves as a backdraft damper on the hot line.

Assembling Ductwork

Round ductwork is usually shipped in cartons containing eight 2-foot or 5-foot sections. Larger-diameter ductwork usually snaps together quite easily, but smaller ducts may require widening the groove (female side) for the entire length of the duct, to make assembling go more smoothly. The best time to cut round ducts to length is before snapping them together, so don't get carried away when putting ducts together.

After the ductwork is snapped together, the sections are fitted together. Insert the male end into the female end and wiggle them together (naturally...). This step will require two people if the ductwork is large. Don't push the two pieces together as far as they will go, but leave a ¼-inch gap between the end of the female joint and the stop collar on the male joint, to allow for expansion and contraction of the metal. Also be sure that the snaptogether seams don't line up; if one pops loose, all the rest of the seams can pop loose with it.

If the pieces of the ductwork refuse to fit together (quite likely), it will be necessary to further crimp the male end. This can be done

Photo 9-2: The box for this large adapter was fabricated by a sheet-metal shop, which saved time and minimized hassles for the owner. Two 12-inch starting collars were attached to the lid with the round elbows to simplify connections. A square motorized damper will fit on the other end of the adapter. Note the liberal use of black silicone for sealing.

by clamping the crimped end with a large pair of needle-nose pliers and twisting them to deepen the groove in the crimp. Do this at five or six spots and the piece will slide more easily into the next one. You can buy a special hand crimper (about $12), which is a very helpful, timesaving tool.

After two sections are assembled, seal the joint with duct tape, and put three self-tapping screws through the tape and both pieces of ductwork. Driving in saber-pointed, hex-head screws with the high-speed gear on an electric drill is the easiest way to accomplish this. Be sure to rub the tape to get it firmly stuck to the ductwork, and be sure the screws penetrate the male (inner) end of the joint. This assembly must be completely airtight.

Flexible duct is easily attached to round sheet-metal duct by inserting the latter several inches into the flexible duct. Secure the joint with a couple wraps of duct tape, and place a wire or large hose clamp over the tape.

After three or four duct sections are assembled, begin hanging them. Be sure that the crimped (male) end of the ducting points in the same direction as the airflow to limit air turbulence and static pressure. Ductwork is easily hung from floor joists with plumber's strap or wire used about every 6 feet.

Elbows

Adjustable elbows are installed, taped and screwed on like the straight sections. They are about the same price as 90-degree elbows but are much easier to use since they can make any angle between 90 and 180 degrees. A good procedure is to put a temporary piece of tape on the elbow, following its curve, after it is adjusted to fix it in place while installing the next duct section. Tape the next straight

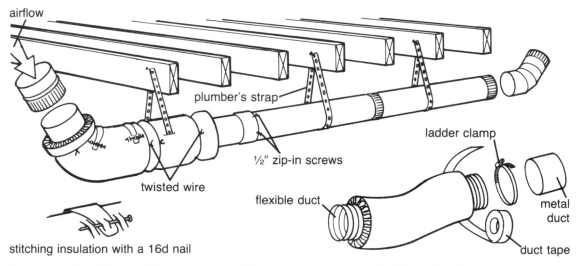

airflow

plumber's strap

ladder clamp

½" zip-in screws

twisted wire

flexible duct

metal duct

stitching insulation with a 16d nail

duct tape

Figure 9-2: Installing and insulating round ductwork is quite simple. The individual sections are snapped together, crimped ends are inserted inside uncrimped ends, and joints are sealed with screws and duct tape to stop air leaks. Ductwork should be supported every 6 feet with plumber's strap. The amount of insulation required around the ductwork will depend upon your particular installation. Flexible duct is attached to round sheet-metal ducting with duct tape and large ladder (hose) clamps.

section in place, but don't secure it with screws at this point because you may have to readjust the elbow. When the entire run of ductwork is completed, come back and put in the screws and tape all the seams in the elbow to stop air leaks. If the joints in your sheet-metal elbows are stiff and difficult to adjust to the desired angles (very likely), spray the inside sparingly with WD 40 or other spray lubricant. A tiny amount of this can seep through to the outside of the duct, but it won't prevent the duct tape from sticking. To adjust large elbows, sit in a cross-legged position with your legs encircling the duct and begin "elbow wrestling." Avoid squashing the elbow or you will never get it into the proper shape. This can be a hassle, but, as ever, persevere.

It is important that your duct system be as airtight as possible. If cold air is pulled into the system upstream of the blower or

before it reaches storage, the output temperatures can fall considerably. Although most people don't do it, it is not a bad idea to go as far as to tape the long, straight seams where your round ductwork sections snap together.

Outlet ducts from the collector should be insulated with fiberglass duct insulation or 3½-inch fiberglass batts until they reach the point of use. If all the heat from your system goes to the crawl space, the "hot" ductwork under your house doesn't need to be insulated since heat loss here isn't critical. Return ducts need to be insulated only if they run outside the heated area.

Insulate round or square sheet-metal ductwork with 1- or 2-inch-thick, 4-foot-wide batts available from your local plumbing and heating contractor. Measure the circumference of the duct, and cut a 4-foot-wide section that is 6 to 9 inches longer than this measurement (8 to 12 inches longer for rectangular duct). Wrap it around the duct so that it is snug but not really tight. Use two or three 16-penny nails in each 4-foot section to stitch or pin this overlap together as in figure 9-2. Put on the next section of insu-

Photo 9-3: Large, adjustable elbows often don't want to budge. Andy Zaugg demonstrates that his elbow wrestling skills are in top form. The trick is to use your legs to hold part of the elbow without deforming it. A little lubrication of the seams doesn't hurt either.

lation, overlapping the first piece by 3 or 4 inches. When the plumber's strap holding up the ductwork is in the way, cut slits in the insulation so that it will fit around the strap, and stitch the seam back together. When all of the insulation is in place, come back and put small wires (precut to the correct length) around the ductwork and insulation every 2 feet to hold it securely in place. Elbows are more difficult to insulate. Make folds in the insulation and use plenty of nails for stitching and three or four wires at each elbow. Some "waste" here is inevitable; just be sure the entire elbow is covered.

If you use foil-faced duct insulation, it will not only be cleaner to work with but you will also be able to use the handy insulation stapler (see chapter 5) for holding the insulation together. This will eliminate the need for nail-stitching and wiring. Always wear a respirator mask and long sleeves when dealing with insulation.

Never run ductwork where it will be exposed to the weather unless it is completely unavoidable. If you must have exposed, rectangular ductwork, it can be insulated on the inside with one or more layers of 1-inch, rigid fiberglass duct liner. The technique is expensive and second-rate, but if you decide to use it, get help and materials from a sheet-metal shop. Another method, recommended by the Small Farm Energy Project, is to enclose insulated round ductwork inside flexible plastic drain tile. They use this setup for portable solar grain driers.

Returns

Building codes and common sense dictate that all of the warm air from a heating system, no matter if it is forced air or solar, travels through some type of ductwork to the point of use. This isn't the case with return air, however, and air that is being returned to the collector can move through unlined and uninsulated races. These races can be simply and inexpensively built by framing-in a corner of the room or one end of a closet and enclosing the race with drywall or paneling. Races for return air should be as airtight as possible, but this isn't a critical consideration unless they're outside.

Outlet races can be built and lined with foil-faced insulation, but since they must be

Photo 9-4: The ductwork for this tilted ground mount runs from the collector manifold through the rim joist and into the crawl space. Note how all joints in the elbows have been taped.

completely airtight, it is usually easier to use insulated sheet-metal ductwork.

A common method of using solar air is to heat the coldest area of a house, such as rooms on the north end, then let the air find its way back to a collector return located in warmer, south-facing rooms. When this is done, it is necessary to cut a couple of inches off the bottom of any interior door that could restrict the flow of air back to the return. Putting grilles on both sides of a hollow-core door also allows free flow of return air. This is important because, if the room is well sealed and air is blown into it, the room will become pressurized, the solar air will take another path, and there won't be an even distribution of air through the house.

Installing the Distribution System

Blowers

Although there are many different types of blowers available, squirrel cage blowers are the type used in almost all solar applications. Unfortunately, at this writing there isn't a single one on the market that comes ready to install without some modification.

The inlet and outlet openings of these blowers are usually odd dimensions and won't hook up to ductwork directly. One of the simplest methods of attaching ductwork to them is by installing sheet-metal caps (pipe plugs) over the openings. Caps are readily available in 1-inch graduations from your sheet-metal shop. They should be slightly larger than the blower opening and are usually smaller than your distribution ductwork so you will need *transitions* to join them to the ductwork. These can often be made on site by squeezing round ductwork into a funnel shape and securing it with screws and silicone for sealing.

To attach the cap to the outlet end of the blower, mark the necessary square opening on the cap with a felt-tipped pen, and drill or punch a hole in the center of this opening. With snips, cut diagonals from the center out to the four corners, and bend them up to create lips. Cut off some of the metal in the center of the cap to make lips 1 or 2 inches wide that will fit over the outside of the square end of the blower. Drill pilot holes through the sheet metal and the blower housing (it's heavy gauge), and screw and silicone the cap in place. Use short (¾-inch) screws and make sure their placement doesn't interfere with the operation of the blower. On the round, inlet end of the blower cut a round hole in the plug, slightly larger than the opening, and screw and silicone the plug directly onto the housing, again taking care that your screws don't interfere with the blower's moving parts.

Installing caps directly onto the blower is the quick and easy method of blower attachment, but the best way is to have transitions made up at the local sheet-metal shop. A gradual transition to and from the blower will help reduce static pressure at these points. A typical transition may be from a 9½-by-7-inch rectangle to a 12-inch circle, which would be difficult to make yourself without special tools. Also, never change the direction of the collar attached to the round inlet of a blower. This can simplify hookups but will foul up the air delivery.

It is a good idea to install *vibration collars* on both sides of the blower to keep the ductwork from vibrating when the blower is running. They are made of two sheet-metal collars with a vinyl or canvas material between them, and they attach quite easily to ductwork using techniques already mentioned. Vibration collars are usually not air-tight at the vinyl seam, so caulk it with silicone. Short sections of flexible duct can also be used as vibration collars.

Many blowers come with several small holes punched through the casing. These serve no purpose that we have been able to figure out, so at this point put small dabs of silicone on them to seal them up.

It is important to mount your blower in an accessible location since the motor bearings need lubricating twice a year. The actual placement of the blower in the outlet ducting makes no difference, so locate it with ease of service in mind. Find the oil port (or ports) on the blower, and make sure that they feed the bearings. It may be necessary to remove the motor and turn it on its axis so that it can be oiled after installation, but it is usually easier to alter the planned placement.

When installing a blower in a crawl space, it is best to set it on cinder blocks. For a quiet system don't attach the blower to floor joists or other structural members that can resonate. Set the cinder blocks firmly in the crawl-space dirt, and set the blower on top of them. Further fastening is probably unnecessary if vibration collars are installed because the blower isn't very heavy and the surrounding ductwork is self-supporting.

If the blower is to be mounted in the attic, hang it from the rafters with several pieces of plumber's strap. Put pieces of rubber (inner tube) on the strap where it touches the blower and the roof rafters to help dampen vibration. If it is necessary to secure a strap to the blower casing, use short screws and drive them where they won't interfere with the movement of the blower blades. If the blower must be mounted directly onto the ceiling joists, put a layer of dense, closed-cell foam (or several layers of cut-up inner tube) underneath the mount. This is important since

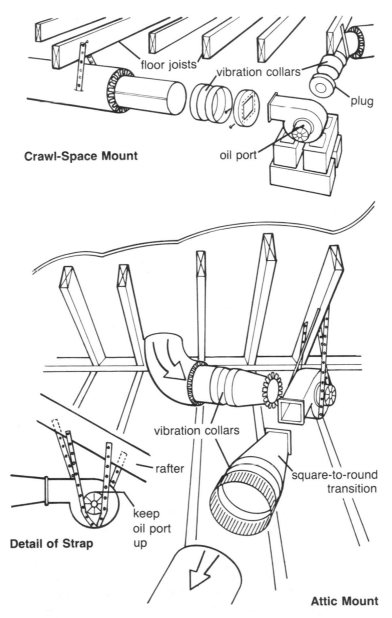

Crawl-Space Mount

floor joists

vibration collars

plug

oil port

Detail of Strap

vibration collars

rafter

keep
oil port
up

square-to-round
transition

Attic Mount

Figure 9-3: Blowers mounted in a crawl space can simply sit on top of cinder blocks. In an attic they can be hung from rafters with padded plumber's strap to reduce vibration and to allow for quiet operation. Vibration collars are also essential for minimizing vibration noises. Gradual transitions to and from the blower, as shown in the attic installation, will help minimize resistance to airflow as the air enters and leaves the blower.

even small blowers vibrate quite a lot and can be very annoying if not mounted properly.

If the blower for your system is belt-driven, there are a few other things to keep in mind:

• Keep the pulleys and the belt clean. Dirt and grease will speed up wear on the belt.

• Loosen the motor mount to put the belt on the pulleys. Never force a belt over the pulleys.

• Align two pulleys by sighting down the "Vs" in them.

• Adjust the tension on the belt so that the belt can deflect about ¾ inch for each foot between the pulleys.

• Run the blower to make sure it rotates in the proper direction. It may be necessary to change the wiring of the motor.

Any type of blower can be insulated with fiberglass batts like the rest of the ductwork, but never put insulation near or over the motor. Blower motors need to be well exposed to cool air to keep them from overheating and burning up. For this reason blowers should never be installed so that the motor is.in the airstream. Even motors rated for high temperatures tend to burn up under these conditions.

Cutting Holes

Cutting holes through existing walls or floors is a necessary job in retrofit situations. "Look twice and cut once" is a good rule to follow because there inevitably will be a water pipe, an electrical wire or a floor joist right in the way of your planned ductwork.

Start small and drill a tiny test hole or drive a long nail through the wall or floor where the ductwork will be located. Have a helper on the other side yell measurements

from this pilot hole to likely obstructions on his side. Enlarge the hole with a large drill bit, and cut out the hole using a reciprocating saw or jigsaw (the easy way) or a keyhole saw (elbow grease needed). You will often encounter nails, especially when cutting through several layers of flooring, so use a wood-and-metal-combination blade in your saw. Be aggressive, but try to be neat. If you don't have a power saw for hole cutting in a particularly awkward location, you may have to resort to drilling out the perimeter with a large wood bit, a crude but effective technique. In most cases, however, it is important that your holes are smooth and the correct size, especially when they are cut through the living room floor. Take your time and do a good job and it will be easier to trim them out or install grilles over them later.

Making a hole in a concrete or masonry wall is another story, and it is difficult to make a clean job of it. Brick veneer on the exterior of a house comes off quite easily with a hammer and masonry chisel and usually doesn't require any structural support after the inner frame wall is exposed. Two or three blocks can also be removed easily from a block wall with a cold chisel if the cores haven't been filled with concrete. Adding some structural support, such as an angle-iron frame, may be necessary in this case.

Poured concrete walls, especially older ones that aren't steel reinforced, are the worst. Avoid punching holes through them at all costs. Unless you have access to elaborate masonry saws, the hole will end up being a complete mess and can weaken the wall.

Beware of randomly cutting holes in the floor or ceiling. Fire codes typically forbid open holes between floors that don't have fusible-link fire dampers installed in them. The idea here is that in case of a fire it is important that different floors be isolated from each other. It is acceptable to have a return register cut through the floor into the crawl space if the return is ducted all the way to the collector, but it isn't acceptable to have an open hole simply dumping air into the crawl space. Call your fire department if you are in doubt.

Installing Floor Registers

Don't cut holes for floor registers until the necessary hardware for their installation is on hand. For each outlet you need a floor register and a *boot* or adapter that makes the transition from the round ductwork to the rectangular opening of the floor register. The square end of the boot is pushed up from the crawl space, and its top edge is screwed or nailed to the end grain on the flooring. Another method is to bend over the top four edges of the boot so that a flange is created

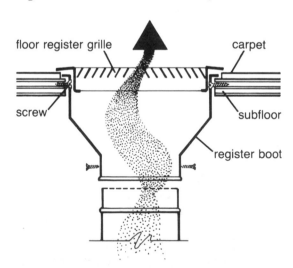

floor register grille   carpet

screw   subfloor

register boot

Figure 9-4: Half-inch flanges are bent on the end of floor register boots after they are pushed through holes in the floor. Then they are attached to the subfloor with screws or nails. A register grille fits inside the boot and over carpeting. It's best to install the boot and register before attaching ductwork to it.

that rests on the floor (not carpet). A flange on the floor register fits inside the boot, and a trim flange (1 to 1½ inches wide) fits over the carpet and any part of the boot that may be showing. It usually isn't desirable to secure the grille when it is placed on top of carpet, but if the grille goes over bare flooring, a couple of small dabs of silicone will hold it in place and still allow for future removal. Hot air can also be delivered into the living area through ceiling grilles, even through grilles in partitions, if the need arises.

Dealing with Dampers

Backdraft and manual dampers are built in rectangular sheet-metal boxes so for installation onto round ductwork, round-to-rectangular transitions are needed. Rectangular ductwork fits together a little differently from round ducting. In standard practice the straight edges are joined with *S-clips* and *cleats:* Two opposite sides are joined with S-clips, and the other two sides are joined with cleats. However, a special technique is required for installing cleats so for our purposes we will deal only with S-clips, which are available at all sheet-metal shops. To use them, you cut 1-inch slits at the corners of the end of the transition and at the corners of the damper box. A ½-inch slit is usually sufficient, and that's what the sheet-metal shop will deliver on specially made boxes, but a 1-inch slit is easier to work with. Cut the S-clips ¼ inch shorter than the edge on which they fit. Slide the S-clips onto the edge of your transition, opposite sides in and out as in figure 9-5. Slowly slide the edges of the damper box into the S-clips, and keep wiggling everything until it is joined tightly. Drill ⅛-inch holes, two to a side, so that they go through the entire assembly, and drive in ¾-inch screws to snug everything up. Seal with duct tape and/or silicone to ensure an airtight seal. This is another job to be done in the workshop, not the attic.

If the box you are installing won't fit together with the transition this way because of a size difference, secure it to the square transition the best way possible. Use plenty of screws and silicone and be sure to tape the joint. The two pieces of metal must overlap at least ⅜ inch everywhere and have screws through them. Keep the box square when in-

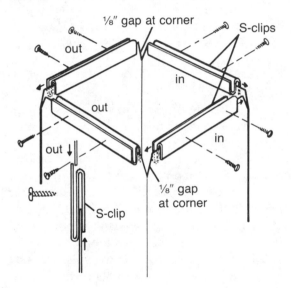

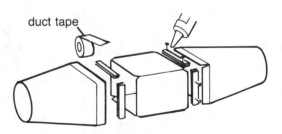

Figure 9-5: Use S-clips to attach rectangular sections of ductwork. Alternate the sides of the clips, in and out on different sides of the duct, and secure the sides with screws driven through predrilled holes. Seal the joint with silicone or duct tape.

stalling it, and check to see that your makeshift installation doesn't interfere with the damper's operation.

Air Filters

The return or inlet duct to an active collector should have an air filter placed somewhere in it. Almost all collector systems have a backdraft damper in the return line, and this may be the easiest place to install an air filter, too. You can use a piece of furnace filter that is oversized for the opening of your backdraft damper. The filter is held in place with a small spring-loaded curtain rod or with a piece of sheet metal cut and bent to act as a spring (see figure 9-6). If your return duct is from the house (direct-use system), you can install a filter grille in the return air register. It is available from a sheet-metal shop or plumbing and heating dealer. Call and ask for available sizes early in the design process.

For larger systems use an air filter box, which is a thin, square box with round ductwork attached to both ends. You can have it made at the sheet-metal shop. Make the box itself airtight. If you anticipate high static pressure in your ductwork, consider ordering oversized filter boxes, which will present less resistance to airflow.

In-and-Out Tabs

In many installations, especially those with exterior manifolds or an air-to-water exchange tank, it is necessary to attach a small piece of duct at a right angle to a large one. The best way to do this is with in-and-out tabs. To make them, you snip 1- to 1½-inch tabs every inch around one end of the smaller duct section. Bend three or four of these tabs out at a 90-degree angle, and insert the rest into the receiving (larger) section. Reach inside the smaller duct and bend the other tabs

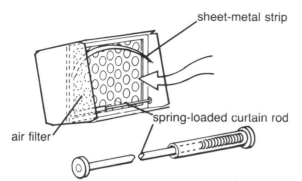

Furnace Filter Installed in Backdraft Damper

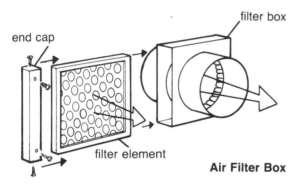

Figure 9-6: Furnace filter material can be held directly in front of the exposed end of a backdraft damper with either a spring-loaded curtain rod or a strip of sheet metal. More complex systems require an air filter box installed at some point in the system.

over against the sides of the large duct. Make sure the inside tabs are tight before screwing the outside tabs to the larger duct. Silicone-seal the outside of the joint. *Starting collars* can be ordered from the sheet-metal shop to accomplish this, but making collars on site is very easy.

Tying into Your Collector

After the distribution ductwork has been run and the blower has been mounted, it is time to tie into the collector. Do this step after the insulating backing is on the collector but before the absorber is installed. The best way to tie-in is with rectangular floor register boots.

Cut a hole the same size as the boot through the back of the collector. Push the boot through the hole. With snips, cut 2 inches into the corner edges of the boot and bend these cut ends back at a 90-degree angle to form 2-inch flanges. Push these flanges into a bed of silicone, and screw them securely through the insulative backing and to the plywood. Silicone-seal the cut edges of the plywood and insulation to prevent air leaks and to ensure that they aren't exposed to high temperatures.

It Won't Fit

The cardinal rule of duct sizing is "keep it large and don't choke it down at any point," but sometimes this isn't easy. Every now and then there is a retrofit installation where the ductwork just won't fit. Let's take the case of trying to run a big duct from a large collector through an 8-inch rim joist, with floor joists 16 inches on center. There isn't enough room between the floor joists to allow the properly sized duct to fit into the crawl space. It may be necessary to run two ducts through adjacent joist spaces and tee them together once they are inside the crawl space. This is easier than rebuilding the floor supports on one end of the house. If choking the ductwork is unavoidable at some point, make gradual transitions up to and away from the problem area.

Building Manifolds

Locating the inlet and outlet vents of a collector is accomplished easily in small, internally manifolded collectors. Planning external manifolds takes more thought, however, because they are large, and the inlet and outlet each has more than one port. The collector must be positioned so there is ample room behind it for external manifolds.

In vertical collector designs it is possible to locate the manifolds inside the stud wall. A specially made, sheet-metal duct fits into this cavity, and short rectangular sleeves (corresponding to the various port locations) are attached to the manifold before it is placed in the wall. The plywood backing for the collector has holes cut for these ports, and the sleeve-manifold assembly is attached to it before the plywood is nailed in place. Poly-isocyanurate foam could be used to insulate this space and serve as the manifold, but proper sealing is very difficult.

Roof installations usually have room in the attic for installing the manifold behind the collector. Holes for the ports are cut through the plywood, and the manifold is hung between roof rafters with plumber's strap.

Installing external manifolds in slanted ground mounts and built-up roof mounts can be more of a problem because the space in the upper and lower corners behind the collector is limited. When designing any of these installations, make careful drawings to make sure everything will fit. You can leave yourself extra room to work by bringing the slanted collector out from the wall or by building a slightly oversized framework for a roof mount.

Sizing

Active collectors must have an unrestricted airflow from the manifold and the airflow cavity. This is accomplished by making the ports about as wide as the baffle spaces they serve. The manifold itself should have approximately the same cross-sectional area as the rest of the ducting. For manifolds with balancing dampers, the total cross-sectional area of the ports should be slightly larger than that of the baffle spaces. For manifolds without dampers it should be slightly smaller. Let's look at the following example.

Collector: 120 ft^2, each manifold has three ports and balancing dampers

Ductwork: 10" round = $\pi (5)^2$ = 78.54 in^2

Manifold: 8" x 10" rectangle = 80 in^2

Ports: three 2" x 14" register boots @ 28 in^2 = 84 in^2

Photo 9-5: When you're using in-and-out tabs to connect two pieces of round ductwork, the slits on the inserted piece need to be of different lengths to conform with the curve of the receiving (larger) duct. Reach inside the piece of duct with the tabs to bend the inside tabs up against the sides of the other duct section.

Working with a Sheet-Metal Shop

The special tools needed for custom sheet-metal work are certainly very handy, but they're also expensive, especially if you're going to be a one-time ductworker. You probably will need some custom pieces made for your solar system: a damper box, an unusual elbow or transition. Instead of struggling to make these yourself, you can visit a sheet-metal shop, where you can get your custom work done properly, neatly and without tears. The success of your visit depends on knowing what you need. This place is a tremendous resource, and the people who work there are typically patient types who will take time with you to make installing your ductwork much easier.

Some shops in your area may already have solar experience or, at least, interest. Check the Yellow Pages for shops that include solar heating in their ad copy, or ask around to see which shops have done solar work. Smaller shops can often be easier to work with and easier when it comes to billing you. If you have a large order, shop around for the best total bid. Prices can vary by as much as 50 percent. Labor costs should be slightly less than material costs for simple items, slightly more for complex ones, but the total cost can be considerably more if you need only one of a kind. Charges for running Pittsburgh joints on absorbers should be between $15 and $25 an hour, and it shouldn't take over 30 minutes. If you buy valley metal for collector flashing, the shop will charge about a dollar a break to make it up to your dimensions.

When you visit the shop to get your bids, bring along detailed drawings of the device you need, label as many dimensions as possible, and show more than one view of the piece. Design your devices to be as simple as possible, and don't worry too much about whether or not it can be built. Sheet-metal workers are skilled craftsmen, and they can figure these things out for you.

Always visit the shop instead of trying to explain things over the telephone. If there is any confusion about what is being built, talk it over. Talk is cheap at this point. Be sure to indicate tolerances since there can be problems in trying to install an 8-inch-square box when the application calls for one that is 8¼ inches.

Sheet-metal workers have little trouble working within a tolerance of ¹⁄₁₆ inch.

Be sure to ask if a stock item can be ordered that will fill your needs. This will make both you and the shop owner happy: You save money; he saves time. If possible, bring the devices you have to the shop. It can be easier, for example, to make the proper transitions for a blower assembly if it's right there in the shop.

Photo 9-6: Nothing beats a helpful sheet-metal man when it comes time to install ductwork. My sheet-metal man, Dennis, helps to simplify my complex designs and has made all kinds of strangely shaped goodies for me. Here he poses with the collectors he built on slow days to heat his shop.

Inlets and outlets to the collector must be long and wide enough for free air movement, or the static pressure will be excessive, which could require the use of a more powerful blower.

Construction Steps

Round ductwork is suitable for building manifolds with few ports, but rectangular ducting is easier to work with if there are several. When there are only two inlet or outlet ports, a simple tee-fitting attached to round ductwork can deliver air to the two lines feeding the collector. Air will then enter through register boots. Balancing dampers will usually be needed in the two different lines to ensure an even flow down each line.

Rectangular manifolds, feeding two or more ports, are easily and inexpensively made from button-lock ducting. This metal duct material comes in pairs of L-shaped sections that snap together to form a rectangular duct. It is standard in many sizes, along with corresponding end caps that also snap on. Eight-by-ten-inch button lock costs about $3 per foot; the end caps $1 each.

The ports into and out of the collector are made by installing floor register boots onto the manifold. In regular heating systems these boots are a transition between round distribution ductwork and rectangular floor registers. A 6-inch-round to 12-by-2-inch-rectangular boot costs about $3. The round end is attached to the manifold, the rectangular end is attached to the collector. Let's look at the construction sequence for building manifolds from button-lock ducting and register boots.

Design the Manifolds

Make detailed drawings that include all measurements: total length, distance be-tween ports, attachment points for the ductwork, etc. Make sure that the boots will extend at least 2 inches past the plywood into the collector when they're installed. If your collector design requires balancing dampers (they're discussed in chapter 4), it may be necessary to install short sections of round ducting between the round end of the boot and the manifold so these dampers will have a place to go.

Snap the Button-Lock Together

Cut the button-lock pieces to length before assembly. It may be necessary to relieve the female side of the joint slightly by twisting a slotted screwdriver in the opening if the male refuses to fit. Put on the end caps. Once again it may be necessary to relieve the female side of the joint. Snap the caps in place and bend over the corners with a hammer to secure them.

If it is necessary to splice two sections of button lock, cut them to length adding 2 inches for overlap. Put on both end caps and snip the overlap ends back on one section so that the two sections telescope together. Secure with screws, silicone and duct tape.

Cut Holes for Register Boots

Hold the round end of the boot against the side of the manifold box where it will be attached, and scribe a circle onto the manifold with a felt-tipped pen. Hold a large slotted screwdriver at an angle in the center of this circle and pound down on the middle of the blade with a hammer to start the hole. Cut out the hole, just inside your line, with aviation snips. Left or right cutters work better than straight cutters. Also cut a hole for the inlet or outlet duct that will connect with the manifold. This hole will usually be on the side opposite the ports.

Attach the Boots

Snip 1-inch slits into the round end of the register boot to make in-and-out tabs spaced at 1 inch. Bend four tabs outward at a right angle. Insert the boot into the hole in the manifold. Reach inside the boot and bend over the rest of the tabs so they are flush with the inside of the manifold. Make sure the rectangular end is oriented properly, and screw the four exposed tabs to the manifold with hex-head screws.

Seal the Manifold

Run a bead of silicone around the joint each boot makes with the manifold. Smear this with your finger, but be careful to avoid burrs that may have resulted from cutting the hole. Seal all seams in the manifold, sides

Figure 9-7: *Above and on opposite page.* Working in the attic and in the crawl space.

Working Where the Sun Never Shines

Building your collector is an enjoyable experience. The neighbors will come over and chat, offering encouragement and advice. The sun will shine and the birds will sing. Of course, a lot of work on solar systems happens where the neighbors never visit, the sun never shines and the birds never sing: in dank, dark crawl spaces and in sometimes freezing, sometimes broiling, attics where the blower and much of the ductwork are usually installed. Before you get too far along with your building plans, look at these places with an eye to what it's going to be like to work in them.

Working in either a crawl space that's 2 feet high and full of spiders or an attic that's 130°F can represent a real challenge to the fortitude of even the most steadfast do-it-yourselfer. No matter what the conditions, planning always makes tough jobs easier.

Before you start, look at all the access holes to these spaces to make sure your hardware will fit through them. Make as many advance connections to your blower, ductwork and mechanical devices as you can, to check for fit.

Some crawl spaces are too small to work in, and if yours won't let you in, you probably shouldn't be designing a crawl-space system. If you can get in, but it's damp in there, the first thing to do is cover the entire crawl space with sheets of clear 6-mil plastic.

and end caps with silicone. Allow the silicone to cure for four to six hours before moving the assembly.

Insulate the Manifolds

Manifolds will inevitably be located outside the living area, and heat loss from them must be avoided. Insulate the cold manifold with 3½-inch batts and the hot end with two layers of 3½-inch batts, overlapping the seams. Secure with bailing wire.

Other Construction Possibilities

Manifolds can be built quite easily using ductboard material. The long, rectangular section is made from large flat sheets. The smaller rectangular ports are made from scraps and attached to the manifold with iron-on tape

---

You might be amazed at how much you can accomplish in a day of crawling around in a short crawl space. Take along a trouble light (with a rough-duty bulb), a flashlight and at least one helper for physical and moral support. It's nice to have company down in the crawl space. If you're all alone, bring a radio for some company. When the music's right or the home team is winning, even the worst conditions are improved. If you find you need some basic (lift, drop, pull, push, gopher, etc.) help, consider hiring an enthusiastic (emphasis on enthusiasm) teenager for the weekend. Convince him it's an educational experience.

Work is best done in the attic on a cloudy, rainy day or in early morning, when it's cool up there. Fiberglass insulation jumps onto bare skin like iron filings onto a magnet, which means you can become an itching fuzz ball in no time. Wear your insulating costume: a stocking cap, a long-sleeved shirt, a respirator mask and goggles. And use a garden rake to clear loose-fill insulation from your work area. Lay boards down over the ceiling joists in this area to walk and kneel on. This is much easier than nervously trying to straddle the joists and helps prevent you from dangling your foot through the hole you just put in the ceiling. If at all possible, avoid hookups in tight spots where the roof pitch dips down. Also, keep your head down to avoid the ends of roofing nails that are inevitably sticking through the sheathing, often in the very same spot you decide to stand up straight and (inadvertently, of course) go one-on-one with a nail point.

Oh yes, once you're back out in the sunshine, and the curious neighbors come over to talk solar, be sure to offer them a flashlight so they can poke their heads through the access hole and appreciate the results of your "invisible" labors.

Design 1

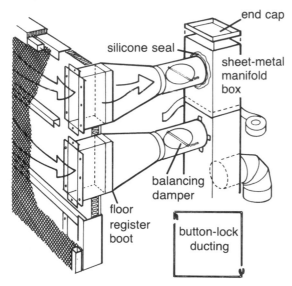

Design 2

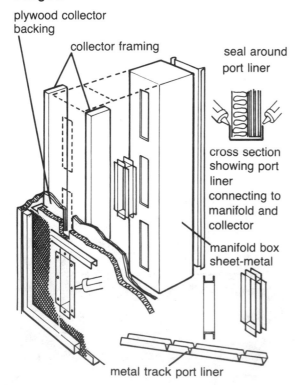

Figure 9-8: This illustration shows two possible manifolding arrangements. In design 1 the collector is fed with floor register boots attached to button-lock ducting. Balancing dampers placed in short duct sections allow regulation of airflow into each baffle space. Design 2 shows the placement of the manifold box directly against the back of the collector. The ports are lined with metal track (or angles) and sealed with silicone before the back panel is installed on the manifold box.

and/or silicone. This type of manifold will cost about the same as button-lock ducting, but it isn't recommended when solar air is delivered directly to the house. The "hot" manifold will require additional insulation outside the ductboard.

Manifolds have been built from plywood attached to the framing members behind the collector. First the manifold is lined with litho plates. All seams are sealed, and then the plywood is nailed over the rafters to form the duct. This method can work, but unfortunately the manifolds of this type that we've seen tend to leak a lot, and we don't recommend this method for any but the most meticulous.

If your manifolds need to be of uncommon shapes or sizes, have them made at the local sheet-metal shop.

Manifold Materials

Button lock is standard in many sizes. Most have an 8-inch dimension on one side, i.e., 8 inches by 10 inches, 8 inches by 12 inches, and up to 8 inches by 24 inches. Many sheet-metal dealers offer prebuilt square and rectangular ductwork in other sizes which may be useful.

Floor register boots are readily available in 6-inch dimensions, i.e., 6 inches round to 2¼ inches by 12 inches or 6 inches to 4 inches by 12 inches. Eight-inch boots may have to be specially ordered. Get the longest rectangular opening available.

Do-It-Yourself Dampers

Building a backdraft or manual damper for your system isn't complicated, but it does require patience and attention to detail to get a truly tight seal. Simple, home-built dampers are adequate for many points in a system, but if you require a very tightly sealing damper that will *completely* stop airflow or isolate part of your system, you'd better look to a commercially manufactured item. Building tightly sealing motorized dampers is out of the range of do-it-yourselfers. They will cost you a fortune, even if you "pay" yourself only $2 an hour, and they will probably leak. If it looks like the static pressure in your planned

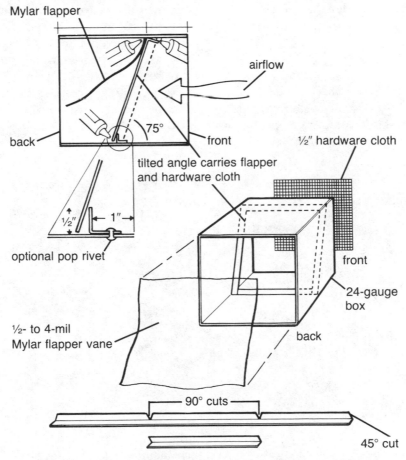

Figure 9-9: When building a backdraft damper, first install the sheet-metal angles in the box, then the hardware cloth and finally the lightweight flapper. Hold the damper up to a light to check for air leaks around the flapper.

ductwork installation could be high, consider installing a slightly oversized damper to help reduce the resistance to flow.

Building a Backdraft Damper

Materials
½" x 1" angle, 24-ga galvanized
½" hardware cloth
24-ga rectangular box (made at sheet-metal shop)
4-mil Tedlar or Mylar for the flap (½ mil for convective dampers)
silicone caulk

Labor
2½ hours (total for the first one)

Almost all active air systems require a backdraft damper at some point in the system. Their construction is different from the convective backdraft dampers discussed in chapter 7 because they must be built with materials that can withstand continuous temperatures of 200°F and a higher airflow rate. Round backdraft dampers are almost impossible to build so we will describe the construction of a square one. You will need to install square-to-round transitions, or oversize the backdraft damper and cap it at both ends in order to tie into round ductwork. The movable flap in this damper is installed at an angle to allow for easier airflow through it and to provide a more positive gravity closure that prevents unwanted back flow when it's mounted in horizontal ductwork. It can also be used in vertical return lines to prevent reverse flow down the ductwork at night. Damper boxes are usually square, but ours will be rectangular so the instruction will be easier to follow.

Photo 9-7: Shown here are a typical home-built backdraft damper (*left*) and a manual damper (*right*). The flapper vane on the backdraft damper has been marked with diagonal lines. The manual damper is shown in the open postion. When the damper is closed, the spring stretches to the other screw to keep the vane closed.

Construction Steps

1)   Order a 24-gauge rectangular box at the sheet-metal shop. The length should be equal to width (12-by-14-inch opening by 14 inches long in our example—see figure 9-10). Also order a length of standard 28-gauge, ½-by-1-inch angle equal to the perimeter dimensions of the box (12″ + 14″ + 12″ + 14″ + 2″ extra = 54″).

2)   Carefully measure the inside dimensions of the box, and cut a piece of ½-by-1-inch angle to fit snugly inside the bottom of the box. This angle will be slightly shorter than the longest dimension of the box (14 inches), and its 1-inch side will rest against the box. The top-and-sides angle is made from a single piece. Cut exact 90-degree notches (use a try square) in the ½-inch leg of the remainder of the angle so it will fold up and just fit inside the top and sides of the box (¹⁄₁₆-inch gap). Insert this angle into the box near the front, and slope the sides forward at about a 75-degree angle. The top angle will be 2 inches from the front of the box, and the bottom about 3½ inches (in our example). It will be necessary to cut the corners off the ends of the 1-inch legs on the side (12-inch) legs in order for them to fit inside the box. Secure the top and side legs to the sides of the box with several small pieces of duct tape. Insert the bottom angle (previously cut) so that its upper (½-inch) leg rests directly behind, not in front of, the 12-inch side legs. Secure the bottom angle with duct tape. At the top of the box there will be a gap where the 1-inch side of the 14-inch leg doesn't lie snugly against the box. Fill this gap in three or four places with silicone. Push silicone under the 12-inch side legs and the 14-inch bottom leg. Make sure the front surface is flat. Allow the silicone to cure overnight. Remove the duct tape and carefully and neatly seal the front edge of the ½-inch angle. Use a tiny bead of silicone and press it into the gap with your finger.

3)   Cut a rectangular piece of hardware cloth (with ½-inch holes) ½ inch smaller than the box dimensions, 11½ inches by 13½ inches. This must be absolutely flat. If it isn't and can't be made to be, cut another piece. Set the damper box on it's front side, insert the hardware cloth on the back (upstream side) of the angles, and attach with seven or eight globs of silicone along each edge. Be generous, but don't goo it past the front of the hardware cloth. Allow the silicone to cure overnight.

4)   Cut a piece of 4-mil Tedlar or Mylar smaller than the inside dimensions of the damper box (11⅝ inches by 13⅝ inches) to act as a flap. Check for fit. Run a small bead of silicone on the front edge of the top angle, and press the Tedlar into place so that it is square in the box and firmly against the upper angle. Next, run another bead of silicone along the top edge of the flap. Point with your finger to get a secure attachment between the flap, the angle and top side of the box. Allow the silicone to cure for four hours.

5)   Test the flap. Hold the box upside down, and make sure the flap doesn't rub on the sides of the box. Carefully trim the flap with scissors if necessary—not too much though. It must overlap all angles at least ¼ inch.

Flappers for backdraft dampers in active systems should be no thicker than 4 mils and can be made from Mylar or Tedlar film. Tedlar is the best choice, but it can be more difficult to obtain than Mylar. Some types of Tedlar cannot be adhered to on one or both sides so be sure to check for this. Tedlar no. 40BG20SE is 4 mil that can be adhered to on both sides. Four-mil Mylar is available from draftsmen or drafting supply stores. The only drawback to using Mylar is that it shouldn't be placed where it sees sunlight, or it will get brittle and fall apart, but this isn't a problem in active systems.

Here are some additional tips to keep in mind when constructing a backdraft damper.

• Galvanized parts that are attached together with silicone should be cleaned with rubbing alcohol prior to assembly.

• If you want to complete the backdraft damper in one day, you can pop-rivet the angles to the sides of the box and solder the hardware cloth in place.

• Avoid racking (distorting) the box when installing it in the ductwork, and make sure the flapper can move freely.

Building a Manual Damper

Materials
¼"-diameter spring (3" long, pulls to 6" fairly easily)
24-ga ductwork
EPDM weatherstripping (available at large

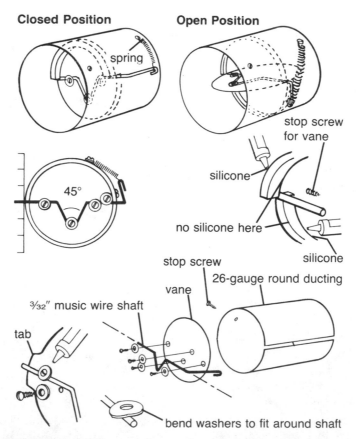

Figure 9-10: Careful positioning of the vane and weatherstripping are important when building a tightly sealing manual damper. Note that the tab on the vane is located directly opposite the joint on the round ducting.

hardware stores; felt weatherstripping can be used but isn't as good)

24- or 22-ga flat sheet metal

$3/32''$ music wire (available at the local hobby store)

sheet-metal screws ($3/4''$ and $1''$)

silicone caulk

$1/4''$ washers

Labor

$2^{1/2}$ hours (for first one)

Manual dampers are usually installed in pairs in larger systems and can be useful in installations that have two points of use, such as a manually operated two-mode system, or in installations requiring summer ventilation. Often one damper in a set must be more tightly sealing than the other, and for this reason it is a good idea to complete one damper before building the other(s). You'll be experienced on the second one and can use the tighter damper in the critical ductwork line. The round manual damper we describe (see figure 9-10) is simply an elaborate stovepipe damper that is easily placed into round ductwork. A tightly fitting round vane acts as the damper and closes against weatherstripping attached to the duct. The vane is held closed with a small spring.

Tightly sealing manual dampers are easier to build in small sizes (6 and 8 inches in diameter) than in larger sizes (10 and 12 inches). If you have ductwork over 12 inches in diameter, your system is probably big enough to justify motorized dampers. Our example is for a 10-inch manual damper.

Construction Steps

1) Cut a round section of heavy-gauge ductwork (the heavier the better) 2 inches longer than its diameter (12 inches long for a 10-inch damper). Snap the duct section together. Next cut out the vane. Use a hammer and nail to punch a dimple in the center of a piece of flat galvanized sheet metal, 24 gauge or heavier. With a compass, scribe a circle with a diameter $1/4$ inch less than the diameter of the ductwork ($9^{3/4}$ inches). Cut out the circle, leaving about $1/4$ inch of material outside of the scribed line and then cut out the vane exactly along the line, leaving a $1/8$-by-$1/2$-inch tab on one edge. Making two cuts is *not* an unnecessary step. It helps keep the vane flat and ensures an accurate second cut. The small tab on the vane will be located opposite the joint in the ducting and will help center the vane.

2) Find a small cardboard box or large wooden block that will fit inside the duct and insert it about halfway in. This is important, so make one if necessary. Set the vane inside the duct and on top of the box, and center it with the tab on the vane opposite the joint in the ductwork. There should be a $1/8$-inch clearance between the vane and the duct all the way around except at the tab. Use a felt-tipped pen to mark the points where the shaft will penetrate the duct. One penetration will be through the seam in the duct, the other directly opposite and centered on the tab on the vane.

3) Make the shaft for the vane from $3/32$-inch (or $1/8$-inch) music wire. Don't use a coat hanger. Bend it as in figure 9-10 using two pairs of pliers or a vise and hammer (it is very stiff) and cut it with a hacksaw. Bend the long leg to be 45 degrees off the plane formed by the shaft and the triangular section. Bend and twist the wire so that the two sides of the shaft are in line with each other. Insert the shaft through the holes so that it lies on top of the vane. Center the vane by inserting a few 6-penny nails in between it and the duct. Trace the shaft's placement onto the vane with a felt-tipped pen, remove the vane and drill four $1/8$-inch holes through it

as in figure 9-10. Make a 30-degree bend on one edge of four ¼-inch washers, and secure the shaft to the vane with four sheet-metal screws driven through the washers. If the shaft isn't tight against the vane, put in more screws and washers. Reposition the shaft if the vane is no longer centered in the duct.

4) The next step is to attach weatherstripping to the side of the duct. The vane will make a compression seal against this material. EPDM weatherstripping should be used, although felt is a second choice. Do not use vinyl weatherstripping. Remove the adhesive (usually standard) from the EPDM weatherstripping by peeling and vigorous rubbing. Do not rely on this adhesive to hold it in place. Cut the weatherstripping exactly to length so that it will butt against the seam in the duct and run around the inside circumference to the opposite shaft penetration (the length equals one-half the circumference minus ¼ inch). Rest the vane on the small cardboard box. Carefully run a bead of silicone on the back side of the EPDM and ease it into place. Once it's in position, push it firmly against the side of the duct, but avoid putting too much pressure on the vane or smearing silicone onto the vane. Allow the silicone to cure for four hours or more. Turn the damper over and place a ½-pound (or less) weight on the "sealed" side of the vane to hold it firmly against the first gasket. Silicone the other gasket in place, remove the weight and push the second gasket down about ⅛ inch all around. Gently close the damper five or six times (tapping motion) to knock the second seal into the proper position. Open the vane and, after the silicone has cured, hold the damper up to a light to check your seal. Don't be afraid to scrape the weatherstripping off one side and replace it if you need a very tightly sealing damper.

Attach one end of the 3-inch spring to the end of the shaft. Pull the spring across the outside of the duct to close the damper. Drive a sheet-metal screw halfway into the duct to act as an attachment for the spring. It should be in a position that allows the spring to be under considerable tension while holding the vane closed without interfering with the vane's operation. Drive a long (1-inch) sheet-metal screw through the side of the duct directly opposite the seam in the duct to act as a stop for the vane when the damper is open. Position another screw to hold the spring when the vane is in the open position. Silicone all screw penetrations into the vane and the duct.

Here are some additional tips on building a manual damper.

• If your installation calls for a very tightly sealing manual damper, have the casing and vane made at a shop from 24- or 22-gauge material.

• If you use felt for the seal, use a fairly narrow piece and back it up with silicone. Don't use vinyl or other low-temperature materials.

• When installing the damper, make sure it is placed in a well-supported, straight section of ductwork. The vane will not seal properly if the duct in which it is placed is racked in any way. Take your time and check its movement after installation.

Guillotine Dampers

Another design for manual dampers is the *guillotine damper*. The vane is made from ¾-inch plywood or heavy-gauge galvanized sheet metal, and the box is made much like the box for an air filter housing. These dampers must be made very carefully, or they're sure to leak, and they also require transitions to connect with round ducts. However, if your

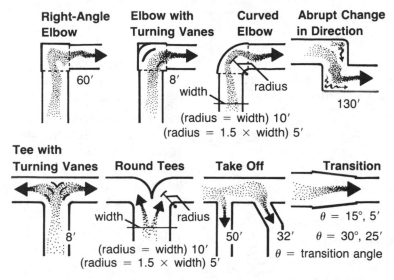

Figure 9-11: Different duct fittings and transitions represent different equivalent lengths of straight ductwork. Abrupt changes of direction can greatly increase the static pressure in your ducting system and should be avoided.

sheet-metal shop has had experience with guillotine dampers, they may be able to help you build a good one.

Testing Dampers

All backdraft and manual dampers will have some leakage, and you can make a simple test to tell how much. Place the damper in the bathtub, and run a full stream of water into the damper. If it doesn't fill up, it is too leaky and could present a problem if your installation demands a tight seal.

If you order a commercially manufactured motorized damper, take it to the bathroom for testing. Plug it in and motor the damper closed; then unplug it and put the motor in a plastic bag. Run a quarter stream of water into the damper and if it doesn't fill up, dry it off, send it back, and order from another company.

Static Pressure: Duct Design and Blower Selection

Proper duct design and blower selection are vital if an active solar air collector is going to operate efficiently. Probably between a third and a half of all homemade air collectors have either ducting that is undersized or a blower that delivers too slow an airflow. These situations both lead to poor system performance.

Calculating or estimating the static pressure (SP) in your system will take only a few minutes if your air delivery system is simple. It becomes more difficult as the complexity of your system increases, and the do-it-yourself builder of a large, very complex system may need to seek professional advice from a mechanical engineer.

Before choosing the blower, you need to know the static pressure of the system. In

TABLE 9-1

Variables That Cause Low or High Static Pressure in Ductwork

Low Static Pressure	High Static Pressure
Large ducts	Small ductwork, undersized
Short duct runs	Long runs of ductwork
Gradual changes in direction of flow	Abrupt changes in direction of flow
Low airflow volume	High airflow volume
Low airflow velocity	High airflow velocity

order to calculate this, you must know the size of ductwork to be used, the equivalent length of the ductwork (actual run, plus additions for elbows and fittings), the volume of air to be moved (in cfm) and the static pressure presented by different components in the system: the collector, the air-to-water heat exchanger and the rock storage.

Static Pressure in Ductwork

Since static pressure is usually described in inches (water gauge, WG) per 100 feet, one of the first things that must be determined is the distance the air moves through the system. Different duct fittings, elbows, tees and takeoffs represent different equivalents of straight ductwork. An abrupt, right-angle elbow represents the same resistance to airflow as would 60 feet of straight ductwork of the same size. A curved elbow will represent an equivalent of only 10 feet of straight ductwork, so curved elbows should be used whenever possible. Figure 9-11 shows some common fittings and their equivalent in duct length. It is immediately obvious that abrupt changes in the ductwork size or direction can greatly increase the equivalent

duct length and should be avoided whenever possible.

Collector Components

Static pressure should be minimized everywhere in your system except where a transfer of heat takes place. These places include the collector, the air-to-water heat exchanger and the rock storage bin. In the collector, for example, air needs to "rub" against the absorber in order to get a good heat transfer from the plate into the moving stream of air, and this requires more air turbulence and a higher resistance to flow than is found in the ductwork. When heat is taken out of the solar air, the same principle applies. There must be some static pressure present in an air-to-water heat exchanger or a rock bin for an effective transfer of heat.

The active collector we described in chapter 8 has an increased static pressure by having a narrow airflow cavity. It is very difficult to calculate the exact static pressure a collector will present to a system, but in this design it is between 0.20 and 0.35 inch WG.

TABLE 9-2

Desired Volume of Air through an Active System

Elevation (ft)	Airflow through Collector (cfm/ft²)
Sea level	2½
2,500	2½ to 3
5,000	3
7,000	4

NOTE: The velocity of the air that is moved through an active system depends on the desired temperature rise through the collector and the altitude of installation. This table shows the airflow needed for a 50-degree temperature rise, which is about right for both water- and space-heating applications.

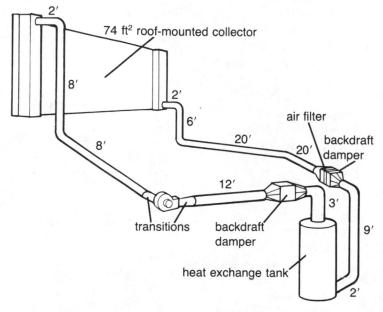

74 ft² roof-mounted collector

Figure 9-12: To get a quick idea of the static pressure that will be present in your system, make a drawing of it. Determine the equivalent length of the ductwork and fittings and the static pressure it will present. Then add the static pressure presented by the other components in your system.

This is a wide range, but there are many factors that influence this figure, including how the air enters and leaves the collector and the type of baffling within the collector. The optimal collector design will have long rectangular boots feeding nearly the entire baffle width and a one-direction flow path. Then, all the static pressure is used to transfer heat, and there is no unnecessary static pressure created by undersized ports and abrupt corners. This design isn't always possible in retrofit situations, and it is sometimes impossible to build a collector with long rectangular manifold ports. Likewise, it may be necessary to have turns in the airflow path in order to have a 20-foot-long flow path. Both of these options are usually satisfactory (see the discussion of collector design considerations in

chapter 4) but will increase the static pressure in the collector and may require the use of a large blower. Proper manifolding is usually more important than having a straight airflow path because turning vanes can be installed inside the baffle spaces to direct airflow and limit unwanted static pressure.

The static pressure presented by the air-to-water heat exchange tank we feature in chapter 10 is approximately 0.22 inch for up to 250 cfm, 0.34 inch for 300 cfm and 0.60 inch or more for 400 cfm. Higher outlet temperatures are achieved with lower airflows. Systems using an 80-gallon tank for domestic water heating work best with a flow rate of 200 to 300 cfm. The static pressure presented by commercially manufactured, in-duct air-to-water heat exchangers will vary greatly with their design, and if you plan on using them, you need to check their static pressure at your desired airflow before designing your ductwork or choosing your blower (see chapter 10).

A complete listing of the static pressure presented by rock boxes is included in chapter 11. If you plan to build one, read chapter 11 before continuing your static-pressure calculation.

Velocity

A final consideration in static pressure calculation is the speed with which the air moves through the system. The maximum air velocity for residential applications is 800 feet per minute (fpm). If air is moved faster than this, turbulence is created within the duct, and the static pressure in the system increases very rapidly. It can also cause annoying "whining" in the ductwork. This noise problem is particularly noticeable in direct-use systems. The maximum air velocity can be increased slightly at higher altitudes because the air there is less dense.

Here is a quick calculation for the static pressure in the simple hot water preheat system in figure 9-12.

Location: Salt Lake City, Utah, elevation approximately 5,000 ft

Collector size: 74 ft² (3 horizontal panes of 46-by-76-inch glass)

Desired airflow: 74 ft² x 3 cfm/ft² = 222 cfm

Ductwork: straight runs = 92 ft + 9 elbows @ 10 ft each = 90 ft (equivalent) = 182 ft total

System components:	inch WG
Collector: straight pass, large manifold ports	= 0.20
Blower transitions: tapered and gradual	= 0.02
Air filter: oversized for ductwork	= 0.07
Backdraft dampers: 2 each @ 0.05 inch WG	= 0.10
Preheat tank/exchanger: @ 222 cfm	= 0.20
Ductwork: from table 9-3, 9″ round (estimate)	= 0.10
Total static pressure	= 0.69

After doing the calculation for your system, your initial reaction may be to try to under-size your ductwork since it seems such a small part of the total static pressure in your system. Avoid this temptation or you will end up with noisy ductwork, and an excessively high static pressure.

You should be able to figure the static pressure in your system to within 10 to 20 percent using this method. If you are completely confused about static pressure, go to your local plumbing and heating contractor. He has a slide-rule-like device for calculating static pressure, and one setting tells the whole story. If you desire a more exact calculation of the static pressure in your ductwork, turn to Appendix 6 where you can see how we came up with our charts.

Blowers

Once you have an idea of the desired airflow in cfm and the static pressure present in your system, you are ready to choose a blower. Let's examine the different types of centrifugal squirrel cage blowers that are suitable for collector applications. The first thing to look at is the blade arrangement on the squirrel cage.

A *forward-curve* blower is the most common and desirable type for use in small installations. The tips of the blades on the squirrel cage are inclined in the direction of rotation. This blower runs very quietly and is the best choice when the static pressure in the system is fairly low (under 0.8 inch WG).

A *backward-inclined* blower has blades that are inclined away from the direction of the flow, so the blower will operate at higher speeds, yet be noisier, than forward-curve blowers. Backward-inclined blowers are more efficient than forward-curve blowers and can be used in systems with high static pressure (between 0.8 and 2.0 inches WG), but in operation they sound like vacuum cleaners unless they are well isolated from occupied areas. These blowers are available only in larger sizes and find their best application in systems with extensive ductwork and/or a rock

TABLE 9-3

Rules of Thumb for Static Pressure in Ductwork Systems

Desired Airflow (cfm)	Total of Straight Runs (ft)	Number of Elbows	Equivalent Length (ft)	Duct Size (inches round)	Static Pressure (inch WG)	Velocity (fpm)
150	50	3	80	6	0.15	780
150	50	6	110	7	0.10	570
250	50	4	90	8	0.10	735
250	100	6	160	9	0.10	580
350	60	6	120	9	0.15	810
350	120	8	200	10	0.14	660
500	60	6	120	12 (10)	0.07 (0.15)	650 (930)
500	120	10	220	12	0.12	650
650	80	6	140	14 (12)	0.07 (0.15)	620 (850)
650	120	10	220	14	0.12	620
800	80	6	140	14	0.08	760
800	120	10	220	14	0.13	760
1,000	80	8	160	16 (14)	0.15 (0.15)	730 (950)
1,000	150	12	270	16	0.13	730
1,300	80	8	160	18 (16)	0.07 (0.11)	750 (950)
1,300	150	12	270	18	0.11	750

NOTE: For purpose of convenience, rectangular ductwork is converted to its equivalent in round ducting. Ductwork that is nearly square has an effective cross-sectional area of approximately 95 percent, so a 10″ × 12″ rectangular duct with an effective area of 114 square inches (120 in^2 × 0.95) is equivalent to a 12-inch round duct (area = πr^2 = 3.14 × 6^2 = 113 in^2). The more the ductwork is out of square the smaller its effective cross-sectional area. If one side of the duct is twice as long as the other, the effective cross-sectional area will be approximately 90 percent of the calculated area, so a 4″ × 8″ duct will have an effective cross-sectional area of 32 in^2 × 0.90 = 29 in^2 and be equivalent to 6-inch round ductwork. If the inside radius of a corner is equal to the diameter of a duct, then a 90-degree elbow is the equivalent of 10 feet of duct (see figure 9-11).

For a more exact calculation of the static pressure in ductwork, see Appendix 6.

storage component.

A *radial-blade* blower has straight blades and limited solar application. It operates at very high static pressures but is noisier than both forward-curve and backward-inclined blowers.

TABLE 9-4	
Static Pressure in Collector Components	
	Inch WG
Collectors and manifolds	
Straight pass (large rectangular ports in manifold)	0.20 to 0.25
Convoluted pass (undersized or round ports)	0.30 to 0.35
Transitions to and from blower	
Gradual and tapered	0.02
Abrupt, not tapered	0.05
Air filters	
Thick filter same size as ductwork	0.15
Thin filter in box larger than ductwork	0.05
Dampers	
Backdraft or motorized, sized to ductwork	0.05
Damper larger than ductwork	0.02
Hot water preheat exchange tanks	
Less than 250 cfm	0.20
Up to 400 cfm	0.30
Rock bins	
5 feet of rock @ 20 fpm	0.10
6 feet of rock @ 30 fpm	0.30
Crawl-space return	
Air has to force its way into return	0.05

These three types of blowers have two basic inlet configurations: single inlet and double inlet. Double-inlet blowers draw air from both sides of the squirrel cage and are useful only in simple installations where you are blowing air into the collector (instead of pulling air out) in a crawl-space system. Single-inlet blowers are the preferred choice for most systems.

The next consideration is how the motor is mounted onto the blower. The two options for single-inlet blowers are direct drive and belt drive. In direct-drive blowers the motor is mounted onto the side of the blower housing, and its shaft drives the squirrel cage directly. Most small, direct-drive blowers are operated by either a shaded-pole motor or a permanent-split-capacitor (PSC) motor. Even though they are more expensive, the PSC blowers are preferred in all cases because they use less power. In belt-driven blowers the motor is mounted on top and away from the blower housing.

With both mounting arrangements, proper ventilation of the motor will prevent overheating and motor burnout. Different blower motors accomplish motor cooling in different ways. With belt-driven blowers the motor is mounted away from the squirrel cage and has its own ventilation system. As long as the motor is mounted outside of the moving stream of hot air and doesn't have an excessive load placed on it, it will run cool. *Never mount any blower motor inside the air-stream.* Surprisingly, motors don't have to work as hard when the static pressure in a system is high. When there is more resistance to airflow, the blower won't be moving as much air. When there is no static pressure, the blower will be moving a lot of air, and the motor will be put under a greater strain. Many belt-driven blowers are designed so that there must be a certain amount of static pressure present in the system or they will over-work, overheat and burn up. Be sure to check for this when ordering a belt-driven blower, and don't test a blower of this type for more than a few minutes before installation in the ductwork.

The motors in many direct-drive blowers, on the other hand, don't have their own ventilation systems but, instead, rely on air movement inside the blowers for cooling. As air moves through the blower and down the ductwork, a small amount of air is pulled through the motor housing, and this cools the motor. The heat generated by the motor is thus added to the air system. These direct-drive blowers, unlike belt drives, cannot tolerate high static pressures. If there is only a small amount of air moving through the blower, the flow through the motor will also be small and insufficient to keep it cool, even though the motor isn't working as hard. For this reason some catalogs listing direct-drive blowers indicate a *cut-off static pressure*, which is the highest recommended static pressure for a given system. If the static pressure in your system is close to this figure for a particular blower, order a larger one. Those that don't have a cutoff listed are self-venti-lating and can tolerate higher static pressures. When choosing a direct-drive blower, also be sure to get one that has a thermal protection breaker. If the motor does get too hot, this breaker will trip and turn off the motor and keep it from burning up.

Belt-driven blowers are the logical choice for two-mode systems where one delivery mode encounters very different static pressure than the other mode. The system can even be designed around the wide range of static pressures that belt-driven blowers will tolerate. For example, if a two-mode system delivers air either into the house or to a domestic water preheat tank, the tank should receive a smaller flow of air than the house. If a belt-driven motor is used in this system, the air-to-water heat exchanger or the duct-work delivering air to it can be "choked down," increasing static pressure and decreasing air delivery. To accomplish this with a direct-drive blower is much more complicated. The blower would need two speeds, the ductwork to the tank would need to be large, and a relay would need to be installed to operate the two speeds on the blower motor.

High-Static-Pressure Blowers

A static pressure over 1.0 inch WG could be considered unusually high in most air-heater systems. Static pressures this high should be avoided, but since they do occur, there are special, high-static-pressure blowers available that can handle up to 2.0 inches WG. Another possibility for solving existing high-static-pressure problems is to place a second, "booster" blower at some point in the system. One blower could blow air into the collector and another pull it out. This would help to speed air delivery and may be more cost-effective than replacing an under-sized blower.

Blower Selection

All blowers are rated to deliver a certain flow (cfm) through a certain resistance (static pressure). Let's use table 9-5 to determine the size of a blower for our example hot water heater in Salt Lake City (see figure 9-12). The desired airflow for this system is 222 cfm at 0.68 inch of static pressure. Number 4C666, a permanent-split-capacitor blower, looks like a good choice. It delivers 200 cfm at 0.7 inch and 300 cfm at 0.5 inch. The next larger blower, no. 4C667, is quite a bit too large so the smaller blower is the best choice.

TABLE 9-5

Selecting a Blower for Your System

Desired Airflow (cfm)	Static Pressure in System (inches WG)	Grainger Order Number and Type
100	0.3	2C647 SP*
	0.5	2C610 SP*
150	0.3	4C446 SP*
	0.5	4C447 SP*
200	0.3	4C447 SP*
	0.5	4C448 SP*
	0.7	4C666 PSC*
300	0.3	4C666 PSC*
	0.5	4C666 PSC*
	0.7	4C667 PSC* or 7C037 DD
	0.3	4C667 PSC*
400	0.5	4C445 SP* or 7C037 DD
	0.7	7C808 BD
	1.0	7C038 DD
500	0.3	7C037 DD
	0.5	4C668 PSC* or 2C946 SP*
	0.7	7C808 BD or 7C651 BD
	1.0	7C039 DD
600	0.3	4C668 PSC* or 2C646 SP*
	0.5	7C808 BD
	0.7	7C810 BD or 4C054 SP*
	1.0	7C039 DD
750	0.5	7C038 DD or 4C054 SP*
	0.7	7C812 BD
	1.0	7C812 (increase rpm)
900	0.5	7C812 BD or 7C039 DD
	0.7	7C647 DD
	1.0	7C554 RB
	1.2	7C630 BI
1,100	0.7	7C652 BD or 7C554 RB
	1.0	7C623 (adjust sheave, i.e., pulley)
	1.2	7C632 BI

SOURCE: *Grainger's Wholesale Net Price Motorbook* (Chicago, Ill.: W. W. Grainger).

NOTES: The following abbreviations are used in this table: BD = belt driven; BI = backward inclined; DD = direct drive; PSC = permanent split capacitor; RB = radial blade; SP = shaded pole.

Few of these blowers deliver the exact volume of air and static pressure required (550 cfm @ 0.25″ WG instead of 500 cfm @ 0.3″ WG, for example). For most installations it is a good idea to order a slightly oversized blower rather than a slightly undersized one.

When comparing blowers, compare motor horsepower or watt usage as well as air delivery and price.

*Cannot tolerate high SP.

Motor Speed Control

The motor pulley on most belt-driven blowers can be adjusted to change the blower speed. This is done by spreading the two halves of a *variable-pitch pulley* to slow the speed or by moving them closer together to increase speed. There isn't a wide range of adjustment here, and this is more of a fine-tuning strategy. To make greater changes you can change the diameter of one of the pulleys. If the motor has a swing (or otherwise adjustable) mount, you won't have to change the belt size.

Many installations can benefit from a blower that operates at more than one speed. Two-speed blowers work well for this function, but they are difficult to locate and not available with a variety of airflow deliveries. The simplest way to regulate blower speed is with a rheostat that lowers the voltage reaching the motor. A standard light-dimmer switch from the hardware store will work just fine for this control, but be sure to get one that is rated higher than the wattage of your blower. Shaded-pole and PSC blowers can use a control of this type, but capacitor-start and three-phase motors cannot.

Automatic two-speed control can be achieved by manually adjusting the dimmer switch or hooking the dimmer switch to a relay that activates the slower speed. In a water-heating/space-heating two-mode system, for example, the same thermostats that control the motorized dampers can be used to activate the relay, slowing the blower speed for water heating. Since the blower has mass attached to it (the squirrel cage), the motor won't start if the voltage delivered is less than 60 percent of that used for full speed. The range of operating speeds available is therefore from 60 percent to full speed.

Air Delivery Checkout

After your system is installed and op-erating, it is a good idea to check the actual amount of air it is moving. Holding your hand in front of the return or an open section of ductwork won't tell you much. Air moving at a velocity of 800 feet per minute will only be moving about 9 miles per hour so you won't feel a great rush of air. The best way to measure airflow is with an *anemometer*. This instrument measures airflow very accurately, but be sure to get instructions on how to get accurate measurements. Take your time and correct for altitude and temperature if necessary.

A simpler way to check airflow is to check the temperature rise through the collector. The difference in the temperature of air entering and leaving a collector should be about 50°F on a sunny day at noon. If it is higher, you may want to increase blower speed. Measure the inlet and outlet temperature with an indoor/outdoor thermometer with a remote sensor or with a bimetal, dial thermometer (about $12 at a heating contractor supply store). Table 9-6 tells what's going on at different temperature differentials. For more precise methods of monitoring of your system, see Appendix 5.

Balancing Airflows

Whenever two or more collectors are connected in parallel so that one blower drives air through all of them, it is important to balance the different airflows so that each collector has roughly the same flow rate per square foot. Thus a blower should pull twice as much air through a 100-square-foot collector as through one that is 50 square feet. The best way to accomplish this is through proper duct sizing. The ducting for a 100-square-foot collector should have twice the cross-sectional area as that for a 50-square-foot collector. Proper duct sizing, however, is not always a sufficiently reliable method of ensuring the proper airflow, and it is always advisable to place adjustable dampers in the different lines to accomplish a balance. In systems with round ductwork, an inexpensive butterfly damper can be placed in each line so that the flows can be balanced. If the flows are balanced, the temperature differentials of all the air streams will be about the same. Indeed, the simplest way to balance the airflow from two or more collectors is to restrict the coolest flows until all flows have the same temperature. Less flow will yield a higher temperature. Your fingertips can sense about a 3°F temperature differential, and you can balance airflows closely enough by feeling the uninsulated ductwork while the system is hot and operating. Be sure to wait a couple of minutes after damping down the cool lines to recheck the temperature, since it will take the system about that long to stabilize.

If it is impossible to reach all the different lines from the same point, an inexpensive indoor/outdoor thermometer with a remote bulb sensor can be used. Punch a small hole in the ductwork and insert the bulb into the airstream of each line.

It is also a good idea to balance the airflow of the distribution ductwork to the house. Once again, this can best be accomplished by proper duct design, but if the runs of ductwork are of different sizes, lengths or static pressures and if different flows to different rooms of the house are desired, in-line dampers or dampered registers may be needed. Butterfly dampers will accomplish the desired balance, but they also increase the overall static pressure in the system (see Appendix 6).

For Further Reference

ASHRAE. *Handbook of Fundamentals.* Atlanta: American Society of Heating, Refrigerating and Air-Conditioning Engineers, 1981. This publication is available from ASHRAE, 1791 Tullie Circle NE, Atlanta, GA 30329.

Meyer, L. A. *Sheet Metal Shop Practice.* 4th ed. Alsip, Ill.: American Technical Publishers, 1976.

TABLE 9-6

Inlet and Outlet Temperature Differentials

Temperature Differential

= 20 to 40°F —collector operating very efficiently, may need to slow the blower speed to get hotter air, unless there is an extensive distribution system (several registers) in which the velocity at each register is relatively low and doesn't cause a chilling effect

= 40 to 60°F —collector operating efficiently, this is about right for space heating retrofits with possibly limited distribution and the need to have a higher temperature rise because of the relatively high velocity at each register

= 50 to 70°F —about right for hot water heating

= 70°F or more —collector is operating inefficiently, need to increase blower speed or install a larger blower

10

SOLAR WATER HEATING

Site-built air-heating collectors that are tilted are a natural for domestic water heating in addition to their primary space heating function. Installation of the subsystem and its controls is simple, and since the collectors can be used throughout the year, the payback on money invested in a water-heating system will be rapid. It is important to note that liquid-medium collectors do perform better for heating water than do air heaters per square foot of collector. But air-heating systems that are sized for space heating nevertheless perform admirably for summer-time water heating because they incorporate a large collector area. If you have limited space available for a collector and no plans for solar space heating, a liquid-based collector system will probably be the better selection. But for do-it-yourself combination space-heating/water-heating installations, a large, air-based collector system is the best choice. We have also found that a smaller (75 to 100 square feet), tilted air-heating collector can in some regions be used cost-effectively to heat water year-round with no space-heating mode. But, again, their cost-effectiveness is not universal; in some cases liquid heating systems will be the best choice. The following design guidelines refer to both space/domestic hot water combinations and domestic hot water-only systems.

Design Guidelines for Air-to-Water Systems

Collectors for hot water systems need to deliver higher temperatures than when the collectors are used for space heating. Air that is 100 to 120°F is adequate for heating a room, but usable hot water temperatures are 120°F and above. It is a good idea to use a selective surface absorber in an air system that includes water heating, especially in the hot half closest to the outlet. Double glazing can be more cost-effective than single glazing, especially where outside temperatures are low. Double glazing is a good idea if you can't build as large a collector array as you would like, because of space limitations or whatever. A smaller, double-glazed collector will perform better than a single-glazed unit of the same size, especially in colder climates. More insulation behind the collector and around ductwork is helpful in minimizing heat losses from the airstream. The airflow rate in the water-heating mode should be lower than it is in the space-heating mode. The blower should be sized to deliver 2 cfm per square foot of collector (at sea level) rather than 2½ cfm, so there will be a higher temperature rise through the collector.

Most importantly, there needs to be a good exchange of heat at the air-to-water heat exchanger so that a maximum amount of heat is taken out of the moving airstream and the air returned to the collector is as cool as possible. Since a collector's efficiency decreases when it operates at high temperatures, it is usually most cost-effective to use solar air to preheat a tank of water to 100 to 120°F and to feed a well-insulated water heater with this preheated water. Figure 10-1 shows two

systems used with air heaters, one with an *in-duct coil-type heat exchanger* and one with a *tank-type heat exchanger*. Both exchangers are types of *air-to-water heater exchangers*.

Since a storage tank is necessary in all water-heating systems, it makes sense to exchange the heat at the tank itself. Tank-type exchangers can offer as much transfer surface as in-duct exchangers; they are easier to build

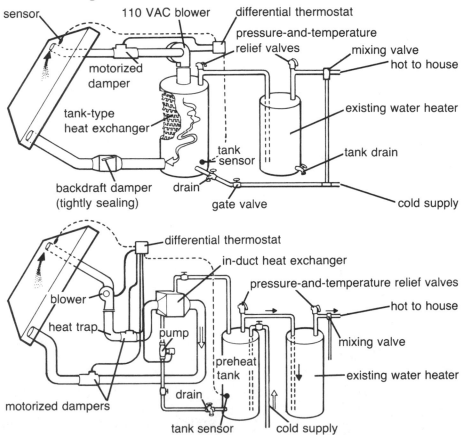

Figure 10-1: Systems that heat water with solar-heated air use an air-to-water heat exchanger. Two types of exchangers are shown: the finned-tank type and the in-duct type. We prefer tank types because they are easier to build and install and because they are simpler, less expensive and more trouble-free than in-duct exchangers. This illustration shows simple valving and plumbing arrangements for both types of systems. A tightly sealing damper is necessary with an in-duct exchanger to prevent cold air from trickling through the ductwork and possibly causing the exchanger to freeze and burst. Never locate an in-duct exchanger directly below collectors; the preferred location is at the top of a vertical duct run.

and are less expensive. No pump is required, as it is with an in-duct exchanger. Plumbing is simplified, and there is little danger of freezing problems.

In-duct exchangers do have their applications, however, especially in cases where it is desirable to exchange heat in a small space near the collector or storage. Well-insulated pipes are run to the storage tank, and there is no need to cut large holes and install ductwork to the tank.

Photo 10-1: A tank-type air-to-water heat exchanger is made easily from an 80-gallon storage tank. After the end caps on the jacket were removed, the jacket was split to remove the insulation to make room for the fins.

Isolating the Exchanger

It is important to isolate the heat exchanger from the collector at night to prevent heat loss. This is especially important when using an in-duct exchanger. Water won't be moving through the exchanger coils at night, and even a tiny trickle of very cold air from the collector can freeze the stagnant water and possibly burst the exchanger. These exchangers should be located high in the system or at the top of a heat trap to prevent this. A tightly sealing motorized damper is also a must.

Tank-type exchangers won't freeze up, but they can lose a lot of heat at night if they aren't properly isolated. A well-built backdraft damper is usually sufficient, but your system may require a motorized damper if there is no heat trap on the ductwork from the collector.

Sizing

The amount of hot water needed by a family will vary somewhat, but a good rule of thumb is 20 gallons per person per day for average use habits. Thus an average family of four would use about 80 gallons of hot water every day. The solar preheat tank for this system should be sized at 80 gallons to provide the existing water heater with all the preheated water the family will need. Families that are very energy conscious (10 to 15 gallons per person per day) can get by with a tank system that is considerably smaller than this. In any case, the tank should be glass or epoxy-glass lined and should last at least as long as the water-heater tank (a new one lasts 10 to 15 years).

A collector for a hot water system needs to be sized for the tank and for the climate. It needs to be large if the installation is in a cloudy, cold location or if the preheat tank is large. The collector can be small if the solar heater is located in a sunny, warm area and the tank is small. Larger collectors (or those with selective surface absorbers) will provide a greater annual percentage of hot water—smaller collectors a smaller percentage. As a rule of thumb, figure that 1 to 1½ square feet of collector will heat 1 gallon of storage. (See chapter 3 for information on sizing a hot water system for your site.)

Using a Large Collector for Heating Water

If your installation has a large, tilted space-heating collector, it can be used very effectively for preheating water in summer, which means shortening the payback period of the system. The low speed on a two-speed blower can be used, or a second, smaller blower can be installed, or a light-dimmer switch can be used to slow the blower speed. Any of these options will provide a slower, higher-temperature flow of air to the air-to-water heat exchanger. If your space-heating collector is smaller than 1½ to 2 square feet of collector per gallon of water in the preheat tank, a properly sized exchanger should be able to take enough heat out of your collector to keep it adequately cool in summer, and the air can be moved in a closed loop with no need for venting to the atmosphere. If the collector is larger than this, it should be fed with outside air and the air coming off the exchanger exhausted to the outside after it heats the water. This will prevent the collector from reaching undesirably high temperatures while still providing plenty of hot water. It doesn't mean that the heat exchanger is inadequate; a large space-heating collector can truly overwhelm (and overheat) the storage tank, and it must be vented to the atmosphere.

In many cases larger collectors (at least 120 square feet) can provide nearly all of a family's summertime hot water needs. The back-up water heater can in some cases be bypassed and turned off for the summer. It is also possible to tie a hot water preheat system into a solar system with storage.

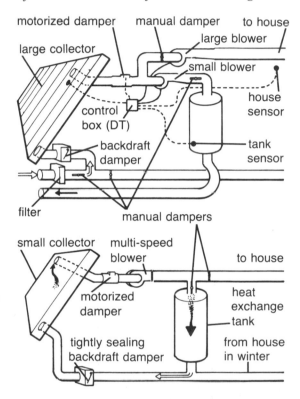

Figure 10-2: When a large collector that is built for space heating is used for heating water in the summer, it is important to prevent collector overheating. This is accomplished by adjusting manual dampers so that the collector is fed with outside air. Wiring is also modified so that a smaller blower operates off a thermostat sensor on the exchange tank. If the collector is small relative to the heat exchange tank, the air can run in a closed loop without overheating the collector. Ideally this system will also have a slower airflow through it during the summer, by changing speeds on the blower. A slower flow rate gives a higher outlet temperature needed for domestic hot water.

One-Tank Systems

Most electric water heaters and hot water storage tanks have top and bottom heating elements, or fittings to allow for them. If the tank-type heat exchanger is made with this type of tank, the solar exchanger is placed on the lower portion of the tank and the bottom heating element can be disconnected. The top electrode remains operable to bring the water to a usable temperature if there isn't enough solar input.

If your existing water heater is wearing out and you want to install a solar water heater, you can replace it with a larger, single tank. This will save space over a two-tank system. There will be less tank surface area to insulate and less heat loss.

Most residences could benefit from a 120-gallon tank unless both the collector and consumption of hot water is small (less than 20 gallons per person). The collector for a single-tank system with a 120-gallon tank should be sized to heat between 80 and 120 gallons of water. Since large gas-fired tanks tend to be pretty expensive, this system requires an electric backup. Two-tank systems are a better choice if your existing gas water heater is in good shape or if your hot water consumption is high (the construction of a one-tank system is discussed later in this chapter).

Tank Placement

The ideal place for a tank is next to the existing water heater, and if you have a large utility room/washroom or basement, you're in business. Plumbing can be kept to a minimum and there will be easy access to control valves and drains. Another good choice is a closet or other storage area near the existing water heater.

If you don't have room in your living area, a well-insulated preheat tank can be placed in a crawl space or attic. Getting a large tank into these locations can be difficult, so plan ahead. A tank in the attic must be placed so that its weight is spread over three to four ceiling joists. Placing it over an interior load-bearing wall is also a good idea. Along with strong supports an attic tank needs a drip pan in case of leaks. The pan has a hose that is run to a drain or outside the house. Don't hesitate to consult a structural engineer if you are in doubt about an attic installation. When it's located in a crawl space, a tank can sit horizontally in a wooden cradle on top of cinder block supports, although it is usually necessary to pour a small concrete pad to support this arrangement since any settling of the tank can cause plumbing leaks. You'll need at least 2½ feet of vertical clearance for the tank and its cradle.

There is a slight advantage in placing the solar tank vertically rather than horizontally, because hotter water can be taken off the top of a vertical tank due to heat stratification in the tank. A horizontal mount can provide preheated water of nearly the same temperature, however, if you use a side port on the tank to take water from the top of the tank.

Thermostatic Controls

A differential thermostat (DT) is the key to getting *hot* water from your preheat tank. A DT turns on the blower and sends solar heat to the heat exchanger (in-duct or tank type) only when the collector is warmer than the tank. In the late afternoon, when the tank is hot and the collector is warm, some usable heat will be lost from the collector, but a DT will ensure that the collector isn't delivering air that would cool the tank at this time. An 18°F on differential and a 5°F off differential thermostat is about right for this thermostat.

There are differential sensors available

that screw directly into a plumbing fitting on the bottom of a tank. If you use a tank-type exchanger, a standard sensor can be used. It should be put in an immersion well low on the tank, or be attached to the outside of the bottom of the tank (cold end) with a glob of silicone, in a location that is easy to reach should the sensor fail. The silicone must not insulate the sensor from the tank, and the sensor must be insulated from the airflow cavity so that it reads the tank temperature, not the air temperature.

More complicated systems, such as a two-mode system, will require an additional thermostat to act as a shutoff when the tank reaches a desired temperature. This is typically a remote bulb thermostat called an *aquastat* (Grainger's no. 2E146). The sensing bulb is placed inside an immersion well that extends into the tank, allowing the thermostat to be replaced without draining the system. The thermostat should be mounted so that it senses the tank temperature near the bottom of the tank. In a two-mode system where water heating is the first priority, this aquastat will trip a relay when the tank is heated up, and solar air will be diverted to another point of use. If water heating is the second priority, this additional thermostat may not be required, but as a safety measure all systems that have water heating as the last priority should have an aquastat installed to turn off the blower if the tank ever reaches 180°F (see Appendix 2 for a wiring schematic for a two-mode water-heating system).

Demand-Type Water Heaters

The best back-up tank for a solar water-heating system may be no tank at all. *Tankless* or *demand-type* water heaters are becoming more popular for back-up water heat-ing. Widely used in Europe and Japan, these electric and gas-fired units heat water only when a hot water tap is opened. They can deliver as much hot water as a conventional tank-type water heater with as little as one-half the fuel consumption.

Demand-type water heaters operate in one of two ways. The type suitable for a solar backup heats water to a preset temperature. If the solar tank is delivering water that's hot enough, this type won't turn on. If the solar preheat tank is cool, the water from it will be raised to the preset temperature with a corresponding fuel savings from the solar input. The type that isn't suitable for solar backup produces a set temperature rise in the water flowing through it. This is fine if the heater is always being fed water from the water main at a constant temperature, but it's not effective when dealing with the wide range of temperatures encountered in water from a solar storage tank.

Using a demand-type water heater for solar backup not only saves fuel, but it also simplifies the solar plumbing. There is no need to include the possibility of bypassing and turning off the back-up heater when solar is supplying all the hot water needs. If the proper type of tankless heater is used, it won't come on if the solar is hot enough.

Demand-type water heaters cost about twice as much as conventional water heaters, but fuel savings, especially when used in conjunction with a solar system, can make this figure look insignificant. One factor contributing to the present high cost of these units is the fact that most brands are manufactured overseas and have limited distribution in North America. In the next few years look for more units to be manufactured domestically and sold at lower prices. A plumbing schematic appears later in this chapter.

Building an Air-to-Water Heat Exchange Tank

A satisfactory and inexpensive air-to-water heat exchanger can be built using an 80- or 82-gallon hot water storage tank. This is a simple project even for those of you who are new to plumbing. A storage tank has much in common with a conventional water heater. It is an insulated tank enclosed in a sheet-metal jacket. To build the exchanger, first remove the jacket and the insulation. The jacket is replaced around the tank, and solar-heated air is blown into the space between the jacket and tank. The air returns to the collector in a closed loop. Efficiency can be increased if fins are attached to the outside of the tank to increase its surface area and promote better heat transfer. In the most effective tank design these fins are formed by attaching bands to the tank and bending out tabs to create a *porcupine* tank.

Materials
82-gal glass-lined water storage tank (without heating elements)
plumbing fittings for the tank
12 to 17 sheet-metal bands, 4″ wide and equal in length to the tank circumference
36 ½″ #8 sheet-metal screws
12 to 17 1¼″ #10 bolts with 2 washers each
silicone (2 tubes)
pressure-and-temperature relief valve (20,000 Btu/hr)

Labor
2 people @ 3 or 4 hours

Construction Steps

Prepare the Tank
Locate and purchase a new 82-gallon, glass-lined hot water storage tank ($270 to $300) without heating elements. A smaller system with less than 80 square feet of collector can utilize a smaller tank. A tall tank will have more surface area for heat transfer than a short, squat tank, so buy the tallest tank available. We suggest buying a new tank rather than using a recycled water heater because you never know how much life an old water-heater tank has left in it. Do not use a galvanized tank; it'll have a much shorter usable life.

Consult the plumbing schematics in this section to plan your fittings. Most storage tanks have plenty of ports to choose from. Remove the jacket and loosen all of the fittings on the tank. (They will be difficult to loosen after the fin bands are attached.) Remove the screws (or drill out the pop rivets) that hold the round jacket end caps on both ends of the tank. Set the tank on its side and pull off the ends. If the tank is insulated with fiberglass batts, pull these out and remove the jacket in one piece. If the tank is insulated with foam, as most new ones are, slit the jacket lengthways using a saber saw with a short (1-inch), metal cutting blade. Dig out the insulation as you peel the jacket off. If foam insulation is stuck to the jacket, tank or end caps, scrape it off with a chisel and use a wire brush to clean the surface.

Before attaching the bands to the tank, lift it up and set it vertically onto the bottom end cap with insulation under the concave bottom of the tank. Center the tank over the end cap, and run a large bead of silicone around the bottom of the tank to seal off any perforations in the end cap. Let the silicone cure for a day or two before moving the tank or putting the jacket back in place. Later, when moving the tank around, don't rest the tank on the edge of the bottom end cap, to avoid disturbing this seal.

Prepare and Install the Bands

If, for example, the tank has 2 inches of insulation between the tank and the jacket, the fin bands required for it should be 4 inches wide. The tabs are cut to extend 1½ inches from the tank. If your tank had thicker original insulation, use wider bands, but be sure and leave a ½-inch gap between the jacket and the tab ends, or the tank will have excessively high static pressure. The bands should be as long as the circumference of the tank plus 2 inches for fastening the ends together at the *attachment tabs* (see figure 10-3).

Install the first band before making the rest so you will get a feel for things. Wrap it around the tank and bend up both ends 1¼ inches at a 90-degree angle to form the attachment tabs. When the band is held tightly around the tank, these tabs should be about ½ inch apart. This is a critical measurement. Remove the band and drill ¼-inch holes in both tabs. The edge of the hole should be about ⅛ inch from the bend in the tab. Cut 1½-inch-deep slits in both sides of the band every inch. Try to make the slits opposite each other. Don't bend out the heat exchange tabs yet. Run a ¼-inch bead of silicone down the center of the band that will be attached to the tank. The silicone eliminates any air gaps between the tank and the bands.

With help from an assistant carefully position the first band on the lower portion of the tank. Pull it tight and have your helper hold it in position while you bolt the two attachment tabs together. Use 1¼-inch #10 bolts with a washer on both sides. Bolt the band on very tight, and then bend up the tabs at alternate 60-degree angles (see figure 10-3). If you have used the proper amount of silicone, a small amount should ooze out around most of the edges. This silicone at-tachment should be as thin as possible, and there should be no air gaps between the band and the tank. Silicone is not a great heat conductor, but it is the best material (at a reasonable price) for attaching the fins. We have attached bands using heat-conducting compounds such as Thermon Cement but have found that it is more difficult to work with. These compounds also don't compress as well as silicone when the bands are tightened, so there is more material between the band and the tank, which results in poor transfer of heat.

Plan ahead when placing bands near tank ports, and cut off or bend over tabs where necessary. Place the bands as closely together as possible (the more the better) while still allowing yourself room to bend up the tabs.

If your system requires an aquastat thermostat, install the "well" for it to a fitting high on the tank. If you are using a plug-type sensor (thermistor) for a DT, install it in a low port on the tank. A regular surface-mount thermistor can be put in an immersion well on the tank or bonded to the tank, near the bottom.

Install the Jacket and Boots

Wrestle the jacket back onto the lower end cap and secure it in one or two places with sheet-metal screws. If you had to split the jacket, cover the joint with a strip of sheet metal or join the two edges with S-clips. Finish attaching the jacket to the bottom cap with screws about every 8 inches.

Cut an 8-inch-round hole near one edge of the top end cap and another in the jacket near the bottom side of the tank. Install short sections of 8-inch duct to these holes, with four tabs outside the jacket and screwed to it and the other tabs bent over the inside. The tabs will be different lengths on the lower boot since you are attaching a round boot to a round tank. Slide the boot through the 8-inch hole on the side of the tank, and mark it with a felt-tipped pen before snipping the tabs so they will fit snugly together.

Place the top end cap on the jacket so that the 8-inch hot inlet extension is more or less opposite the lower, outlet extension. Se-

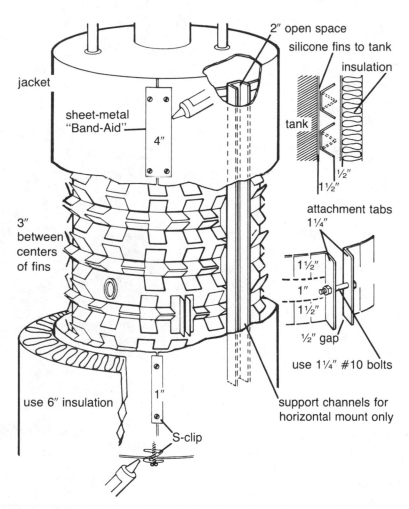

Figure 10-3: The finned bands that form the tank-type exchanger must fit snugly around the tank. Bolts connecting the attachment tabs clamp the band on very tightly. The fins are bent out after the band is in place. You can join the cut edge in the jacket by overlapping pieces of sheet metal or with S-clips. Seal with silicone to make the jacket airtight.

cure the top end cap to the jacket with screws about every 8 inches. Don't silicone-seal seams at this point.

If you are using an air sensor probe for your differential thermostat, put it in the immersion well, or attach it to the tank with a glob of silicone at this point. There should be an access hole through the jacket around a fitting at the bottom of the tank.

Pressure-Test, Seal and Insulate the Tank

Set the tank on some bricks where it will be finally located. Plumb the tank to a water supply and fill the tank. As the storage tank fills, the air in it will force its way into the existing water heater. The air will bubble up from the bottom of the water heater and blow out an open hot tap in the house. This may stir up the sediment in the bottom of the water heater so your hot water will be brown for a couple of hours until the sediment settles. This can be avoided if a fitting at the top of the storage tank can be opened while filling, to bleed off the air.

Once the tank is filled and pressurized, check for water leaks. Then silicone-seal the seams at both end caps and the vertical seam in the jacket, if it was split. Cut sheet-metal covers for all unused holes through the jacket, and silicone and screw these into place. Seal around all plumbing penetrations. The jacket must be *completely* airtight. The tank must also be very well insulated. Wrap 6-inch insulation around and on top of the tank and secure with wires. Also insulate around the bricks under the tank. Make sure the pressure-and-temperature relief valve is long enough so that an overflow of water won't soak the insulation.

After the tank is insulated, it may be desirable (but not necessary, for appearance only) to place a jacket over the insulation. This can

be fabricated on a flat surface from litho plates or galvanized sheet metal. Allow for a 2-inch overlap on the sides, and build the lid by cutting tabs in a round, flat sheet. Another possibility for covering the insulation is to wrap it in aluminized Mylar or polyethylene. Hold this in place with wire, and silicone or staple the seams together.

Horizontal Tanks

If your tank will lie horizontally, leave the tabs against the tank in two vertical rows 18 inches apart. Securely silicone six short

Photo 10-2: Round boots are attached to the top end cap and to the jacket low on the tank. The boots are placed to be more or less opposite each other, and all holes in the jacket are covered with sheet-metal scraps and sealed with silicone. This exchanger was built using a water-filled coil of $\frac{5}{8}$-inch copper pipe spiraling around the tank inside the jacket. In our tests it didn't perform as well as the fin type.

sections of 2-by-2-by-⅛-inch channel over the tabs and to the tank after all of the tabs are installed. Use duct tape to hold the channel in place until the silicone cures. These sections will rest on the cradle that supports the tank so plan their placement carefully.

To mount a horizontal tank, build a wooden cradle from 2 x 8s or 2 x 10s that supports the tank in three places and that extends halfway around the circumference of the tank. Set the tank in the cradle, and fill any gaps between the tank and cradle with silicone before filling the tank with water. If the tank and cradle are placed in the attic, locate them so their weight (80 gallons equals 700 pounds) is distributed evenly over a load-bearing wall or several joists. When placing the tank in the crawl space, set the cradle on a 3-inch reinforced concrete pad.

Dribblers

When building a horizontal tank, it is a good idea to build a *dribbler* for the cold incoming line. This is a capped, 3-foot section of copper pipe with holes drilled in one side. It connects with the incoming cold line located at the bottom of the tank. The holes in this extension are pointed at the bottom of the tank when installed so that cold water entering the tank doesn't stir up the entire tank and reduce temperature stratification. The dribbler is made by reaming out the ridge on a ½- to ¾-inch copper flush bushing so a ½-inch pipe will slide through it. This ar-

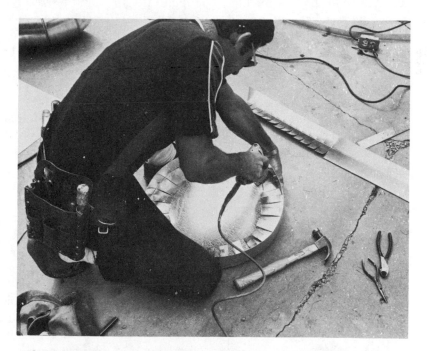

Photo 10-3: If your tank doesn't come with a jacket suitable for enclosing the fins, you will have to build one from 30-gauge galvanized sheet metal or litho plates. The end caps are screwed or pop-riveted together and sealed with silicone.

rangement is soldered inside a ¾-inch fitting which attaches to the tank. Drill enough equally spaced ⅛-inch holes to equal the area of the pipe diameter, and tighten the tank fitting until the holes point downward. If the only suitable fitting for the cold, incoming line on a horizontal tank is in through the center of the bottom end cap, the dribbler can have a downward jog in it so that water enters the tank on the bottom (see figure 10-4).

Building a One-Tank System

One-tank systems can be built using a tank-type exchanger or an in-duct exchanger. A tank exchanger can be built on the tank by cutting the jacket two-thirds of the way up the tank. (Look out for electrical wires inside the jacket that run between the heating elements.) Remove the lower jacket section and attach the bands to the tank, leaving a 4-inch space top and bottom to act as a manifold for distributing air around the tank. The upper third of the jacket is left in place with insulation underneath because the top heating element will still be active. The hot air duct is located opposite the cold return to the collector, and a band of sheet metal seals the horizontal joint in the jacket.

TABLE 10-1
Static Pressure in Tank-Type Exchangers

Airflow (cfm)	Static Pressure (inch WG)
250	0.22
300	0.34
400	0.60

NOTE: The figures apply to a tank with fins ½ inch away from the jacket.

When using an in-duct exchanger, dump the solar-heated water into the middle of the tank or slightly above this point, depending upon the fittings on your tank. Use a dribbler that points downward on this line, to help preserve stratification in the tank.

In-Duct Exchanger Systems

These systems require a pump wired in parallel with the collector blower. When the blower comes on, water is circulated through the exchanger. Both are controlled by a plug-type sensor from a differential thermostat. This sensor is mounted on a fitting low on the storage tank.

When ordering an in-duct exchanger (which should cost from $200 to $400), the coil and pump must be sized for the airflow. For best efficiency look for a coil that is rated for a flow rate of 2 gallons per minute through the exchanger. You will need to specify the airflow from your collector (cfm) as well as the average temperature of the air delivered to the coil. This temperature will vary greatly depending on your system but should be between 120 and 160°F. You will also need to specify the temperature of the water going into the coil. Since this also varies greatly, the temperature of your cold tap water is a good choice. Considering all of these variables, it is a good idea to get help from a local plumbing-and-heating contractor who is interested in your project. He can also give you tips on installing the exchanger.

You need to make sure that the exchanger you use is designed to heat cold water from hot air. Some folks have used standard tube-in-fin sections used in hot water baseboard heating. Unfortunately, they are designed to heat air, not water, and are therefore water-to-air (not air-to-water) exchangers. Coils of this type work, but not as well as ones specifically designed for air-to-water applications.

Other Tanks

The idea for the porcupine tank came to us from W. Scott Morris of Sante Fe, New Mexico, who has used tanks of this type, but with longer fins, for air-to-water heat exchangers in convective (passive) solar air installations. We have also built other tank-type exchangers for active systems using vertical fins and tanks in which the air was forced to

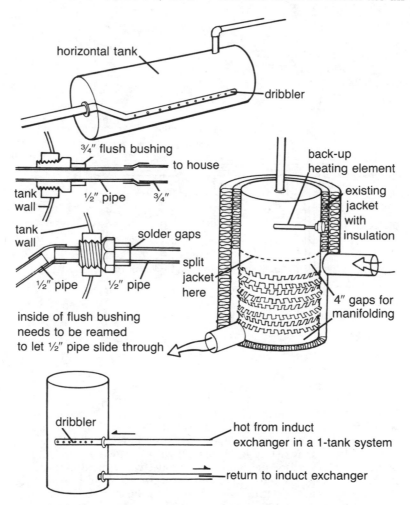

Figure 10-4: In one-tank systems the jacket on a large tank is cut one-half to two-thirds of the way up. The finned band is attached to the lower portion of the tank, and the heating element at the top of the tank is left in place as a backup. Dribblers help preserve temperature stratification in one-tank systems and horizontally mounted exchange tanks. They are made by reaming out the inside of a ½- to ¾-inch bushing so that a ½-inch pipe will slide through it.

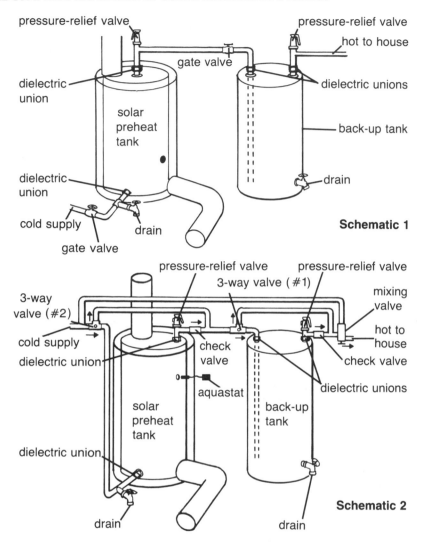

Figure 10-5: Shown here are two plumbing schematics for air-to-water, tank-type heat exchangers. Schematic 1 shows the simplest setup possible. Cold water enters the solar preheat tank which feeds the back-up tank (existing water heater) with warmed water. Both tanks require pressure-and-temperature relief valves, and dielectric unions are used between copper and galvanized pipe to prevent galvanic corrosion. The entire arrangement must be shut down to replace or service either tank. Schematic 2 shows a more elaborate arrangement in which a three-way valve (#1) allows the back-up heater to be bypassed in the summer. Another three-way valve (#2) allows for bypassing the solar tank for replacement or repairs. Two check valves control the direction of flow. An aquastat controls dampers that divert air away from the preheat tank when it is hot enough. A tempering valve adds cold water to the outgoing hot water when it is above 130°F and should never be omitted unless the collector is very small in relation to the tank.

spiral the tank. We found these were more difficult to build and didn't perform as well as the porcupine tank exchanger. We have also built tanks where a water-filled, spiral coil of ¾-inch copper pipe inside the jacket transferred heat to the tank by a thermosiphoning effect. Steve Baer of Zomeworks has tested this type of exchanger and found it superior to many finned-type tanks, but our tests have shown the porcupine-type exchanger to be superior to the coil type. Its effectiveness is due to the tight attachment of the fin bands to the tank. Ideally, the fins would be soldered or welded to the outside of the tank, but this is impossible on existing tanks because it would destroy the glass lining inside.

Plumbing the Heat Exchange Tank to the Back-Up Heater

Materials
copper pipe
fittings, valves
flux
pipe joint compound
solder (50-50)
Tools
cylindrical wire brush
long 1″ drill bit (optional)
2 pipe wrenches
propane torch
reamer
steel wool or emery cloth
tube cutter

Once your heat exchange tank is built and set in place, the next consideration is plumbing it between a cold water line and the existing hot water heater. Plumbing work, like electrical work, is often best left to a professional, but it can be done successfully by a careful amateur. Rather than give you complete step-by-step details, our purpose

here is to give you an idea of what should be included in the system and to caution you about tricky and easily overlooked areas. There are many good do-it-yourself books available on plumbing if you need the basics spelled out.

Either copper or plastic pipe can be used for the tie-in. We discuss copper pipe in more detail because it is a better choice for potable water, and since most houses have copper plumbing, you will probably have to tie-in to copper pipe at some point in your system.

When designing your plumbing system, keep all pipe runs to a minimum. This won't be a problem if you locate the solar preheat tank near the water heater. Slope your pipes at least ¼ inch for each foot of run, and put a stop and waste valve at the lowest point in the system to be able to drain the pipes. Make sure you can isolate either tank, and place the valves where they are easily accessible.

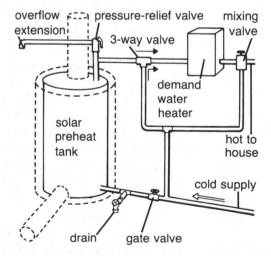

Figure 10-6: This plumbing schematic shows an arrangement for tying a solar heat exchange tank into a demand-type (tankless) water heater. The tempering valve is set above the output temperature of the demand-type heater.

Consider using three-way valves in suitable locations. They are easier to install than two gate valves, cost about the same and help simplify the plumbing.

Connections between galvanized pipe or fittings (usually at the preheat tank) and copper pipe or brass fittings must be made with dielectric unions (or plastic adapters) to prevent galvanic corrosion from taking place. These are readily available at most plumbing suppliers. Be sure to disassemble valves before soldering them. This prevents damage to the rubber O-rings. Gate valves don't need to be taken apart but should be soldered with the valves tightly closed. Install a pressure-and-temperature relief valve on both the solar preheat tank and the existing hot water heater within 6 inches of the top of the tank.

Working with Copper Pipe

Three-quarter-inch copper pipe is a good choice for plumbing your tank. Cut the pipe with a tube cutter and ream out the burr that is created inside the cut. Clean the inside of the female fittings and the outside of male fittings with a cylindrical wire brush and emery cloth. Brush flux on the male sections only and assemble several sections of pipe before soldering. Heat both sides of a joint with a propane torch until the solder (50-50) melts and is drawn completely into the joint. Clean the hot joint with a wet rag. Copper pipe should be supported every 4 feet with copper plumber's strap and should be well insulated.

If you find yourself having problems and yet are still determined to do it yourself, take your measurements, drawings and fittings to the local hardware store. There is undoubtedly a plumbing expert there who can help you figure things out.

Sources for Supplies

In-Duct Heat Exchangers

Climate Control Refrigeration-Coil Division
a unit of Snyder General Corp.
602 Sunnyvale Dr.
Wilmington, NC 28403
Phone: (919) 791-8510

The Coil Co., Inc.
125 S. Front St.
Colwyn, PA 19023
Phone: (215) 461-6100

Davenport Hydronics Inc.
805 S. Inglewood Ave.
Inglewood, CA 90301
Phone: (213) 673-4026

Plumbing Materials

All of the plumbing materials you need will be readily available at any large hardware store, although dielectric unions may have to be ordered, depending upon common plumbing practices in your area.

Tank-Type Heat Exchangers

We know of only one company that offers tank-type, air-to-water heat exchangers. These tanks, prewired with a differential thermostat and blower, are ready to install. We haven't installed any of these tanks because they are quite expensive ($1,400 for a 75-gallon tank), but they look like a good item. Contact the following for more information:

Park Energy Co.
Star Route, Box 9
Jackson, WY 83001
Phone: Park Energy requests that consumers write for information.

11

THERMAL STORAGE FOR LARGE SYSTEMS

Systems with large collector area (over 200 square feet) can often benefit from the controlled heat delivery available from a thermal storage component. But before you consider storage, you need to determine first if your collector is large enough to require it (see chapter 3 for a discussion of system sizing). You will also need to take a hard look at where the storage component can be located since it will be very large and very heavy.

The first design decision to make regarding thermal storage is what material to use for storing the solar heat. An ideal storage material is one that is readily available and inexpensive, quickly absorbs and releases heat, holds a lot of heat per unit of weight or volume and doesn't outgas, collect dust or deteriorate over time. Both rocks and water, the two most commonly used storage materials, satisfy these requirements. A rock thermal storage bin is usually used with air-heating collectors, while liquid-medium collectors usually use a large water tank for storage. Because water has about 2½ times the heat capacity of rock, the storage tank in a liquid system can be less than half the size of a rock bin to store the same amount of heat with the same temperature rise. This, however, is the only advantage water storage can claim over rock storage.

An ideal heat storage material allows the heat stored in it to stratify, that is, to be separated into horizontal layers. An understanding of heat stratification is the basis for designing an effective rock thermal storage bin. In rock storage systems the airflow from the collector heats the rocks from top to bottom. The top of the rock bin receives the hottest air from the collector, and because of convection within the rocks, it remains the hottest. When heat is removed from rock storage, it is taken out by air flowing from the bottom of the bin to the top. The warmest rocks on top of the bin are exposed to air that has already been preheated by the cooler rocks below; thus 130°F air can be taken from a rock bin whose average temperature is 100°F and which contains 70°F rocks in the cool (lower) end. Since the collector is always being fed with relatively cool air from the bottom of the bin, it operates at a high efficiency. If, for example, late in the afternoon a system delivers collector air cooler than 130°F, it will lower the temperature of the hottest rocks on top of storage but still add heat to the bin.

In order to make the best use of stratification in a rock box, you must properly size and proportion it, and it must provide for the best transfer of heat going into and out of the box. Yet despite their large size, rock boxes are desirable means of storing heat. Air flows quite easily through them so an oversized blower isn't required to get an adequate airflow. They are also clean. The low-velocity air moving through the bin won't pick up the tiny bits of sand that are likely to form through years of expansion and contraction of the rocks, and this won't be blown into the living area.

The only possible problem that can develop in a rock bin is the formation of mildew or mold. This problem is easily avoided if these steps are taken: The rock should be cleaned of dirt and other sediment before you buy it; after you take delivery (have the rock poured onto a plastic sheet), let the rock air dry for two to four days before transferring it to the bin. The rock bin itself must be well sealed from ground moisture; and the temperature of the rocks shouldn't be allowed to fall below the dew point (the temperature at which moisture will condense out of the air and onto the rocks). Mildew is usually a problem only in poorly designed rock beds that store surplus heat from a humid greenhouse. We haven't heard of it being a problem when a rock bed is used in an air-heater system, which delivers a relatively dry heat.

Some System Design Options

Systems with thermal storage typically have three operating modes: the collector heating the house, the collector heating the rock bin, and the rock bin heating the house.

Direct space heating from the collector is always the first priority. In some systems the back-up heater is tied into the solar system, creating a fourth mode. A domestic water-heating subsystem can also add another mode, which will be discussed later in the chapter.

The controls necessary for proper air delivery in the different modes will vary depending upon whether one or two blowers operate the system. One-blower systems require less electricity to operate, but it is often difficult to get the proper airflow for each mode when using only one blower. Two-blower systems allow for easier control of airflow rates and are usually easier to retrofit.

Let's look first at a few simple space-heating systems. Our examples will show schematics for systems that use individual dampers to control air movement, rather than systems using an *air handler* unit that has all the dampers mounted in one box. Air handlers will be discussed later in this chapter.

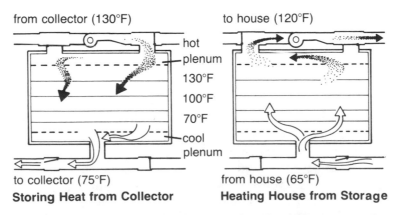

Storing Heat from Collector

Heating House from Storage

Figure 11-1: It is important to use enough rock in a heat storage bin so that the rocks on the bottom of the bin remain cool relative to the temperature of the rocks at the top of the bin. When the temperature is stratified in the bin, the collector can operate efficiently in the collector-to-storage mode, and higher-temperature air can be taken from the bin in the storage-to-house mode.

Two-Blower, Three-Mode System

In most retrofit applications a three-mode system is the recommended setup for delivering solar heat to the house. In this system the existing heating system is kept separate from the solar system and not integrated into

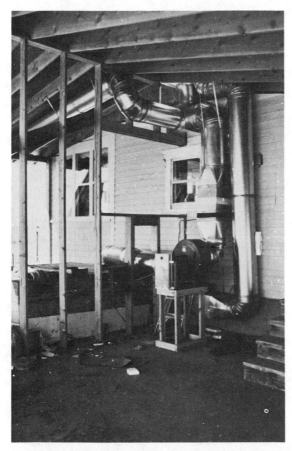

Photo 11-1: The blower, ductwork and other components for a large three-mode system can take up a lot of room, so plan for it in your design. Ducts from two collectors tie together at the top of this photo. Solar-heated air then passes through a motorized damper to a large belt-driven blower and then is blown into the top of the rock bin. The return to the upper collector is the vertical duct on the right. This entire arrangement was later framed-in and insulated.

the solar controls. This is desirable because, if the control system is "down" for any reason, you will still have a heat source. No complicated modifications to the furnace are required, and the solar ductwork can be designed to perform the way you want it to. A three-mode system is the only choice for forced-air solar heat delivery if the back-up heating system is anything other than a forced-air furnace.

A three-mode system is by far the simplest approach for do-it-yourself retrofits with storage. (Figure 11-2 shows an airflow schematic for a three-mode system.) The system has two blowers, one that takes heat from the collector and one that delivers it to the house. (These two functions can also be accomplished with one blower, much like the one-blower, four-mode system discussed later in the chapter.)

Let's follow the control logic of a three-mode system. In the first mode the collector heats the house. A two-stage house thermostat says the house needs heat. When the differential thermostat sensors say the collector is warmer than the house, the two motorized dampers open, allowing the collector blower and the delivery blower to deliver heat to the house.

In the second mode the collector heats the rock bin. The house thermostat says the house doesn't need heat so damper 2 closes and the delivery blower shuts off. The DT sensors say the collector is warmer than storage. Damper number 1 opens, and the collector blower turns on to deliver heat to storage.

In the third mode, heat from storage goes to the house. The house thermostat calls for heat. The plenum thermostat says the storage is warmer than its set-point (typically between 90 and 100°F). Damper number 2 opens, and the delivery blower turns on. The DT

2-Blower, 3-Mode System

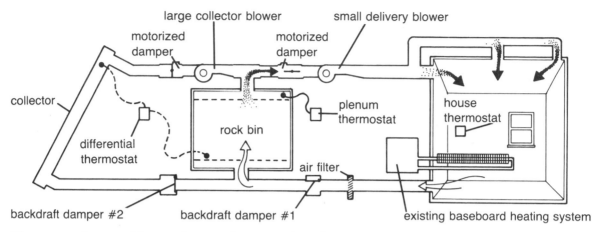

Figure 11-2: In a two-blower, three-mode system the solar system is separate from the existing heating system (shown here as a hot water baseboard heater). The auxiliary heater provides an independent, fourth mode of operation. This arrangement, shown in the storage-to-house mode, is a likely choice for many retrofit applications where the existing heating system isn't forced air.

sensors say the collector is cooler than the house so damper number 1 is closed and no air is drawn from the collector.

There is also an auxiliary, or nonsolar, mode. The back-up furnace comes on when the house cools enough to activate the second stage of the house thermostat. Damper number 1 is closed, and the collector blower is off.

The backdraft dampers allow air to flow in only one direction, like check valves in a water line. Backdraft damper number 1 prevents air from being pushed into the house during the second mode, collector to storage, and backdraft damper number 2 is needed to prevent cold air in the collector from entering the rock bin at night.

Relays are used with the various thermostatic controls, and putting it all together to accomplish these steps can get complicated. Consult the wiring schematic for this system in Appendix 2 for their placement and operation.

Two-Blower, Four-Mode System

If you integrate a solar system with an existing forced-air furnace, a two-blower, four-mode system could be a good choice. Controls are more complicated than those for a two-blower, three-mode, but you can use the existing ductwork to distribute heat to the house. You can use the heating elements in the existing furnace, but don't be tempted to use your existing furnace blower to deliver solar heat to the house. Its motor is probably mounted in the return airstream and has been designed to be cooled by return air from the house. Hot solar air from the collector or storage will cause it to overheat and burn up.

This four-mode works just like the two-blower, three-mode except that the back-up heating delivery system shares its controls with the solar system, to create a fourth mode. Proper control wiring in this system is very important because if the delivery blower quits, or doesn't operate properly, the system can't provide heat whether or not the solar is "hot."

Consideration must also be given to balancing airflow rates through the ductwork as we will see later.

One-Blower, Four-Mode System

Most solar air systems in new construction use a single blower to move air in all modes of operation. The back-up heat is provided by an electric in-duct heater, and as in the two-blower, four-mode system, all return air travels through storage so that even if the rocks are below 90°F, they will contribute some heat to the air.

Flow Rates

Each mode of operation in the systems we have discussed has an optimal airflow rate, so ideally a well-designed system could include several blower speeds. All systems will ideally move more air in the collector-to-storage mode than in the house delivery modes. It is important not to draw the air through the rock bin too rapidly in the storage-to-house mode to avoid draftiness when the air enters the house and to get the highest-temperature heat available from the rocks.

In a two-blower, three- or four-mode system optimal airflow rates are obtained by using a large blower as the collector blower and a smaller blower as the house blower. In a two-blower, four-mode system the collector-to-house and storage-to-house modes should have a fairly slow airflow. (In the collector-to-house mode both blowers will be operating, and some airflow will be delivered to storage.) In the auxiliary (nonsolar) mode a higher flow rate may be required to keep the electric in-duct heater from overheating. Check the manufacturer's specifications if you plan to use an electric in-duct back-up heater.

One-blower systems can be built to have two blower speeds: a high speed for the collector-to-storage and auxiliary-to-house modes and a slower speed for the collector-to-house and storage-to-house modes.

Regulating the static pressure in the system is another way to control air delivery. To state it simply, if the collector, house ductwork and rock bin each present approximately the same static pressure to the blower, then each of the three solar modes will have the same airflow rate at the same blower speed. The size of the rocks and the dimensions of the rock bin may be chosen so that the storage presents the desired static pressure in relation to the collector. Restrictions in the size of the house ductwork can yield a similar effect. These are good points to be aware of, but don't be overly concerned with flow rates. With some thought and good guesswork, one or two speeds may be used without sacrificing efficiency in almost any simple storage system. In the simple two-blower, three-mode systems we usually retrofit, we simply size the collector blower to the square footage of the collector and the delivery blower to deliver about ½ cfm (at sea level) for each square foot of house to be heated, and everything works out fine.

Air Handlers

The approach we generally recommend for do-it-yourselfers is to build their control package from component blowers, dampers and thermostats. This, however, isn't the only way that air delivery can be controlled. *Air handlers* are prefabricated units that contain all the necessary blowers, dampers and control circuitry. Many are also available with a heat exchange coil for heating domestic water. Provided there is space available for an air handler, it can be easier to install than component controls. We prefer using components, however, because they can be installed for about half the price of an air handler.

Components are also more versatile because you can put them where you want them, rather than run all of the ductwork to a central air handler. Air handlers can also create more static pressure in a system than would components. Also, component dampers are, as a rule, more tightly sealing than the dampers in most air handlers now on the market.

Air handlers come with a prewired control box. This can be an advantage if you can use the manufacturer's control schematic. When using component controls, you must order a control box or build your own. It's best to purchase a control box from the company that supplies you with dampers. You will need to send them your airflow schematic, and they usually require you to fill out a form that covers air delivery and controls needed for each mode. This should be easy to figure out and will help you get a better idea of exactly how your system will operate. Plan on some back-and-forth correspondence if you are trying something new. Most control boxes for component systems are easy to install, and you don't have to fool with relays and such. You just hook some wires together.

Designing a Rock Box

In most cases a rock storage bin should be sized for the collector area and not for the house heat needs. The idea is to collect all the available heat from the collector and then use it up within a reasonable length of time while maintaining a usable temperature in the hottest rocks. If you build a huge rock bin for a small collector, hoping to store heat for three days of cloudy weather, the rocks will never heat up to a usable delivery temperature. Economics usually dictate building a solar system that provides no more than 70 percent of your annual heating needs, so the rock

bin should be designed so that the heat within it is exhausted within one or two cold, cloudy days. Some form of back-up heating will always be required in a cost-effective system.

Calculations involving the amount of heat a collector can deliver per square foot and the amount of heat rocks can store per pound indicate that a ratio of ½ to ¾ cubic foot of rock per square foot of collector is optimal. Rock used for solar storage must have about 30 percent voids (open spaces) and will weigh about 100 pounds per cubic foot, so the weight of the rock bin will be between 50 and 75 pounds of rock per square foot of collector. If the amount of rock is less than ½ cubic foot per square foot of collector, all of the available heat won't be removed from the airstream, and return air to the collector will be too warm. An oversized rock bin, on the other hand, won't readily heat up to a useful temperature. Since the collector is sized for the house and the local climate, and the storage is sized for the collector, climatic variations

---

Rules of Thumb for Designing a Rock Box

- Size of bin: ½ to ¾ cubic feet (50 to 75 pounds) of rock per square foot of collector
- Size of rock: ¾- to 1½-inch diameter
- Direction of flow through rocks: vertical rather than horizontal
- Length of flow: 5 to 7 feet
- Plenums (air cavities): 8 to 12 inches top and bottom

If you follow these rules, your bin will have the following:

- Flow rate through rocks: 20 to 30 feet per minute
- Static pressure: 0.15 to 0.25 inch WG

don't affect rock bin sizing directly.

Several factors work together to create an even and thorough transfer of heat from the airstream to the rocks. It is important that the air move slowly through the bin and that it pass through a large surface area of rock at the top of the bin. If air is sped over a small surface area, heat transfer will not be as complete. The ideal velocity is 20 to 30 fpm, so the surface area on top of a rock bin for a 300-square-foot collector delivering 750 cfm (300 ft² x 2.5 cfm/ft² of collector, see chapter 4) will need to be between 25 and 30 square feet (750 cfm ÷ 25 fpm = 30; 750 ÷ 30 = 25) or about 5 feet by 5 feet or 5 feet by 6 feet.

The rock box needs to provide an adequate resistance to the flow of air so that the air doesn't channel down through the box in three or four places, creating hot spots and uneven heat distribution. Even heat distribution can be ensured by making sure the bin doesn't have any places where the air can cheat and flow too easily (as can happen in a bin with horizontal flow). The desired static pressure in a rock bin is between 0.15 inch and 0.25 inch WG. Air will tend to take the easiest path if the static pressure is less than 0.15 inch WG, and an unnecessary load will be placed on the blower if it is over 0.25 inch WG. The right static pressure can be obtained by juggling three factors: air velocity through the bin (which we already determined will need to be quite low), the size of the individual rocks and the length of the airflow path through the bin.

The recommended size for the rock is between ¾ inch and 1½ inches in diameter. Rocks smaller than this size will have more surface area for heat transfer, but they will also present too much resistance to the airflow. Larger rocks will require a longer airflow path (over 7 feet from inlet to outlet) in order to create the right static pressure in the bin.

Our final consideration in getting the heat out of the air and into the rocks is the length of the flow, which is determined by the amount of rock and the surface area of the rock at the top of the bin. The flow through the storage bin should always be vertical, that is, from top to bottom, and not horizontal. Boxes with a horizontal flow are more difficult to seal. Air will channelize through the rocks, and proper temperature stratification will be difficult to obtain. Bins with a horizontal flow have an advantage in that they have a lower profile and will fit into places that are too short for a vertical flow. U-shaped bins (see figure 11-3) are a satisfactory compromise if space is limited. The flow through the rocks is still vertical, and stratification will be good in the hot half (and not so good in the cool half). In a U-shaped bin all of the ductwork can be attached at the top of the bin, which may also help to simplify installation. Rock bins with a single vertical flow perform slightly better than U-shaped bins, however, and should be used whenever possible.

Often one of the main considerations in retrofit situations is making a rock bin fit into a basement or other room while still allowing for a sufficiently long airflow path and for plenty of space at the top and bottom of the bin for free airflow and even heat distribution. Figure 11-3 shows some different-sized vertical-flow rock bins for a 300-square-foot collector that fit into an 8-foot-high space while still allowing 1-foot-deep plenums at the top and bottom.

Cold Return from the Top

In some installations, especially those with a buried rock box, there is no access to

Rock Bin Sizing Calculations

A properly designed bin should be able to store one sunny day's worth of heat output from the collector, with a 50 to 70°F temperature rise in the rocks. The return air to the collector is about 70°F, and air from the collector averages about 130°F (radiant floor heating systems can tolerate only a 10°F temperature rise, from about 60 to 70°F, so the storage must be five times larger than for a rock bin). On a sunny winter day 1 square foot of collector will deliver about 780 Btu:

280 Btu/hr (insolation) x 7 hrs x
40% (collector efficiency) = 784 Btu

Rock weighs about 100 pounds per cubic foot (140 lbs of rock/ft³ with 30% voids = 98 lbs of rock/ft³) and has a specific heat of 0.21. To determine the number of Btu's stored, the calculation is:

weight of   specific   temperature
rock in lb × heat in   × rise in °F
           Btu/(lb)(°F)

For about 780 Btu this is:
25 lbs x 0.21 x 149°F temperature rise
50 lbs x 0.21 x 74°F temperature rise
75 lbs x 0.21 x 49°F temperature rise
100 lbs x 0.21 x 37°F temperature rise

So 50 to 75 pounds (or ½ to ¾ cubic foot) of rock per square foot of collector is about the right amount.

Let's look at an example of rock bin sizing for a 300-square-foot collector.

Collector airflow in cfm: 300 ft²
× 2.5 cfm/ft² of collector = 750 cfm

Desired velocity through rocks: 25 fpm

Surface area of rocks: 750 cfm
÷ 25 fpm = 30 ft²

Amount of rocks: 300 ft² collector × 0.5 ft³
rock/ft² of collector = 150 ft³ of rock

Length of flow: 150 ft³ ÷ 30 ft²
= 5 ft of flow through rocks

From table 11-1 we can estimate that the average static pressure for 5 feet of flow at 25 fpm through ¾- to 1½-inch-diameter mixed rock is 0.18 inch WG, which is within our limits (between 0.15 inch and 0.25 inch WG).

TABLE 11-1

Static Pressure in Rock Bin Storage

Rock Depth (ft)	Rock Size (in)	Face Velocity Across Rock Bin (static pressure in inch WG)		
		20 fpm	25 fpm	30 fpm
5	¾	0.14	0.23	0.34
	1½	0.06	0.12	0.17
5½	¾	0.16	0.24	0.37
	1½	0.07	0.13	0.19
6	¾	0.17	0.27	0.40
	1½	0.07	0.14	0.20

the bottom of the box, and the cold return duct must be inside the box. This is no problem if the cold duct is sturdy enough that the rock doesn't crush it and if there is adequate open space provided at the bottom of the bin for good air distribution.

Heavy metal culvert of the proper diameter is the best material to use for this application. Some bins of this type have been built using transite (underground) ducting or regular heavy-gauge, sheet-metal ductwork. Transite ducting is an asbestos material that is very hard to work with, and sheet metal could collapse, so we don't recommend either of these materials. When you run a duct down through the center of the gravel, it is a good idea to install a pyramid-shaped turning vane under the duct to direct airflow at the bottom of the bin. It is not necessary to insulate this duct.

Placement of the Rock Box

Sizing calculations indicate that the rock bin for a large collector must be huge. It can easily take up half of the space in a bedroom or your entire utility room. Many people try to find a place for it in the garage or build a separate structure for it outside the living space. This isn't a good idea. Even the best-insulated rock bins will lose a lot of heat. If they are located outside the living area, the heat that is lost can't be used. On the other hand, the heat lost from a rock bin located in a basement or cellar will be useful in heating the floor above. This can make a great deal of difference in overall system performance. If your house doesn't have a basement, you may have to sacrifice some living area or limit the size of the collector.

If you think you might have to put your bin in the garage or in a small addition, we should probably say a few words about it. In the first place, you can't use too much insulation. The sides and bottom of the bin should be insulated to R-40 and the top to R-60. Ductwork into and out of storage should be insulated with at least 3½ inches of fiberglass. The rock box and ductwork must also be *completely* airtight since any leaks will greatly reduce performance. Use plenty of silicone and remember that fiberglass insulation, no matter how thickly installed, doesn't stop air leaks.

Using the Back Bedroom

If you have a wood-frame floor and no basement and decide to use a first-floor room for a storage bin, you are in for a major project. The floor will have to be removed and the bin set on the crawl-space dirt. First cut out the flooring, leaving the floor joists in place. Then pour two foundations, one for the wooden stemwall that will be the new support for the remaining parts of the floor joists, and a separate pad for the rock box. Build the stemwalls that are necessary to support the floor joists before cutting the joists out. Allow plenty of room for ducting and air handling equipment.

Storage Options

Direct Radiation from Storage

The simplest air delivery scheme from storage is no airflow at all. If your heat storage bin can be placed in a central location inside your dwelling, you may be able to take advantage of direct radiation from it for your heating needs. Radiant heat sources are very desirable, especially when dealing with the relatively low-grade temperatures that solar collectors deliver. For such a system to work

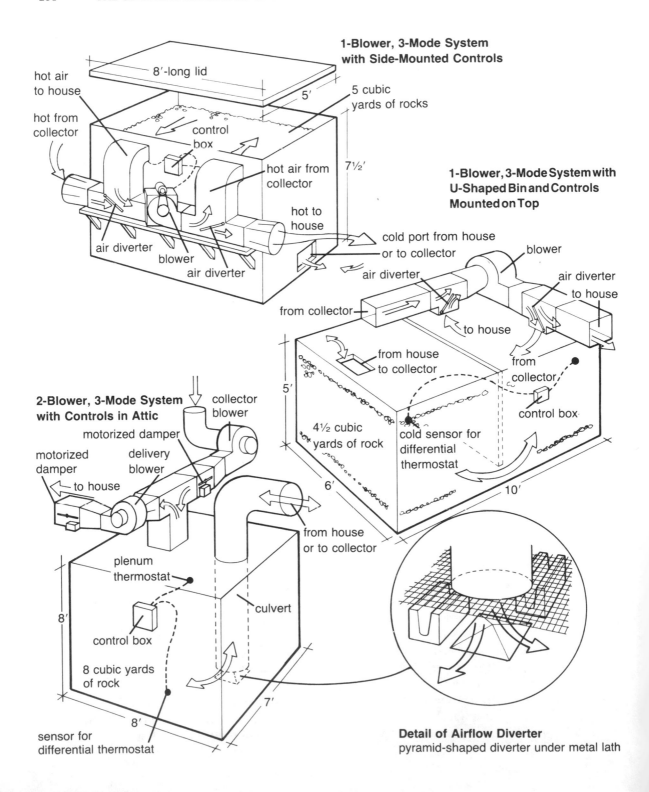

1-Blower, 3-Mode System with Side-Mounted Controls

8'-long lid

hot air to house

hot from collector

control box

5'

5 cubic yards of rocks

hot air from collector

7½'

hot to house

air diverter

blower

air diverter

cold port from house or to collector

air diverter

1-Blower, 3-Mode System with U-Shaped Bin and Controls Mounted on Top

blower

air diverter to house

from collector

from house to collector

from collector

control box

5'

4½ cubic yards of rock

cold sensor for differential thermostat

6'

10'

2-Blower, 3-Mode System with Controls in Attic

collector blower

motorized damper

delivery blower

motorized damper

to house

plenum thermostat

culvert

control box

from house or to collector

8'

8 cubic yards of rock

8'

7'

sensor for differential thermostat

Detail of Airflow Diverter
pyramid-shaped diverter under metal lath

Figure 11-3: *On opposite page.* Be sure to allow yourself plenty of room to install the devices that control airflow to and from a rock bin. Of many different possibilities, three good arrangements are shown here.

properly, the storage must be centrally located so that it can "shine" its radiant heat into several rooms. The outside of the box should also be made of a material, such as masonry or concrete, which is a good radiator.

Twenty or 30 installations built in New Mexico have used radiant heat storage. These systems use a thermosiphon or passive air collector to deliver heat to a rock bed under a concrete slab floor or into interior mass walls, which are essentially heavy-mesh wire cages filled with sized rock and plastered on both sides. These systems have performed very well and should work just as well with an active system. An active system would certainly be easier to design and retrofit.

Since the mass wall will be a radiant heat source, stratification in the storage won't be required. Solar air can be blown into the bottom of the wall and up through the rock pile. Natural convection in the wall will tend to keep the entire wall the same temperature. Since this type of storage won't have any distribution system from the storage to the house, controls are greatly simplified and it would be a natural storage system for a two-mode system. The mass wall would store heat when the immediate heating demands of the living area had been satisfied. A high-limit thermostatic control might be needed to prevent overheating of the mass. Direct radiation systems need to be sized at least two times larger than conventional rock boxes, to minimize temperature swings.

Slab-Floor Storage

In a new construction application another way to use direct radiation from storage is with a slab-floor storage system. Solar-heated air is blown into an insulated gravel bed located under a slab floor. Controls are very simple for this system. A differential thermostat turns on the blower when the collector is warmer than storage, and backdraft or motorized dampers prevent nighttime heat loss from the storage.

Sizing storage for a system like this is very different from rock bin sizing. A rock box is designed for a temperature rise of 50°F or more, from 90 to 140°F, for example. It really doesn't matter because its temperature doesn't directly affect comfort levels in the house. To be effective and comfortable, a slab-floor storage should have only a 10°F temperature rise, ideally between 60 and 70°F. If the floor is cooler than 60°F, the house will be cold; if it is warmer than 72°F, the house will be uncomfortably hot. This limited temperature range indicates that the storage mass must be larger than a rock bin if each is charged by the same size collector. A rule of thumb for slab-floor storage is to use at least 4 to 5 cubic feet of rock per square foot of collector. It is much better to over-size storage in a system like this than to under-size it.

The real key to the success of slab-floor systems is designing a distribution system that will heat all parts of the storage and not produce hot spots. Construction options for these systems will be discussed later in this chapter.

Bricks

Other materials are quite suitable for storing heat. Bricks are a good choice, especially broken ones. Bricks have about the same heat capacity per unit weight as rock (0.20 versus 0.25; this means, however, that bricks will require more volume than rock to provide an equivalent heat storage capacity) and can be used in places where a rock bin may be undesirable. Bricks may be stacked and don't need structural walls to contain them.

One possibility is to use the framed-in closet space underneath a staircase as this storage area. Bricks are stacked in this centrally located structure, and vertical baffles (made from paneling or drywall) are placed between the stacks to direct the air and make a longer airflow path in order to take all of the heat out of the solar air. This storage area will serve as a radiant heat source only, and no heated air will be blown out of it, so heat stratification isn't important. The drywall is removed from the sides of the stairway, and they are plastered with masonry to act as a better heat radiator. The bricks are stacked tightly against the plastered sides of the closet and against the baffles so that the air flows through spaces between the bricks.

When bricks are used as a storage material, it is desirable to have about 30 percent voids (spaces) between them, so broken bricks are more suitable than whole ones. Take your time in stacking them, and make sure there are no easy paths for the air to flow through. Most large brick factories have a huge stack of inexpensive broken culls.

It is also important that the bricks have adequate support beneath them. Never stack bricks directly on a wood floor, and make sure that a concrete floor will support their weight.

Water Containers

One-gallon jugs filled with water are another possibility for a storage medium. They can store more heat for their weight than bricks or rocks, but transferring the heat into them is a problem because of their small surface area. These jugs won't be able to support their own weight over time, so they should be independently supported. One enterprising do-it-yourselfer in Pueblo, Colorado, had the local Coca-Cola plant can several dozen cases of water in cans for his storage system. These cans present more surface area for heat transfer than gallon jugs, but they were very expensive, and he probably would have been better off with bricks.

Water jugs find their best application in a system with a large storage that contains low-grade heat since it is difficult to stack the jugs to produce stratification. Placing water jugs in an insulated box underneath a mobile home is one such case. If you use water containers for storage, design the storage bin so that the containers can be removed if they start to leak.

Phase-Change Materials

Last on our list of options for suitable heat storage materials are *phase-change* materials, of which the most commonly used are *eutectic salts*. These salts are able to store a great deal of heat in a relatively small space because they change state, from solid to liquid, as they are heated. Any substance will absorb or release a great deal of heat at the point at which it melts or freezes, but eutectic salts are designed to maximize this effect. Eutectic salts make this change between 90 and 120°F (collector output temperatures). At first glance they seem ideal for solar heat storage because they can contain the same amount of heat as a rock bin in only about one-fifth to one-tenth the space. However, phase-change materials have several drawbacks. In the first place, eutectic salts will always melt and freeze at one temperature, say 90°F. The solid salts mixture will be heated to 90°F by the collector and take on a lot of heat at this temperature until they melt. They may be heated further to 140°F but won't store much heat between 90 and 140°F. Since they have taken on almost all of their heat at 90°F, they must release it all at 90°F, and the entire storage will be at this temperature most of the time. Stratification can't take place, and thus phase-change storage operates much like a water storage tank. If the salts are designed to melt at 120°F, the stored heat is at a more usable temperature, but the collector must furnish air at a minimum of 120°F to store heat, and collector heat output can be reduced.

Phase-change materials are so handy and compact that we could probably overlook the fact that they don't allow for stratification and use them anyway if it weren't for two other drawbacks—they have an unknown life span and they are very expensive. All storage salts are held in suspension, and they are prone to fall out of suspension after repeated cycling (freezing and melting). Once they separate, they lose their capacity for storing heat. Various companies have designed salts which they claim overcome this problem, but until they accumulate many years of test data from actual installations, the service life of phase-change materials will still be open to question. Since eutectic salts are very expensive, a phase-change storage system can easily be twice the price of a rock storage system that will hold the same amount of heat.

Storage and Hot Water Preheat

If a solar installation has a collector large enough to justify controlled heat storage, it is certainly large enough to justify the installation of a domestic water heater. The system can then be used in the winter when the space-heating needs have been satisfied and during the summer months. In many cases a

water heater can eliminate the need for power venting of a large, tilted collector in the summer.

Placing the tank on its own system is a reasonable setup. The hot water system has its own independent differential thermostat which controls a small blower. This blower delivers heat from the storage bin to the air-to-water heat exchanger through an extra hole cut into the side of the top plenum. The blower moves air over the tank, and the air is returned to the bottom of storage through the return ductwork. An optional motorized damper prevents airflow around the tank when the storage component is delivering heat to the house at night. Solar heat is delivered to the tank whenever the top plenum in the bin is warmer than the water tank, regardless of what the rest of the system is doing. The extra controls (and blower) for this system are inexpensive and simpler to install than many other delivery arrangements. A small heat exchange tank (40 gallons) can be used since it is tied into the large, heated mass of rocks in the bin and there is plenty of heat available for hot water in the evening. The blower moving air around the tank can also be small (250 cfm at a static pressure of 0.4 inch WG) and still provide for a good heat exchange at the tank. Summer heat losses to the house from a poorly insulated storage bin could be a problem with this system.

Another option is to divert solar air into an air-to-water heat exchanger before it reaches storage. This also requires a separate blower and controls. In the summer, when the large blower delivering heat to storage isn't operating, the smaller blower in the hot water loop will ventilate the collector (possibly using outside air) and provide very hot air to the exchanger tank. The rock bin won't be heated, and the large (80-gallon) solar water tank will get very hot. For water preheating in the winter, a thermostat in the rocks can turn off the large storage blower and activate (with a relay) the hot water loop when the middle of the rock bin reaches a set temperature (for example, 110°F). This setting can be changed as more or less heat is needed for space heating.

Another possibility is to build a two-blower, four-mode system that uses the collector blower to heat hot water. The system's first priority is heating the house, the second is heating water and the third is heating the rock storage. An additional motorized damper (air diverter) and several relays are needed to accomplish this. If the ductwork leading to the exchanger is restricted, a smaller flow of hotter air will result. This may be desirable in the summer, but be sure to use a non-overloading, belt-driven blower that can operate through a wide range of static pressures. A large tank should be used, and high temperatures are possible with this system, especially in the summer. The storage isn't heated in the summer, which can be an advantage. This setup requires a two-blower delivery.

A final choice is to bury the tank in the top of the rock bin. Controls are simplified, and the system is self-regulating through the seasons, but the major problem with buried tanks is servicing them. The preheat tank has about the same life expectancy as a water heater (10 to 15 years), and if the tank eventually starts to leak into the rock storage, you have real problems. The rocks within the bin expand and contract, and this can also cause problems. The tank will wiggle around a little bit every day, which plays havoc with the plumbing fittings. We have even heard of a system using copper pipes buried in the rocks for a heat exchanger where the shifting rocks rubbed holes in the pipes after several years. In short, the best place for an air-to-water heat exchanger is outside the storage bin.

Passive Solar Plus Active Storage

In an active system with storage, the rock bin is sized to store whatever output isn't used during the day. When an active system is combined with passive solar heating features, it is usually desirable to over-size the active storage because less "active" heat will be needed during the day and more will need to be delivered to storage. This is especially important if the passive and active systems are heating the same zone. If they are heating different parts of a house, this consideration isn't critical. Similarly, storage systems having a water-heating mode can be slightly undersized since some of the output goes for water heating.

A Super Crawl-Space-Storage System

This is a new idea on the air-heating scene, but it has so many good design features going for it that we couldn't help including it. A small collector that heats a crawl space is a proven, effective way to add solar heat to a house. Heat losses from the crawl space can be greater than from direct-use systems, but lower delivery temperatures from the collector are usable. Small systems have adequate storage in the floor and crawl-space dirt, but what if you want to build a large collector? There just isn't enough available mass in a crawl space to store the heat from a large collector.

Our solution is to use the crawl space for rock storage (see figure 11-4). Lay 2 inches of foil-faced polyisocyanurate insulation on the crawl-space dirt. Down the center of this insulation, lay round or square sheet-metal ducting with holes drilled or punched along both sides near the top. Cover the insulation on both sides of the ducting with clean gravel to a depth of 12 to 18 inches. Use hardware cloth bent in a U-shape to keep the gravel from plugging up the holes in the ductwork. Cover the top of this arrangement with several layers of 6-mil polyethylene, and put ½ inch of sand or dirt on top of the plastic, to hold it firmly on top of the gravel.

Solar air from the collector enters the distribution duct at one end of the bed, and since the other end of the duct is plugged,

the air is fed out the holes and flows horizontally through the gravel. After leaving the bed, the solar air enters the insulated crawl space and returns to the collector. The rocks nearest the distribution duct will be the warmest, but there will be little stratification in the rock bed due to the horizontal airflow. This isn't an important consideration since heat won't be taken out of the bed and delivered to the house by a flow of air.

This gravel bed should follow the same sizing guidelines as a regular rock bin storage. There should be ½ to ¾ cubic foot of rock per square foot of collector (130 lbs of rock/ft^3 with 30% voids = 100 lbs/ft^3). The rock should be ¾- to 1½-inch-diameter washed

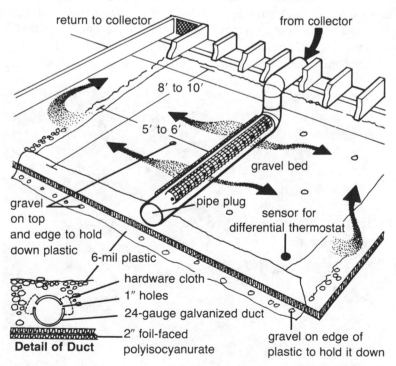

Figure 11-4: If you have a good-sized crawl space, you can store the heat from a large collector in a gravel bed located there. This simple system eliminates the need for the many controls that most systems with storage require.

river rock, and there should be a 5- to 7-foot airflow through rocks. For a 200-square-foot collector with 100 cubic feet of storage, two possible configurations are: 12 inches deep by 10 feet long by 10 feet wide (two 5-foot flows through the rocks) or 12 inches deep by 8½ feet long by 12 feet wide (two 6-foot flows).

Locate the gravel bed under the rooms of the house that are used the most and that require the most heat, since the mass of the gravel will take the heat out of a large collector and deliver radiant heat to the rooms above. In some cases it is desirable to put 1 inch of polyisocyanurate insulation on the sides of the bed (from the floor down to the crawl-space dirt) to further isolate the heat storage. Any heat losses from storage are to the insulated crawl space and therefore aren't losses. Electrical controls for this system are simple—just a blower and a differential ther-

mostat. Place the thermostat sensor near the cold edge of the bed. This is a far cry from the numerous relays, motorized dampers and extras thermostats needed in a conventional three- or four-mode system. Since no air enters the house, no distribution ductwork to the house is needed. If daytime delivery *is* desired, it can be accomplished easily with a simple two-mode control system.

This idea is a natural if you are moving an old house to an existing foundation, but this rarely happens and you are asking, How do you get all that gravel into the crawl space? Moving 3 to 6 cubic yards of gravel into the crawl space through the access hole is pretty well out of the question. Locate a closet or a carpeted corner of one room that will be over the heat storage area and peel back the carpet. Cut a 2-by-2-foot hole through the floor so that two edges of the hole break on floor joists. Dump the rocks down the hole on top of the

Photo 11-2: Locating this wood-frame rock bin in the cellar under a 60-year-old house required a considerable amount of excavation, which was done 5 gallons at a time. The corners of this box are a little overbuilt but this is good insurance. Note the cold-duct port framed into the lower right corner of the bin.

insulation 5 gallons at a time. Use a sturdy rake or hoe to form the mound. Make sure your new "trap door" closes in a tight seal. If this procedure seems difficult, imagine tearing up the entire floor and pouring a pad and footing for a rock storage box or for a mass wall to intercept direct gain in a passive system. Keep in mind the saying of a well-known world leader, "If you want omelets, you've got to break eggs."

Building a Thermal Storage Rock Box

There are several ways a rock bin can be built. In new construction a concrete box and its supporting slab can be poured when the rest of the foundation is poured. There also

---

Rules of Thumb for Constructing Rock Bins Made of Septic Tanks

• Order the stock item that is closest to your needs. Tell the manufacturer where you need "knock outs" for ductwork to be located so he can pour a tank to fit your needs. Also specify anchor bolts at the top of the bin, and build the lid from wood.
• Septic tanks must be buried to the top of the rocks before billing.
• Tar the outside of the tank before setting it into the hole. Dig an oversized hole, and insulate the bottom before setting the tank into place.
• After setting the tank into place, put in the remaining rigid insulation and backfill.

Rules of Thumb for Constructing Poured Concrete Bins

• Walls must be at least 8 inches thick if unsupported by dirt.
• Use ½-inch rebar horizontally and vertically, 8 inches on center.
• Form the bottom inlet port so that the ductwork can slide through it without the framing being removed. Attach and seal this ductwork from the inside of the box.
• Pour the bottom of the bin on several layers of 6-mil polyethylene to act as a vapor barrier. Put insulation under the pour.
• Insulate the outside of the bin with rigid insulation. Use two layers, overlapping seams, or put insulation inside the forms before pouring.
• Seal the joint between the vertical walls and the concrete pad with polyurethane caulk.
• Place foundation bolts in the top of the wall to accommodate a 2 x 6 plate. Seal this plate in place with polyurethane caulk and cover it with litho plates.
• Use cross ties on walls over 8 feet long and on unburied bins. Use large, heavy washers instead of channels to hold the wall in place.

---

have been many successful installations in new construction where a new septic tank (which costs from $400 to $700) was used as a storage bin. A crane is used to set it into place in a large crawl space. Since a septic tank is designed to hold liquids, it is a very sturdy prefabricated unit after it is buried. A wooden rock bin is the best choice in most retrofitted systems because it is easier for the do-it-yourselfer to fabricate than a poured concrete box. With this in mind we will give step-by-step construction details on this type of bin.

The foundation for a rock bin must be sturdy enough to support the great weight of the rocks it will hold. The inside of the box must be completely airtight, and all materials used in it must be able to withstand temperatures of 200°F.

Proper lining and sealing of the bin cannot be overemphasized. Even small air leaks will severely degrade the system's performance, and any large gaps could become an entry for insects or rodents.

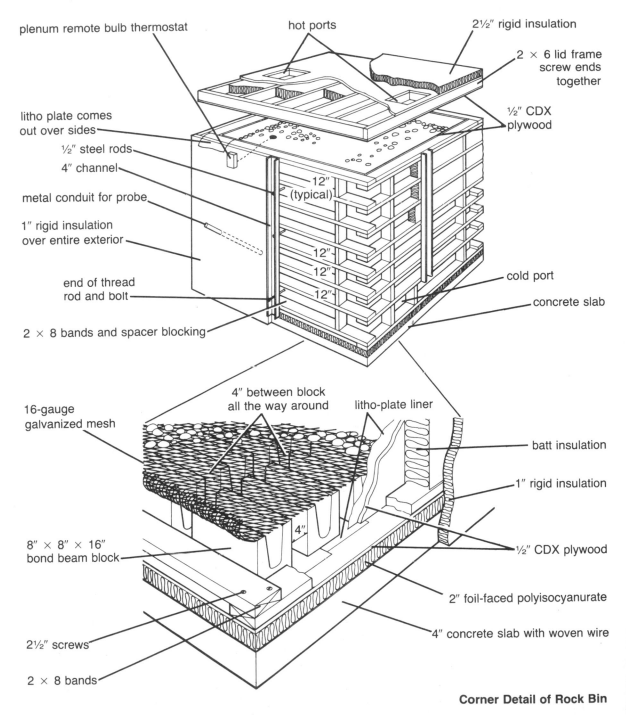

plenum remote bulb thermostat

hot ports

2½" rigid insulation

2 × 6 lid frame screw ends together

½" CDX plywood

litho plate comes out over sides

½" steel rods

4" channel

metal conduit for probe

1" rigid insulation over entire exterior

end of thread rod and bolt

2 × 8 bands and spacer blocking

12" (typical)

12"

12"

12"

cold port

concrete slab

16-gauge galvanized mesh

4" between block all the way around

litho-plate liner

batt insulation

1" rigid insulation

8" × 8" × 16" bond beam block

4"

½" CDX plywood

2" foil-faced polyisocyanurate

2½" screws

4" concrete slab with woven wire

2 × 8 bands

Corner Detail of Rock Bin

Figure 11-5: Although they are easy to construct, rock bins must be built like the proverbial brick outhouse. They sit on top of a reinforced concrete slab. Wooden bands supported with 4-inch channel and metal rods hold the bin together. The cool air plenum at the bottom of the bin is created by laying heavy-gauge metal lath over bond beam blocks before filling the bin with ¾- to 1½-inch washed gravel.

Rocks

Before we get into construction specifics, let's look further at some of the materials that go into a rock bin. Rock used for heat storage must be very clean, dry and properly sized.

Properly sized rock will pass through a 1½-inch screen but not through a ¾-inch screen. Round river rock is a much better choice than crushed rock since it doesn't pack together and the airflow through it is much more uni-

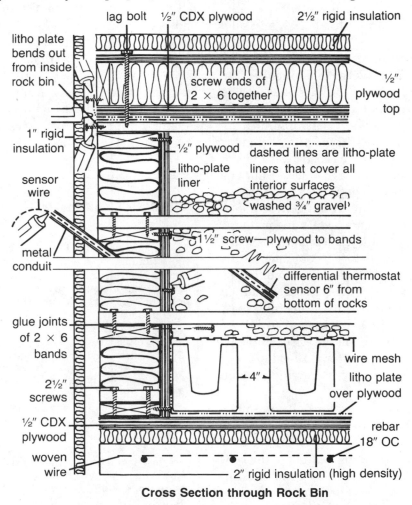

Cross Section through Rock Bin

Figure 11-6: The inside of the lid for a rock bin is lined with litho plate or galvanized sheet metal. The lid is attached to the vertical walls with long lag bolts. This seam is sealed with silicone to make the box completely airtight. The sensor for the differential thermostat runs through metal conduit and is placed in the rocks, 6 inches above the metal lath at the bottom of the bin. All interior surfaces in the bin are lined with metal, and all overlaps are sealed with silicone.

formly distributed. Crushed rock can be used if it is the only kind available, but it must be uniform in size to permit adequate airflow.

You should have the rock washed at the gravel pit where you buy it rather than try to do it in your back yard (never wash it after it has been loaded into the bin). Many sand and gravel companies offer *solar rock*, which is just what you need, or they can deliver a load of oversized gravel at a reasonable price. If there is any doubt at all about what you are getting, drive out to the pit and examine the rock before having it delivered. If there are a few 2-inch rocks in the pile, it's no big deal as long as the rock has been washed and contains no fines (such as sand). It would be a very good idea to perform a *void test* to make sure that you are in fact getting the 30 percent void volume needed for adequate airflow through the bin. It's a simple task: Fill a 5- or 10-gallon (or larger) container with a sample of the actual rock you intend to buy. Shake the bucket so the rock settles and then fill the container with water, keeping track of how many gallons are needed to fill the container to overflowing. If, for example, it took 3 gallons of water for a 10-gallon container, you've got a 30 percent void volume in your rock.

When it comes time for delivery, make sure that the bed of the truck delivering the rock is clean and dry before it is loaded. If there is dirt or sand in the bottom of the bed, it will end up on your pile of rocks. Also, buy the driest rock available. If the rock is very damp when loaded into the bin, you can plan on having water in your ductwork the first day solar heat is delivered to it. Solar rock should cost from $7 to $10 per cubic yard, plus delivery, but even if it costs three times this amount, it is a much better deal than washing rocks by hand, which is an incredible job to be avoided at all costs.

Metal Lath

The expanded metal lath used to make the lower air plenum needs to be sturdy enough to support the rocks and must have a mesh size small enough so that no rocks fall through. It should be 16 gauge or heavier and have three or more holes per inch. One-half-inch, 13-gauge galvanized standard expanded metal lath is a good choice. Metal lath can be purchased from building-materials stores and metal suppliers. It must be galvanized in order to be a permanent material. Although there is very little moisture present in a rock bin that could cause rusting,

---

Rules of Thumb for Constructing U-Shaped Bins

• The center divider must be able to withstand temperatures to 200°F and must make an airtight seal at the top and sides. Build this divider out of ¾-inch plywood insulated with polyisocyanurate insulation and covered with litho plates.

• Bond beam blocks and wire mesh should extend across the entire bottom of the box. The center divider should extend down to the level of the mesh at the top of the blocks.

• Attach and seal the divider to the sides of the box after the mesh is in place. Screw 2 x 2 bracing on both sides of the divider to secure it. Make sure that the top of the divider is ½ inch above the bottom of the top framework.

• A 2 x 4 will rest on top of the divider and be attached to both sides of the box. Cut out a ½-by-¾-inch groove in the bottom side of the 2 x 4 to fit over and seal the top of the divider.

• Crawl inside the box and seal the joint between the divider and the 2 x 4 with silicone, and run litho plates from the divider up over the 2 x 4.

• Fill the bin with rocks slowly to avoid disturbing the divider.

• Build the lid in two sections to ensure an airtight fit to the 2 x 4.

the bin should be built to last the lifetime of the house. Galvanized lath will help to guarantee this.

Bond Beam Blocks

A *bond beam block* is a standard item that is used on the top course of masonry wall construction. These blocks are typically laid with the open "U" facing upward. This cavity is filled with concrete and rebar to form the bond beam. These 8-by-8-by-16-inch blocks are strong enough to support the weight of the rocks and still allow for good air distribution through the bottom of the bin. Don't use regular concrete blocks laid on their sides, and don't use bricks, to support the wire mesh. Bond beam blocks cost slightly more than regular, solid blocks, about 90¢ to $1.10 each.

Insulation

The insulation used in rock storage containers must be able to withstand temperatures up to 200°F, and for this reason extruded polystyrene (Styrofoam) and urethane foam are unacceptable. Polyisocyanurate foam can be used because it is a higher-temperature material, but it needs to be isolated from the rocks and all heated parts of the bin by noncombustible materials. Fiberglass batts are also a good choice for insulation, though they must be placed outside the sealed interior of the rock bin.

Construction Steps

Plan Ahead

Before starting construction of the rock bin, double-check all of your drawings. Make sure there is adequate room at the top and bottom of the bin for ductwork and controls. Be generous! There may be enough room for the devices, but is there enough room for you to get into the space and install them? Our

experience in putting controls on rock bins that other people have built indicates there almost never is.

Pour the Concrete Pad for the Box

The concrete pad that supports a bin should be at least 4 inches (preferably 6 inches) thick, with ½-inch rebar around the perimeter and 2 feet on center through the pad. It should extend 1 foot beyond the entire outside perimeter of the rock bin and should be thicker on the perimeter. If this pad is poured as part of a slab-on-grade or basement floor, install ½-inch insulation board expansion joints between it and the rest of the floor so that it "floats" independently from the rest of the slab. In slab-on-grade retrofits it is a good idea, but not absolutely essential, to rent a large masonry saw and cut a joint around the rock box. After pouring a pad, be sure to wait at least two weeks before filling the bin with rocks, to allow the concrete to attain maximum strength.

Some builders recommend attaching the bottom framework of the bin to the pad with foundation bolts (lag bolts and masonry shields in retrofits). We feel this is an unnecessary step. Since the rock box is massive and the weight in it pushes straight downward, it certainly won't ever slip off the pad.

Assemble the Horizontal Frames

Continuous horizontal frames act as bands to hold the rock bin in shape. These frames can be made of 2 x 6s for smaller bins, but 2 x 8s should be used when sides are longer than 6 feet (up to 10 feet).

Cut the 2 x 8s to the desired length. If the interior of your box is a cube, they will all be the same length and will equal the length of each side plus 15 inches to allow for overlap on both ends. On a flat, level surface (such as a driveway) lay out and snap chalk lines to represent the outside of the rock box. Make sure the diagonal measurements are equal. Set the boards over this pattern, precisely overlapping the edges. Predrill five or six ⅛-inch holes through the overlap. Put a generous smear of glue between the beards, and screw them together with several 2½-inch drywall screws. Once the first frame is assembled, carefully move it to one side to allow the glue to dry. Avoid racking the framework, and keep it square. Assemble the remainder of the horizontal bands. You will need a band every 12 inches and one for the top and bottom of the box. For example, a box that is 8 feet high will need nine frames. The top and bottom frames will have additional 2 x 8s added to opposite sides to create a flat continuous surface. Cut blocks to fit between the frames. You will need four 10½-inch-long blocks for each frame except at the top and bottom. Here they will be 9¾ inches long. Two extra blocks will also be needed at the bottom, to frame-in the cold inlet port.

Assemble the Frames

Lay the bottom frame on a flat surface, with the continuous flat surface facing upward. Cut ½-inch CDX plywood to fit over this frame. Make sure it is square, and screw

---

Rules of Thumb for Constructing Buried Bins

• Do not bury the bin if you have a high water table.
• The entire bin must be completely waterproofed on the outside. Use several layers of 6-mil polyethylene and tar-based waterproofing.
• Use concrete construction. Wood construction is not suitable for buried bins.

---

the plywood to the frame with 1½-inch drywall screws.

Cut 2½-inch foil-faced polyisocyanurate insulation to fit under this first frame and set it on the pad. Set the bottom frame assembly on top of the insulation, plywood down, and toenail four of the 9¾-inch blocks to this frame, 12 inches from each corner, with 16-penny nails. Have a friend stand on the plywood to keep things from bouncing around. Set the next frame in place on top of the blocks, and nail through the frame into the block below. Nail two more 9¾-inch blocks in place to form the cold inlet port into the bottom of the box. Continue to put in the rest of the frames and the spacer blocks up to the top of the box. It will be a bit wobbly and seem quite fragile. Don't be overly concerned about it at this point because the plywood will stiffen everything up. Make sure the frames are square and in line with each other. There should be no gaps between the spacer blocks and frames. Two or three temporary vertical supports may be helpful. The last frame that will form the top of the box will have additional 2 x 8s nailed flat on opposite sides to make the top a flat surface.

Line the Sides of the Box with Plywood

Cut ½-inch CDX (exterior) plywood to the height of the top frame. Start at one corner. Lay the plywood flat against the inside of the box, and push it into the corner to line up all of the frames. Put several 1½-inch drywall screws through the plywood and into the frames to hold the plywood in place. The plywood should be installed so the face grain is perpendicular to the frames. Install the next piece of plywood in the opposite corner of the box. If the frames are not exactly square, it will show up at this point. It may be necessary to put shims between the plywood and the frames if there is a gap of over ¼ inch at any point. Cut and install the rest of the plywood, screwing it to each frame with 1½-inch screws, 16 inches on center. Cut out the cold inlet port in the appropriate sheet of plywood, and screw its edges to the vertical 2 x 8 blocking.

Line and Seal the Inside of the Box

The inside of the rock box must be lined with a high-temperature material and sealed with silicone caulk so that it is *completely* airtight. This lining can be 30-gauge galvanized sheet metal, ½-inch aluminum-faced drywall or litho plate. Litho plate is cheaper than sheet metal and easier to seal than drywall, so we will use it in our example.

Attach the litho plate to the ½-inch plywood with ⁷⁄₁₆-inch drywall screws (or ½-inch hex heads). Seal all overlaps with a generous bead of silicone, and cover all screw heads with silicone. Line the cold port with litho plate at this time, bringing the aluminum plate all the way through the port and overlapping it onto the front surface of the 2 x 8 framework. Also run the litho plate over the top frame on the box. Screw this at the side of the box, not the top. This lining must completely cover the inside of the box and be airtight. If you line the box with foil-faced drywall, attach it with 1-inch drywall screws and seal each screw head and butt joint with silicone. Do not use foil-faced insulation of any type for the innermost layer of the rock box because expansion and contraction of the rocks will crush it.

Set in Bond Beam Blocks and Wire Mesh

Bond beam blocks support the wire mesh and allow air to flow freely from the cold port to the entire bottom surface of the rocks.

Place 8-by-8-by-16-inch bond beam blocks on the bottom of the bin. Set the blocks with the open "U" facing upward, and place them so that air can flow freely to and from all parts of the bin from the cold port. A 4-inch spacing around all sides of each block will help accomplish this. At the sides of the box, set the blocks snugly against the litho plate, to support the edge of the mesh. Screen a sample of gravel with the mesh before putting it in place to make sure no gravel settles through. Cover the bond beam blocks with heavy-gauge (16-gauge) galvanized mesh. Bend the mesh so that it runs at least 12 inches up each side of the box. Overlap adjoining pieces about 4 inches, and wire them securely together every 2 feet. A small scrap of plywood placed on top of the first piece of mesh, for you to stand on, will make it easier to put in the rest of the mesh.

Install Cross Ties

Metal cross ties and vertical braces are your insurance that the rock bin won't ever "blow-out." They must be used on any side 6 feet or longer. There is a lot of movement within the rocks as they heat and cool every day—movement that can weaken even the most sturdily built box after several years.

Four cross ties of ½-inch steel rod and four pieces of 4-inch channel will be required for most boxes. Oblong boxes may require only one set. The steel rod can be purchased, cut and threaded at a local welding shop.

Cut the channel to the height of the box, and drill two holes through each piece of channel dividing it into thirds. Mark the side of the box where the rods will penetrate, and drill slightly oversized holes through the sides of the box. Insert the rods through the box and bolt the channel in place, flat side against the frames. Begin tightening the bolts (use lock washers) and shim any gaps larger than ¼ inch between the channels and the frames. Continue tightening the bolts until the sides of the box have been compressed inward about ⅛ inch. Seal the holes in the box with silicone.

Install Thermostat Probes

Most thermal storage systems will need two thermostat probes inserted in the rock box: a differential thermostat sensor placed low in the rocks and the bulb of a remote bulb thermostat placed on top of the rocks, in the hot plenum.

To install the sensor for the DT, drill a 1-inch hole in the side of the box, 6 inches above the bond beam blocks. Insert a 2-foot piece of rigid ¾-inch conduit horizontally into the box. The end outside the box should extend 1 inch past the frames. The conduit allows you to poke a probe into the center of the rock box and remove it if it malfunctions. When filling the box with rocks, gently cover the conduit and seal its penetration through the side of the box with silicone. Push the sensor to the end of the conduit, and seal the outside with a silicone cap. Conduit can also be run down from the top of the box when it's being filled, if this is easier. Make sure the sensor is located near the center of the box, about 6 inches from the bottom of the gravel.

Insert the remote bulb of the plenum thermostat into the hot plenum of the rock box by drilling a hole through the side of the box. Seal the hole with silicone. This should be done after the bin is filled with rocks but before the lid is secured in place.

Fill the Box

Before filling the box, do a trial-run installation with the ductwork and air handling devices. Move them into place to make sure they will fit and can be installed after the 6-inch lid and 2 inches of foam insulation are in place on top of the box.

Then fill the box with gravel to within 8 to 10 inches of the top. Dump the gravel into the rock box slowly to avoid stretching the metal lath and moving the box around or damaging its sides. If you are filling your box from a dump truck (which is the easy way), build a temporary plywood funnel and *slowly* dump and shovel the gravel into the box. If

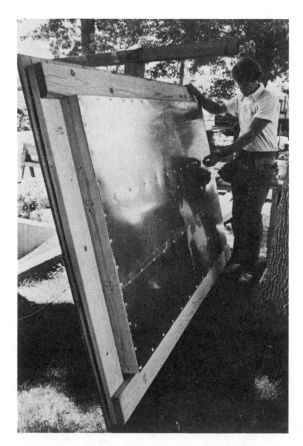

Photo 11-3: In planning a rock bin, be sure to leave yourself plenty of room for installing the lid and the ductwork blower, damper and other components.

you are loading the box by hand with 5-gallon buckets, call up the neighbors for help.

Frame a Lid for the Box

The lid for the rock box must be well insulated and have an airtight seal to the sides of the box. Frame the lid for the box from fir 2 x 6s and allow framed-in holes for ductwork. This lid will have the same outside dimensions as the horizontal frames and will be covered with ½-inch CDX plywood top and bottom. Insulate the lid with 6-inch fiberglass batts. The bottom of the lid and the ports for ductwork are lined with litho plate and sealed with silicone. Bring the litho over the side of the lid, and don't place screws where the lid will rest on the top frame.

Attach the lid to the top frame of the box with 8-inch lag bolts (with washers) spaced 16 inches. Predrill holes through the lid for the lag bolts, and bed the lid in a large bead of silicone. Use an extra glob of silicone where each lag bolt attaches to the top framework. After the lid has been secured, run a bead of silicone around the outside joint between the two layers of litho plate, to further ensure the seal of the lid to the box.

Insulate the Sides and Top of the Box

The amount of insulation required around a storage box will depend largely on where it is located. It should be well insulated (R-20) if it's located in or under a living area and superinsulated (R-30 to R-60) if located elsewhere.

Staple fiberglass batts to the outside of the box between the horizontal frames. Over the framework additional foil-faced rigid insulation can be added. Two inches or more of insulation should be placed on the top of the rock box after the ductwork attachments have been made.

If the sides of the rock box are exposed to a living area, it will be necessary to furrout the sides of the box with 2 x 4s or 2 x 2s to bring the sides of the box out past the channel.

Attach Ductwork to the Rock Box

Cut and bend flanges on the ductwork that will attach to the outside of the box, and screw these flanges through the litho plates and into the wooden framework below. Use plenty of silicone to achieve an airtight seal.

Designing and Building a Slab-on-Gravel Storage

In a slab-floor system there must be five times as much mass as there is in a conventional rock bin. This will allow a very low air velocity through the rocks of 4 to 5 feet per minute, about one-fifth the velocity present in a conventional bin. Hot air from the collector will move so slowly that it will exchange most of its heat where the air first comes into contact with the gravel in the rock bed. This can result in hot spots in the slab, especially if there is a large distance between the "hot" distribution ductwork from the collector and the "cold" return ductwork. The gravel and the slab above it will tend to distribute heat away from hot spots, but the system should be designed to avoid them. Keeping the hot and cold feeder ducts within 4 feet of each other will help minimize hot spots in shallow, 12-inch-deep gravel beds. In beds 2 feet deep or more, there will be a more even distribution of heat, even with longer airflows through the rocks. Another possibility is to design the system so that the hot manifold dumps hot air into the gravel bed under the center of the room located above and not worry about hot spots.

Since the velocity of the air moving through the gravel is low, there will be almost no static pressure in the rocks. For this reason just about any size rock, from ¾-inch pea gravel to 2-inch oversized rock, can be used in the bed without affecting heat transfer. The gravel must be well washed and free of fines, of course. The length of flow through the rocks is important only in preventing hot spots, and any flow between 2 feet (¾-inch gravel) and 6 feet (any size of gravel) will work well.

The mass in a slab storage is sized to allow for a 10°F temperature rise in the course of a day's heat delivery. Because of this, very

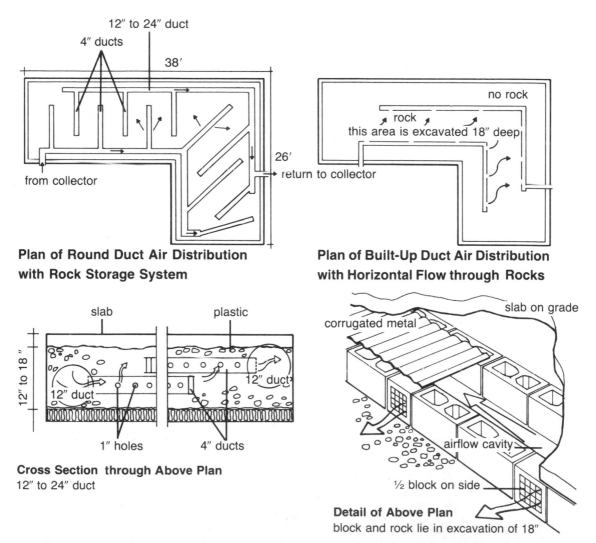

Plan of Round Duct Air Distribution with Rock Storage System

Plan of Built-Up Duct Air Distribution with Horizontal Flow through Rocks

Cross Section through Above Plan
12″ to 24″ duct

Detail of Above Plan
block and rock lie in excavation of 18″

Figure 11-7: This illustration shows two of the many variations possible in a slab-floor heating system. In the system on the left, heat is distributed throughout the floor through heavy-gauge galvanized manifolds and PVC pipe. The system on the right heats only the central portion of the floor, and cinder blocks are used to form the manifolds. In both systems air flows horizontally through the rocks.

Rules of Thumb for Constructing Slab-on-Gravel Storage

- Install vapor barriers of 6-mil polyethylene under the gravel and between the gravel and slab; the storage must be completely waterproof.
- Insulate under the gravel and on the sides of the bed with 2-inch rigid, waterproof insulation inside the vapor barrier.
- Include the slab in the calculation for the amount of mass in the storage. Concrete weighs 120 pounds per cubic foot.
- Pour the slab 3 inches thick; use "remesh" reinforcement.
- Use large PVC pipe for distribution lines and smaller PVC for individual feeder lines. Cap the PVC and drill small holes to add up to less than the cross-sectional area of the pipe.
- Over-size the manifolds. Cold manifolds can be lined with plywood if the storage is completely watertight. Hot manifolds must be airtight.
- Manifolds can also be built using cinder blocks laid on their sides, with galvanized wire mesh holding back the gravel.
- Control the blower with a differential thermostat. Place the storage probe in the gravel near the cold return before pouring the slab. Place the probe inside conduit to allow for removal if necessary.
- Place the hot distribution lines at the bottom of the gravel, and place the cold return ductwork at the top of the gravel directly under the slab.

---

low collector output temperatures are still very usable. Since collectors operate more efficiently at lower temperatures, the collector in a slab storage system can be designed to deliver a great deal of air at a fairly low temperature (80 to 100°F). The collector will probably want a slightly larger airflow channel or a shorter flow path and 1½ to 2 times as much air delivery to accomplish this (see Appendix 4 on collector efficiency). Deliv-

ering cooler air will also help even out potential hot spots on the slab.

Even air distribution to all parts of the slab is very important in this storage system. Since there is almost no static pressure in the gravel, air will tend to take the shortest paths from the hot to the cold ductwork. Designing the distribution and return legs to be opposite each other and of equal length to all parts of storage will prevent this. Over-sizing the manifold in relation to the feeder lines and under-sizing the holes drilled in the feeder lines will ensure that each line is slightly pressurized and receives the same flow.

We don't know of any slab-on-gravel systems that follow all of our design criteria, but all of the ones we've seen work fine. Use plenty of clean gravel, insure good distribution to all parts of the bed, and move a lot of air through the collector and you won't have any problems.

Sources for Supplies

Air Handlers

Component Systems

Heliotrope General manufactures differential thermostats and motorized dampers and builds custom control boxes for systems with multiple delivery modes. Many of their airflow schematics appear overly complicated, but their prices and quality are good.

Heliotrope General, Inc.
3733 Kenora Dr.
Spring Valley, CA 92077
Phone: (714) 460-3930

Hot Stuff Controls offers a complete line of motorized dampers and control boxes for two-, three- and four-mode systems. This small company caters to do-it-yourselfers and can custom-build dampers or controls for active air systems at reasonable prices.

Hot Stuff Controls Inc.
P.O. Box 306
406 Walnut St.
La Jara, CO 81140
Phone: (303) 274-4069

Natural Power builds custom control boxes for do-it-yourself two-, three- and four-mode systems. They don't offer dampers but are a good source for controls.

Natural Power, Inc.
Francestown Turnpike
New Boston, NH 03070
Phone: (603) 487-5512

Prefabricated Air Handlers
Contemporary Systems' No Frills 1200 air handler has been used successfully in many site-built systems with multiple delivery modes.

Contemporary Systems, Inc.
Route 2
Walpole, NH 03608
Phone: (603) 756-4769

Rota Flow 2, made by Delta H Systems, with or without an integral air-to-water heat exchange coil, is a good item for use in systems with multiple delivery modes. Delta H is currently developing an air diverter damper which also looks good.

Delta H Systems, Inc.
16970 Rd. 36
Sterling, CO 80751
Phone: (303) 522-4300

Other sources for air handling equipment include the following:

Helio Thermics, Inc.
1070 Orion St.
Donaldson Industrial Park
Greenville, SC 29605
Phone: (803) 299-1300

Park Energy Co.
Star Route, Box 9
Jackson, WY 83001
Phone: Park Energy requests that consumers write for information

Solar Development, Inc.
11799 E. 30th Ave.
Aurora, CO 80010
Phone: (303) 343-8154

Solaron Corp.
Solar Energy Systems
1885 W. Dartmouth Ave.
Englewood, CO 80110
Phone: (303) 762-1500

Tritec Solar Industries, Inc.
711 Florida Rd.
P.O. Box 3145
Durango, CO 81301
Phone: (303) 247-8497

12

SOME SHINING EXAMPLES

Throughout this book we have concentrated on the various components that go into making an entire collector system. In this final chapter we look at some working installations to give you a better idea about how all the bits and pieces fit together.

In preparing this book, we examined dozens of site-built collector systems and were amazed at both the innovative approaches taken by owner-builders and the tremendous variations among successful systems. We could easily fill a large volume outlining interesting systems, but we have narrowed our choices to these few examples which only begin to reflect the wide variety of possible solar heating applications. Several of the systems in this chapter are installations we have had a hand in, either in design or construction. Most do-it-yourselfers could easily build retrofit projects similar to ones presented here. However, since proper design is very site specific, it is important to stretch your imagination when planning your own system. The shining examples presented here should help you to do just that.

The Gonzales Retrofit: Two Two-Modes

We're very enthusiastic about two-mode delivery systems in this book because we believe they deliver more usable heat from a mid-sized active collector at a better price

than any other system. Even without a storage component the Gonzales system provides so much heat that for all intents and purposes it's really their primary heating system.

This system consists of a pair of two-mode systems. The three wall-mounted collectors were built first. They use one blower to deliver heat first to the house and then to the crawl space. The roof-mounted collector was added later to heat their water first, the crawl space second.

To optimize the second mode of both systems, the foundation around the crawl space was insulated with 6-inch fiberglass batts. Six-mil plastic also was laid over the dirt to form a vapor barrier to limit evaporative cooling by ground moisture.

The Wall-Mounted Collectors

Two small collectors were mounted on the brick-veneer wall below south-facing windows, and a large collector was mounted on the exterior of a stone fireplace between the windows (photo 12-1). Since this two-mode system is used only for space heating, vertical mounting was a correct choice. The collectors happen to be located behind a concrete driveway, and reflection from the driveway probably increases heat gain by about 10 percent (see chapter 4 for information on reflective gain). The vertical collector area totals about 126 square feet (equal to about 7 percent of the 1,700 square feet of heated liv-

Photo 12-1: The four collectors on the Gonzales home operate in two independent two-mode systems. The three vertical collectors deliver solar-heated air either directly into the house or into an insulated crawl space. The roof-mounted collector heats water as its first priority and also delivers supplemental heat to the crawl space during the winter months.

ing space). Although the three collectors are of different sizes, each was designed to accommodate 34-by-76-inch tempered glass panes and each was built with a 1⅝-inch airflow cavity.

Collector 1

The west collector is glazed with two panes of tempered glass. The plywood backing was fastened to brick veneer and is 1 inch wider on the bottom and sides and 1¼ inches wider at the top than the collector frame. This was done to accommodate 1 inch of perimeter insulation and flashing and to provide a slope for the flashing at the top.

The airflow path through the collector is longer than optimal (26 feet) but is still acceptable. The air flows across, up and back

across and is then pulled down to the outlet ductwork. The hot and cold boots fit onto 6-inch-round duct and enter the crawl space through holes cut through the rim joist. A short piece of 2 x 6 was cut to fit under the central glass joint to act as a brace for the collector frame and both panes of glass while the silicone between the glass and the collector frame cured. No other bottom supports are used.

Collector 2

The large central collector was built from a scaffold and has a 21-foot airflow path. Three vertical baffles and a turning baffle direct airflow and support the absorber plate. The side baffles split the inlet and outlet holes to provide for a more even air distribution inside the collector (see figure 12-1). Elbows, torpedo boots and specially made sheet-metal ducts made hookup to the collector much easier. The inlet and outlet ducts (8-inch) were well-insulated beneath the stone facing. The absorber plate seams are horizontal in this collector. The collector ended up being slightly out of square, and the uppermost pane sticks out about ⅜ inch beyond the frame at one

end, but this presents no real problem since there is another inch of perimeter insulation to cover the glass edge.

Collector 3

The small eastern collector was glazed with a single 34-by-76-inch pane of tempered glass. Baffles screwed through the ductboard insulation into the plywood form a zigzag

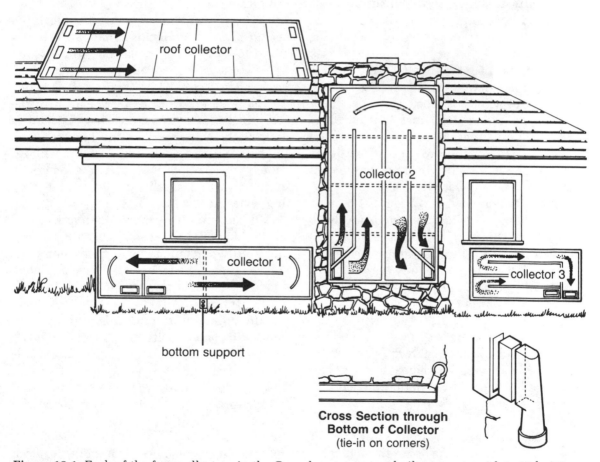

Figure 12-1: Each of the four collectors in the Gonzales system was built to accommodate 34-by-76-inch panes of tempered glass, and each used a different baffling pattern to achieve a 20-foot (or more) flow of air behind the absorber. The ductwork for the three lower collectors runs directly into the crawl space through the rim joist. The ductwork for the roof mount runs through the attic before dropping through the house to the crawl space. The detail shows the tie-in to the large central collector.

pattern that directs the air through a 20-foot flow path. Hot air is drawn off the top of the collector. The inlet and outlet ducts are 4 inches in diameter.

The three vertical collectors present an interesting problem: how to pull air from all three with one blower. This was accomplished by connecting the 4-, 6- and 8-inch outlet ducts to a round, 14-inch, sheet-metal plug (see figure 12-2). Each duct was balanced with a damper so that it supplied the blower with the appropriate airflow rate. (See chapter 9 for information on balancing airflow.)

From the 14-inch plug the hot air enters a 14-to-10-inch reducer (made at a local sheet-metal shop) which connects with the 9-inch opening on the squirrel cage blower. Plugs and vibration collars were installed on both ends of the 815-cfm blower (actual air delivery was about 450 cfm) to simplify the connections. The blower itself sits on cinder blocks in the crawl space. From the blower, the air enters a 12-inch duct and travels 30 feet to a motorized damper (air diverter), which directs the air either to the north end of the crawl space or through the ductwork to the north end of the house. When the house calls for heat, air is sent through three 7-inch ducts that deliver heat to bedrooms on the north end of the house. These lines are the same size and approximately the same length so they didn't need any adjustments to balance the flow. All of the outlet ductwork in the crawl space is insulated with 1½-inch fiberglass.

After entering the bedrooms, the solar air travels back through the house through open doors and reenters the crawl space by way of a floor register at the south end of the house (see figure 12-2).

If the house doesn't need heat, hot air is directed by the air diverter to the north end of the crawl space. It drifts back to the return ductwork at the south end of the house. The return duct has an air filter and a backdraft damper and connects with a 14-by-14-by-18-inch, sheet-metal plenum from which run 4-, 6- and 8-inch uninsulated return ducts to the different collectors. To say the least, the ductwork for this system is about as complex as it can get for a two-mode system.

The Roof Collector

The roof-mounted collector is featured as our example collector in chapter 8. It was sized to carry seven panes of 34-by-76-inch solar glass and is equal in size to the three vertical collectors (126 square feet). A selective surface absorber plate was used in the hot third of the collector. This collector is only slightly oversized for the 80-gallon tank it heats. There was no room for the tank inside the house so it was set in a wood cradle in the crawl space, directly below the existing water heater.

In summer the collector delivers heated air to this tank in a closed loop and provides nearly 100 percent of the Gonzales' hot water needs. Valves were installed to completely bypass the existing water heater during this part of the year.

In winter this two-mode system heats domestic water as its first priority. An adjustable thermostat on the preheat tank sends heat to the crawl space when the tank has reached the desired temperature. In the coldest months of winter, when a lot of space heat is needed, the tank setting is fairly low (100°F). In the spring and fall, when there is less need for space heat, the thermostat setting is raised to 120°F so that more solar energy is used for heating water. In summer the thermostat is set up to 180°F. The collector operates at peak efficiency during the winter because return air to the collector is fairly cool in both modes.

In the summer the hot return air from the tank means less efficient collector operation but high water temperature.

Ductwork for this system is 10 inches in diameter. Both outlet and inlet ducts run through the attic for a short distance. The outlet duct drops through a utility room and through the floor to the blower and heat exchange tank or to the crawl space. A return (inlet) duct allows crawl-space air to return when the system is in the crawl-space mode and has a backdraft damper mounted in it. The crawl-space return duct joins the return duct from the heat exchanger. The air then

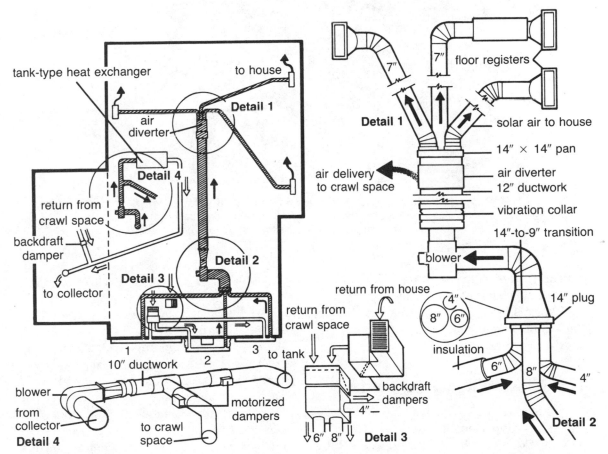

Figure 12-2: This illustration shows the ductwork layout for the Gonzales system. Details 1, 2, and 3 relate to the space-heating two-mode system, detail 4 to the hot water system. Hot ducts are shaded, returns are white. Detail 1 shows the air delivery to either the crawl space or the back bedrooms. Detail 2 is of the ductwork "octopus" and transition to the blower. Returns to the vertical collectors are shown in detail 3. Detail 4 shows the damper arrangement in the crawl space for the roof-mounted, two-mode water-heating system.

Photo 12-2: Ruby Nissley and Mary Kratz built this simple crawl-space heater on their home with help from a local church group. The collector was completed with one Saturday and five evenings of work.

returns to the collector through a duct run through the back of the closet. This return ductwork is insulated with 2 inches of fiberglass to minimize heat loss to the house during the summer.

The Nissley/Kratz Retrofit

This crawl-space system was built by a self-help group from the United Church in La Jara, Colorado. It was installed on Ruby Nissley and Mary Kratz's house with help from four neighbors, one of whom was an experienced collector builder. "I like the concept of the workshop idea where one person is in charge and everyone else can learn," says Mary. "I feel a real sense of accomplishment in what we did, along with the good feeling of getting help from others."

The collector for this system was built directly on a south-facing, 3½-inch stud wall. Solar heat is blown into an insulated crawl space under the west half of the house. The major frustration encountered during construction was caused by limited access to the crawl space. The floor joists were only slightly more than a foot above grade, making crawling into the space nearly impossible. The solution was to assemble the ductwork one section at a time and push it into the crawl space.

This system performs very well for wintertime space heating. The forced-air furnace which heats the west end of the house is set at 60°F and comes on only in the morning or during spells of cloudy weather. The solar-heated part of the house is much warmer than it has been in past winters, as evidenced by Ruby's wintertime piano playing. Ruby says, "I'm taking piano lessons and I practice in the evening. With the furnace temperature set at 60°F, my fingers would be stiff and cold after an hour's practice. But since we put in the collector, the fingers are warm."

Ruby and Mary also like the quiet operation of their system. The blower is mounted on cinder blocks in the crawl space and operates much more quietly than the two existing forced-air furnaces (which the owners would like to eliminate).

Construction Details

The construction of this system was very straightforward. The existing lap siding and ½-inch Celotex sheathing was removed from the wall, and ½-inch CDX plywood was nailed in place. The 3½-inch fiberglass insulation already in the frame wall was left in place to insulate the back of the collector except on the eastern end of the collector. Here a 3½-by-14-inch sheet-metal duct was placed inside the cavity between two studs. This outlet duct takes hot air from the top of the collector down to a hole cut through the rim

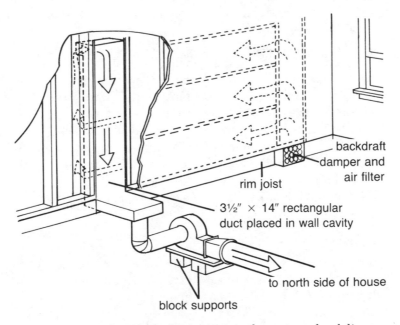

backdraft
damper and
air filter

rim joist

3½" × 14" rectangular
duct placed in wall cavity

to north side of house

block supports

Figure 12-3: In the Nissley/Kratz system the rectangular delivery duct was put in the stud wall cavity before the plywood collector backing was nailed in place. A heat trap is created in this duct at night, minimizing convective heat loss. The backdraft damper for the return was installed in a hole cut in the rim joist.

joist and into the crawl space (see figure 12-3). Building codes required the stud wall duct instead of allowing heated air to pass through an unlined race.

The ½-inch plywood backing was covered with recycled litho plate. Metal track (1⅝ inch) was used around the perimeter and to form the internal manifolds and baffles. This collector has a 25-foot airflow path through a 1⅝-inch air cavity. Five horizontal baffles were installed on top of the litho-plate backing. They were spaced so that they act as supports for the seams of the litho-plate absorber. The internal manifolds on both ends of the collector were connected by single inlet and outlet ports and were formed by installing short sections of track vertically at the ends of the horizontal baffles. The openings into each baffle space were sized so their total cross-sectional area was less than that of the internal manifold, to ensure even distribution (see information on manifold sizing in chapter 9). The blower was hooked to the 3½-by-14-inch outlet duct with special adapters. It was mounted on cinder blocks directly behind the crawl-space access hole and then connected to the 8-inch distribution duct that ran to the north end of the house. The air return from the crawl space is simply a hole cut through the rim joist with a 8-by-10-inch backdraft damper mounted inside it.

Construction of this collector went on for one Saturday and five evenings. The materials cost $800 (1980 prices) or $6.35 per square foot of collector. According to Ruby and Mary the two most difficult jobs in building this system were working in the tiny crawl space and installing the glass. "The glass was bulky and big, and it was good to have help installing it," said Mary. The owners have no reservations at all about their system and wouldn't hesitate to build another one on their

home if they had room for more. Did building the collector make them more energy conscious? To this Mary replied, "Paying high gas bills makes you energy conscious. We put in our system in response to those bills."

The Etherton Retrofit

Bill Etherton, a retired building contractor, installed an air heater on his workshop near Alamosa, Colorado, in 1980. It is a good example of how site-built systems are easily customized to fit an available space. The collector was built with wood for the frame and baffles, ductboard for insulation, corrugated aluminum for the absorber and ¼-inch plate glass for the glazing. Bill already had the lumber and the glass, and he paid $226 for the rest of the materials and the controls.

The ductwork for this system is quite simple. The collector is fed by a 5-inch duct which runs up to the collector from near the floor. The blower and a remote bulb thermostat are located near the peak of the workshop's cathedral ceiling. A large propeller fan is located near the blower outlet and is operated by a wall-mounted switch. It helps to move solar-heated air down from the peak of the roof.

The airflow cavity in the collector is 4 inches deep and has metal lath placed in waves inside it, behind the absorber. The 42-foot airflow path is longer than optimal but works all right with the oversized airflow cavity (two 20-foot flows would have been a much better design approach). Gaskets were placed between the corrugated absorber and the baffles, to make the air travel the full length of the cavity. This installation illustrates good use of recycled ¼-inch plate glass. It was installed in relatively small sections and, un-

Photo 12-3: This inexpensive, custom-built collector provides most of the daytime heating needs of Bill Etherton's workshop.

like tempered glass, could be cut at an angle for the top of the collector.

This 56-square-foot collector does a good job of providing daytime heat to the 24-by-50-foot shop, which is well sealed but poorly insulated. The shop is heated up to 60 to 70°F on sunny winter days. A small woodstove provides back-up heat on cloudy days, and the fuel oil stove that was once used hasn't been turned on since the collector was installed.

Like most builders of direct-use systems, Bill really misses his collector on cloudy days. He is very enthusiastic about his low-cost system, and after two winters in use he says that the collector doesn't owe him a dime.

Dick Ward's System: Retrofitting a Warehouse

The warehouse of Dick Ward, Alamosa, Colorado's Schlitz distributor, is a fine example of a commercial structure using a site-built collector for space heating. A beer warehouse must be kept cool (between 40 and 50°F) but never allowed to reach freezing temperatures. A 192-square-foot, vertical collector on the warehouse helps maintain these temperatures through the winter. The collector provides all the daytime needs for the 5,000-square-foot warehouse every sunny day by delivering a steady flow of 80 to 90°F air, which warms the entire building to 50°F.

Nighttime back-up heating is required, but Dick Ward's savings are now nearly half of his usual natural-gas consumption. This contractor-built system will pay for its total cost of $1,900 (1980 prices) in three or four winters even if tax credits aren't claimed.

A large main duct and two branch ducts located in the ceiling distribute solar heat to the two large rooms of the warehouse, where two large, slowly moving ceiling fans push the solar-heated air toward the floor. Air returns to the collector through ducting which runs to the bottom of the collector. This setup keeps temperatures even from the floor to the ceiling. The system can be turned off by a manual switch if it overheats the warehouse but this rarely happens.

The collector is glazed with Tedlar-coated FRP. ALthough it is guaranteed for 20 years, it will probably need to be replaced after 7 or 8 years. The absorber is corrugated aluminum, and the framework is 1⅝-inch metal track. The back of the collector is insulated with ductboard. Since low delivery temperatures from this collector are usable, it has a very high rate of airflow through it, and the 1⅝-inch airflow cavity isn't too large for the 24-foot airflow path.

Photo 12-4: This collector is very small compared with the huge warehouse it heats. The system operates effectively, however, because the collector is designed to deliver a great deal of low-grade heat.

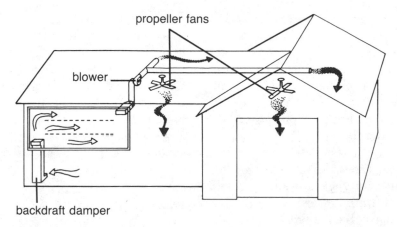

Figure 12-4: Large propeller fans circulate solar-heated air through the two large rooms of the Ward warehouse. The return to the collector runs down to the floor, to promote even air distribution.

Silas Kopf's TAP

Several years ago when Silas Kopf and his family began house-hunting in Northampton, Massachusetts, they looked for a home with good solar exposure. They got what they wanted in 1980, and in 1982 they completed their thermosiphoning collector system (photo 12-5). Originally Silas had hoped to build an attached greenhouse, but he decided instead to go with a collector system that would provide more heat at a lower initial cost.

The three collectors that make up this system were owner built with assistance from a hands-on workshop organized by the Energy Education Center at the University of Massachusetts. All three units are sealed backpass designs with 1½-inch-deep airflow cavities. The large central collector operates convectively, but the two smaller units on each side are fan driven due to obstructions in the wall cavity. The wood framework was designed to accommodate tempered glass units and was mounted on 1-by-10-inch wood

siding after shingle siding was removed. Ductboard was used for the collector backing to seal and insulate the airflow from the wall, and black-painted aluminum siding was used for the absorber. One-mil Teflon was used for the inner glazing, and standard-sized 26½-by-94-inch tempered glass was used for the outer glazing for each collector. Getting the Teflon to lie flat on the adhesive transfer tape was one of the major challenges in building this system.

Louvered grilles trim out the vent openings inside the house and make for an attractive, unobstrusive installation. The upper vents can be closed in summer to prevent undesirable heating of the living area. Superlight ripstop nylon is attached over the vents at the bottom of the collector to act as the backdraft damper flaps. This material opens well when the collector is operating, but Silas has noticed a few small leaks of cold air at night and plans to replace the flapper vanes with ½-mil Tedlar to achieve tighter seals.

During construction, obstructions in the frame wall were discovered so full-sized vents couldn't be used in the two smaller collectors. Instead of running these collectors convectively, Silas had two 55-cfm computer fans installed to force air through them. This works well and delivers a good amount of heat, but Silas questions the time-effectiveness of this procedure. Wiring the thermostat and fans was a major undertaking.

Glass slippage is the one major problem with this system. The panes of tempered glass are installed over Butyl glazing tape and held in place with aluminum flashing at the top and sides of the collector without support beneath the glass. This method hasn't proven to be adequate, and bottom supports will have to be added.

The materials cost for this system in 1981 was $1,350. The owner estimates that it delivers enough heat to replace about 130 gallons of fuel oil during a typical New England winter. After claiming a 60 percent combined state and federal solar tax credit, he will recover the initial cost in three or four years, which clearly makes it a good investment. The Kopfs are presently working on their next solar retrofit project: a greenhouse/sunspace for the southeast corner of their house.

Ron Loser's Black-Roof Collector

In 1979 when Ron Loser began to think about building his new house in Alamosa, Colorado, he heard about Roy Moore's sys-

Photo 12-5: This site-built retrofit on the Kopf house features a large thermosiphoning air panel and two smaller, fan-assisted units. All use corrugated aluminum for the absorber and are double glazed with 1-mil Teflon and tempered glass.

tems in New Mexico, where heat is taken from behind a black-painted metal roof. It seemed like a good idea so he designed a supertight, well-insulated house in which the shingle roof is the collector and the slab floor is the thermal storage. This is a simple, inexpensive and sensible idea, not to mention the fact that a good chunk of the expense of the house is tax deductible under energy credits. Ron is very pleased with this system and estimates that he quickly recovered its cost in just three years.

There are actually two separate solar systems in this house: the black-roof collector and a passive Trombe wall. Let's take a closer look at the roof collector. Although the south-facing roof looks rather conventional, it is really two roofs. The 12-inch roof rafters were sheathed with ½-inch plywood before 2 x 4s were nailed across the roof on 2-foot centers and all around the perimeter. A 2-foot gap alternates at the end of each run of 2 x 4s to create a 180-foot flow path up the entire roof (see figure 12-5). A second layer of ½-inch plywood was nailed over the 2 x 4s. The plywood - and - 2 x 4 sandwich was sealed with caulk to create an airtight channel for the airflow. Roofing felt and black shingles were added with 100 pounds of long roofing nails to promote heat transfer, and the 400-square-foot collector and roof were complete. The north-facing roof is of conventional design and construction.

The slab-floor storage consists of 15 cubic yards (60 tons) of concrete. An 8-inch-diameter duct from the roof feeds four 5-inch ducts under the slab, which connect with an 8-inch return (inlet) duct. A motorized damper and a backdraft damper prevent convective heat losses through the ductwork at night.

An 815-cfm blower, controlled by a differential thermostat (9°F cut-in, 3°F cut-off) and a timer, moves air through the system.

Ron wired the timer ($6) in series with the DT because he felt the blower was running more than necessary to raise the slab temperature to within 3°F of the roof temperature. A 15°F cut-in and a 5°F cut-off DT setting would be a better way to help with this.

Generally the blower comes on much later in the day and runs later into the afternoon than would a blower for a collector. The shingles heat up more slowly than an absorber plate and, acting almost like a low-grade storage, hold their heat longer into the afternoon. On sunny winter days outlet air from the collector is between 90 and 110°F.

How well does an unglazed collector perform in adverse weather conditions? Not too badly, as long as the sun is shining (even a little) and the wind isn't howling. If it is

Photo 12-6: The entire south-facing roof on the Loser house is a huge, low-efficiency collector. Since solar heat is stored in a slab under the house, the low collector outlet temperatures are very usable.

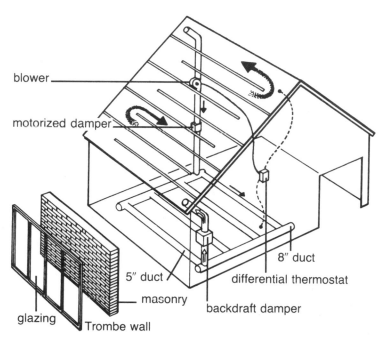

blower

motorized damper

8″ duct

5″ duct

differential thermostat

masonry

backdraft damper

glazing

Trombe wall

Figure 12-5: Solar-heated air flows in a serpentine pattern between the two roofs on the Loser house and is delivered to sturdy ductwork embedded in the slab floor. A passive Trombe wall occupies much of the south wall.

very windy and cold, the system won't come on, but on calm, sunny days below 0°F the blower can run for several hours. Light breezes have little effect on performance, but strong winds will keep the blower off. During these periods, as long as the sun shines, the Trombe wall adds heat to the house. This type of system works even better in milder climates. In areas with hot summers it would be easy to vent the roof convectively or actively during the day and to cool the storage slab actively at night.

This system has a few shortcomings. The 180-foot flow through the roof is too long, and four 45-foot flows would undoubtedly be more effective since the air is probably saturated with heat after about 40 feet. An

airflow through a rock bin, rather than through ducts, would produce better heat transfer to the slab.

In winter, morning temperatures inside the house are consistently 58 to 60°F, and evening temperatures, including the Trombe wall output, are consistently 70 to 74°F. The morning temperatures may seem a little low, but remember we are talking about radiant heat, which is more comfortable at lower air temperatures. Along with the radiant heating, another key to the success of this system is its size, 400 square feet. When a collector is that large, it doesn't have to be all that efficient to do the job, particularly when the house is very tight and well insulated and has a minimum of nonsouth window area.

APPENDIX 1
SUN PATH CHARTS

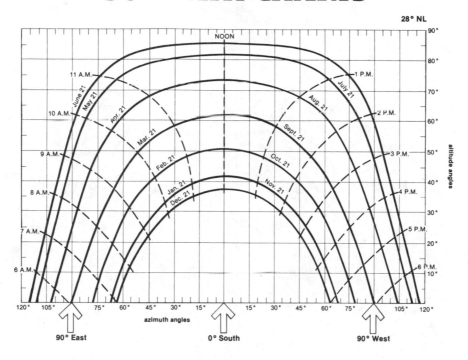

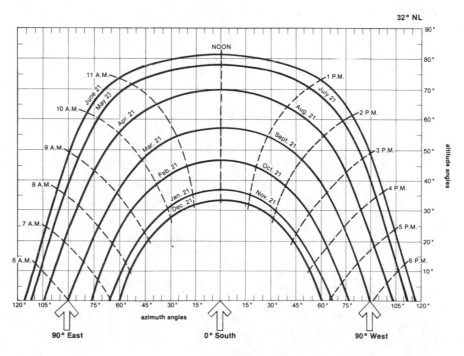

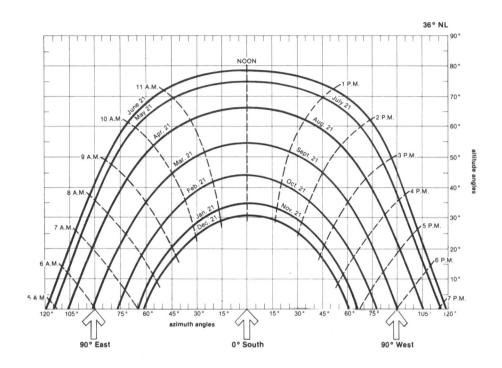

36° NL

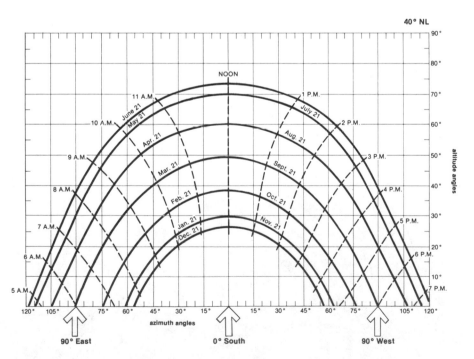

40° NL

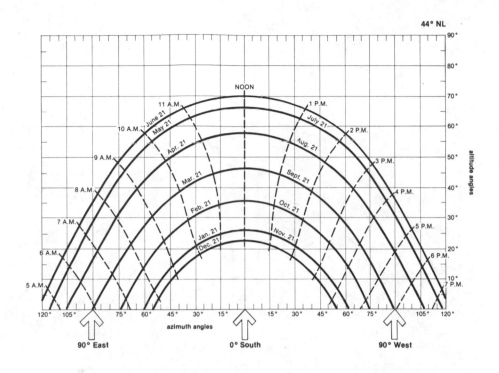

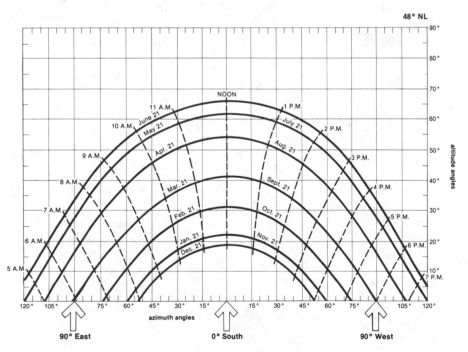

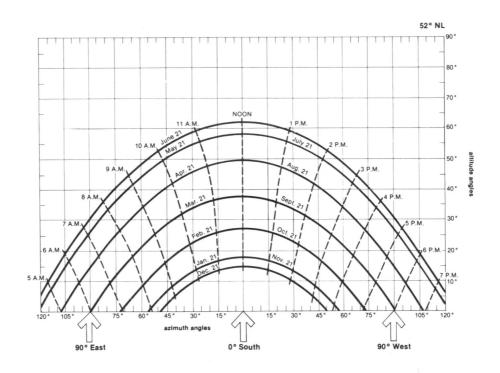

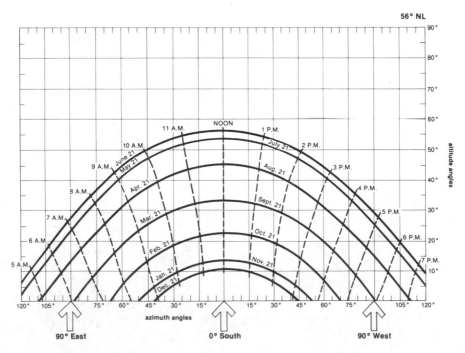

Figures A-1 through A-8: *Pages 297–300.* These sun path charts are for latitudes from 28 to 56 degrees north latitude (NL) at 4-degree intervals. The vertical axis indicates altitude while the horizontal axis shows azimuth angles. South refers to true south. See chapter 4 for an example of how these charts are used to study the movement of the sun at a particular site.
SOURCE: Adapted from Edward Mazria, *The Passive Solar Energy Book* (Emmaus, Pa.: Rodale Press, 1979), with the permission of Rodale Press.

APPENDIX 2
WIRING SCHEMATICS

Wiring the controls for a multi-mode system can be a big undertaking even for professional builders and journeyman electricians. If you are going to wire your own control system, you will need to plan it carefully and check it out with an electrician to make sure it is legal and safe. But if you have never heard of a DPDT relay and don't know what the squiggles mean in the schematics, you should probably enlist expert help. Take your schematics to different electricians and explain all you can about how your system operates.

The schematics that follow represent electrical systems that have been used successfully in air-heater systems. A parts list appears with each schematic. In some cases a heavier- or lighter-duty relay or a fuse with a different amperage may be needed, depending on the current drawn by your blower motor.

Once you know what your system is going to be like and you have located all the parts (relays, fuse holder, thermostats, transformer, etc.), get a knock-out junction box big enough to hold it all. In most control systems there will be some low-voltage components, and to conform with the code, all low-voltage wires must be insulated to 600 volts to be in the same box as 110-volt wires. The wire used to connect sensors to a differential thermostat or a 24-volt house thermostat probably won't have this insulation, so to have them inside the box you can install, inside the main box, a smaller box, in which all the low-voltage connections are made. If any of the low-voltage wiring must enter the main box, short pieces of heavily insulated wire should be used.

More wiring tips: Most relays come prepared for use with crimp connectors. We prefer these connectors, but we never crimp a connector onto solid wire that is larger than 16 gauge. Wherever solid wires come into the control box (i.e., from the blower), you should use Wire Nuts and "pigtails" of stranded wire to connect them to the relevant relay. When using crimp-on connectors, always use suitable crimpers; never use pliers.

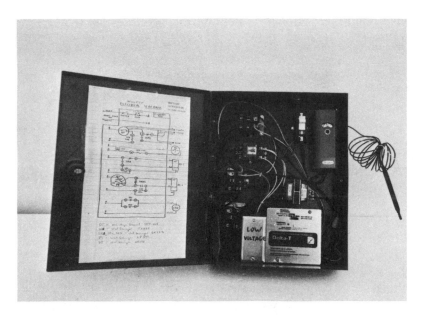

Photo A-1: This is a control box for a four-mode system. It contains a remote bulb thermostat (to the hot plenum in storage), a differential thermostat, four relays, a transformer, a fuse and a box for low-voltage connections. A spacious layout and a copy of the wiring schematic make checking out this section easy.

Once you have the parts and schematic, make a couple of copies of your schematic, and as you install each part, make note of your progress by coloring over the lines on the schematic with a highlighting marker.

Neatness in electrical systems contributes to safety. A clean and spacious layout will help you get the wiring right the first time and make it easier to find any mistakes that you do make.

Electrical Components and Terminology

For those of you who need expert help with the controls for your system, we will now discuss the meaning of the terms and parts used in our control schematics so that you will be knowledgeable enough to deal effectively with those who help you.

If your system has more than one mode of operation, it will probably need at least one relay. Relays are activated by thermostats (or other relays) and, in turn, control the blowers and dampers. A relay is a switch just like a light switch except that a relay's switch is not manual but automatic. Relays use both low-voltage (24 V) and house voltage (120 V) inputs (e.g., from a thermostat) to control a house voltage component (e.g., a blower or motorized damper).

The switch part of a relay is described according to the number and kind of contacts it has. Each set of contacts is called a pole, and each pole may be one of three kinds: single throw, normally open; single throw, normally closed; or double throw.

The single-throw, normally closed type has one contact set that is closed (current can flow) until a control voltage is applied to the relay, which causes the contacts to open (no current flows). In the normally open type the contacts remain open until the relay is energized, at which time they close. The double-throw set of contacts has three kinds of connections: the common, the normally open and the normally closed. A

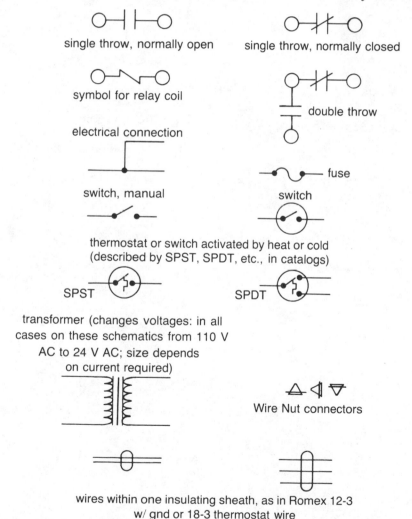

single throw, normally open   single throw, normally closed

symbol for relay coil

double throw

electrical connection

switch, manual   fuse

switch

thermostat or switch activated by heat or cold
(described by SPST, SPDT, etc., in catalogs)

SPST   SPDT

transformer (changes voltages: in all
cases on these schematics from 110 V
AC to 24 V AC; size depends
on current required)

Wire Nut connectors

wires within one insulating sheath, as in Romex 12-3
w/ gnd or 18-3 thermostat wire

Figure A-9: Symbols used in wiring schematics.

relay may have several sets of contacts (poles) and is named accordingly (*double pole single throw*, DPST and *triple pole double throw*, 3PDT). The pin numbers on the relays that we specify are shown in the wiring schematics.

Abbreviations Used in Wiring Sections

AC = alternating current

AD = air diverter

amp = ampere

BD = backdraft damper

blk = black

C blower = collector blower

cfm = cubic feet per minute

CT = collector thermostat

DHW = domestic hot water

DPDT = double pole double throw

DPST = double pole single throw

DT = differential thermostat

1st = first

FS = fan switch

ga = gauge

gnd = ground

H blower = house blower

hp = horsepower

HT = house thermostat

MD = motorized damper (MD1, MD2, etc.)

no. = number

PSC = permanent split capacitor

PT = plenum thermostat

rpm's = revolutions per minute

RT = rock bin thermostat

2d = second

SPDT = single pole double throw

SPST = single pole single throw

SR = solar relay (SRA, SRB, SRB1, etc.)

ST = solar thermostat

T blower = third (small) blower

3PDT = triple pole double throw

trans = transformer, 24 volt

TT = tank thermostat

V = volt

WG = water gauge

Schematic: A Two-Mode System for Domestic Water and Crawl-Space Heating

These controls were used on the sloped, roof-mounted system at the Gonzales residence (see chapter 12). The first priority in this system is domestic water heating, and the second is heating the crawl space. One differential thermostat is used for both modes by having an extra sensor and controlling the two cold sensors with the relay SRB (see figure A-10). If the water tank is heated, the DT compares the temperature of the collector to that of the crawl space and sends solar heat there.

Let's walk through the modes of operation and see what each one means in the schematic. In the water-heating mode the tank thermostat (TT) is on because the tank isn't fully heated. (This is a heating thermostat, and its set of contacts are closed until it reaches the set temperature.) TT energizes solar relay B (SRB), which causes the DT to compare the collector temperature to the tank temperature by closing the normally open contact (9,6). Because the collector is hot, the DT is on, energizing solar relay A (SRA). Thus SRA contact set (9,6) closes to power the blower. Both SRA and SRB are energized in this mode so the power for the first motorized damper (MD1) goes from SRA pin 7 to SRA pin 4 to SRB pin 7 to SRB pin 4 to the opening side of MD1. Power for the second motorized damper (MD2) goes from SRA pin 8 to SRA pin 5 to SRB pin 8 to SRB pin 5 to the closing side of MD2.

In the crawl-space heating mode the tank thermostat (TT) is off (contacts open) because the tank has heated up to the TT set-point. Thus SRB is no longer energized, but because the collector is still hot, DT is on and SRA is energized. DT is now comparing the collector to the crawl space because SRB is off. SRA is still powering the blower on. MD1 is closed as power goes through SRA pins 7 and 4, through SRB pins 7 and 1, to the closing side of MD1. MD2 is open as power goes through SRA pins 8 and 5, through SRB pins 8 and 2, to the opening side of MD2.

In the off mode DT is off, SRA is off, the blower is off and both MD1 and MD2 are in the closed position.

This same schematic could be used differently if desired. For example, if the first-priority mode were to heat the living space, TT would be replaced with a one-stage house thermostat or the first stage of a two-stage house thermostat. The second stage of the two-stage thermostat would activate the furnace. In this case a 24-volt transformer would be used to make the SRB coil circuit a 24-volt circuit, and SRB would be a 24-volt relay. Since the second priority in this case could be to heat water, the two cold DT sensors would need to be switched to prevent overheating the tank in the second mode. A tank thermostat would still be used and

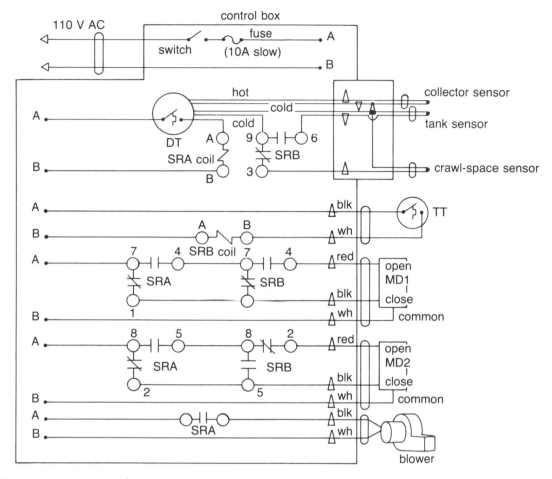

Figure A-10: Wiring schematic for a two-mode system for domestic water and crawl-space heating.

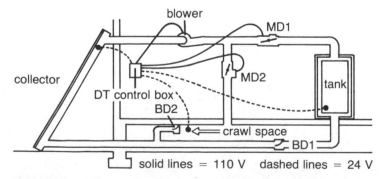

Figure A-11: Airflow schematic for a two-mode system for domestic water and crawl-space heating.

would be wired in series with the house thermostat (replacing TT in the schematic), to turn the air back to the house should the tank reach 200°F.

Parts List for Two-Mode System: Water and Space Heating

DT: (differential thermostat) compares collector temperatures to tank temperature or crawl-space temperature; e.g., Heliotrope General's no. DTT-80A with extra (third) sensor no. SAS-3 and fixed differentials of 18°F on, 8°F off

TT: (tank thermostat) turns heat to crawl space if tank is up to set-point; e.g., W. W. Grainger's no. 2E146

Blower: depends on size of system (see chapter 9)

MD1 and MD2: (motorized dampers 1 and 2) depend on size of ductwork; e.g., Hot Stuff's no. MD-2-10, 10″ diameter

BD1 and BD2: (backdraft dampers 1 and 2) depend on size of ductwork; e.g., Hot Stuff's no. BD-1-10, 10″ x 10″, or Round BD-2-10

SRA and SRB: (solar relays A and B) 3PDT (triple pole double throw), 120-volt coils; e.g., W. W. Grainger's no. 5X841, each with socket no. 5X853

NOTE: If the blower is too large for the contacts of no. 5X841, then use DPDT (double pole double throw) no. 5X838 and no. 3X745, with coils wired in parallel, and no. 3X745 replacing contact set (9,6) in SRA (for the blower).

TABLE A-1

Logic Chart for Two-Mode System
for Domestic Water and Crawl-Space Heating

Mode	DT	TT	Blower	MD1	MD2	BD1	BD2
Collector to tank	on	on	on	open	closed	open	closed
Collector to crawl space	on	off	on	closed	open	closed	open
Off	off		off	closed	closed	closed	closed

NOTE: DT = differential thermostat; TT = tank thermostat; MD1 and MD2 = motorized dampers 1 and 2; BD1 and BD2 = backdraft dampers 1 and 2.

Fuse: depends on blower current—slow-blow in any case; Fusetron's SSU holder w/ switch

Control box: Hoff's AHE, 15″ x 12″ x 4″ hinged-cover knock-out box or equivalent

Wire: 14-ga, 600-volt stranded copper, or larger if fuse is more than 15 amp

Wire terminals: crimp-on; e.g., W. W. Grainger's no. 4X293

TABLE A-2	
Approximate Material Cost for Parts for Two-Mode System for Domestic Water and Crawl-Space Heating	
DT	$ 70
TT	40
Blower	70
MD1, MD2 ($120 each)	240
BD1, BD2 ($30 each)	60
SRA, SRB w/ sockets	40
Fuse & holder	10
Box	12
Terminals & wire	10
Total	$552

NOTE: DT = differential thermostat; TT = tank thermostat; MD1 and MD2 = motorized dampers 1 and 2; BD1 and BD2 = backdraft dampers 1 and 2; SRA and SRB = solar relays A and B.

Schematic: A Two-Mode System for Space Heating

The controls for this system are used in the Gonzales vertical collector system (chapter 12.) The first mode is heating the living space, and the second is heating the crawl space. It uses a single air diverter rather than two motorized dampers to control air delivery. This simplifies controls but requires that there be a heat trap in the duct system. The single-stage solar house thermostat is independent of the house furnace because in this system the furnace is not used enough to warrant tying the two together. They could be connected by using the first stage of a two-stage thermostat for solar heating and the second stage for getting heat from the back-up furnace. Doing so would eliminate the transformer required in this schematic since the solar relay B (SRB) could be operated by the transformer that is part of the existing furnace (see figure A-14). Notice that the schematic calls for a switch in the house. This vertical collector is made of materials which will tolerate stagnation temperatures so the Gonzales family was provided with a means to turn the system completely off during the summer. A remote bulb thermostat, rather than a DT, controls the system because solar air is delivered directly into the house.

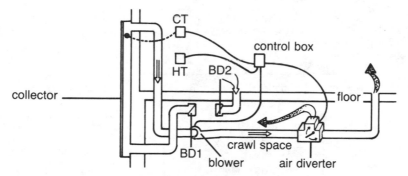

Figure A-12: Airflow schematic for a two-mode system for space heating.

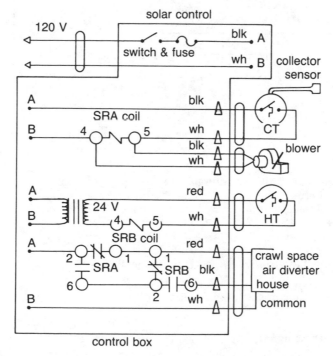

Figure A-13: Wiring schematic for a two-mode system for space heating.

TABLE A-3

Logic Chart for Two-Mode System for Space Heating

Mode	CT	HT	Blower	Air Diverter	BD1	BD2
Collector to house	on	on	on	to house	open	open
Collector to crawl space	on	off	on	to crawl space	open	closed
Off	off		off	to house	closed	closed

NOTE: CT = collector thermostat; HT = house thermostat; BD1 and BD2 = backdraft dampers 1 and 2.

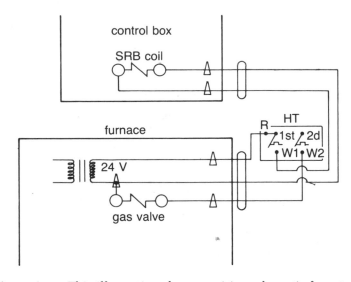

Figure A-14: This illustration shows a wiring schematic for a two-stage thermostat. The R, W1 and W2 shown in the HT (house thermostat) refer to an industry convention by which the terminals of all 24-volt heating thermostats are labeled.

Parts List for Two-Mode, Space-Heating System

CT: (collector thermostat), e.g., Honeywell's no. L6008C1206 (set at 100°F)

HT: (house thermostat), e.g., W. W. Grainger's no. 2E096

Blower: depends on size of system (see chapter 9)

Air diverter: e.g., Hot Stuff's no. AD-2-14, 110-volt air diverter with 14"
x 14" openings

BD1 and BD2: (backdraft dampers 1 and 2), e.g., Hot Stuff's no. BD-1-14,
14" x 14", or Round BD-2-12

SRA: (solar relay A) SPDT (single pole double throw), 120-volt coil; e.g.,
W. W. Grainger's no. 6X563

SRB: (solar relay B) SPDT, 24-volt coil; e.g., W. W. Grainger's no. 6X562

Transformer: e.g., W. W. Grainger's no. 4X743.
NOTE: If W. W. Grainger's no. 2E158 (a line voltage thermostat) is used
for the HT, no transformer is needed and SRB may be Grainger's no.6X563.

Fuse: depends on blower; use Fusetron's SSU with slow-blow fuse

Control box: Hoff's AHE, 15" x 12" x 4" hinged-cover knock-out box or
equivalent

Wire: 14-ga, 600-volt stranded copper, or larger if fuse is more than 15
amp

Wire terminals: crimp-on; e.g., W. W. Grainger's no. 4X293

Schematic: Three- and Four-Mode Systems with Two Blowers

The main difference between the two-blower system described here
and a one-blower system is that different blowers are used for different
modes. The three-mode system shown is the more likely choice for a
retrofit application, with the furnace being ducted in parallel with the
solar ducting. A four-mode system is the more likely arrangement in new
construction, with the house blower moving both the solar and the auxiliary heat. If the auxiliary heating system in new construction were electric or hot water baseboard rather than forced air, a three-mode would be
the likely choice.

The house thermostat (HT) is shown as a two-stage thermostat. Of
course, this need not be the case since two separate thermostats could be
used; one for the solar system (set at a slightly higher temperature) and
one for the furnace. The circuits presented could also be made simpler if
a different kind of damper were used. If the dampers were 110-volt devices

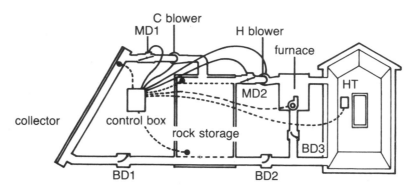

Figure A-15: Airflow schematic for a three-mode system with two blowers.

TABLE A-4

Logic Chart for Three-Mode System with Two Blowers

Mode	DT	PT	HT	Collector Blower	MD1	House Blower	MD2	Furnace
Collector to house	on	on	1st	on	open	on	open	off
Collector to storage	on		off	on	open	off	closed	off
Storage to house	off	on	1st	off	closed	on	open	off
Auxiliary furnace			2d				closed	on
Off	off		off	off	closed	off	closed	off

NOTE: DT = differential thermostat; PT = plenum thermostat; HT = house thermostat; MD1 and MD2 = motorized dampers 1 and 2.

that powered open and sprang closed, they could be wired in parallel with the blowers, and the second pole of each solar relay A and B (SRA and SRB) could be eliminated.

If the auxiliary heat is electric, there is more electrical schematic which we have not shown. There would be a high-power circuit separately controlled by a circuit breaker and unrelated to what we have shown except that it would be controlled by the application of power to what we show as the gas valve or relay coil.

Parts List for Two-Blower, Three- and Four-Mode Systems

DT: (differential thermostat) compares collector temperatures with lower rock bin temperatures and activates collector blower (C blower) and

MD1; e.g., Heliotrope General's no. DTT-80A, with a fixed 18°F to 8°F differential

HT: (house thermostat) 2-stage, 24-volt heating thermostat; e.g., Honeywell's no. T874F with subbase no. Q674D 1040

The first stage of HT turns on the solar heat if PT (plenum thermostat) indicates that there is heat available, and the second stage turns on the

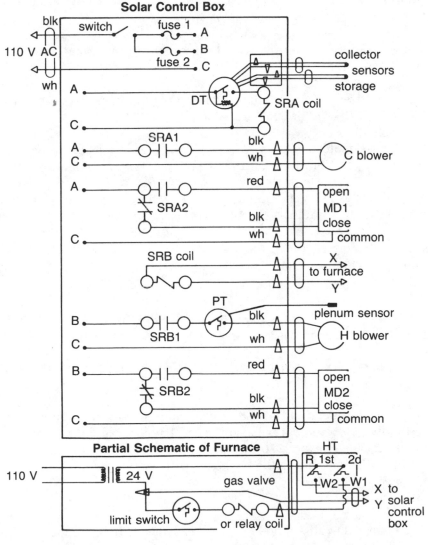

Figure A-16: Wiring schematic for a three-mode system with two blowers.

auxiliary heat, provided that the damper MD2 opens fully (a safety feature on the four-mode only).

PT: (plenum thermostat) detects presence of usable heat in solar system and prevents H blower from coming on unless such heat is present; e.g., W. W. Grainger's no. 2E399 or Johnson Controls' Penn A19ABC-4 or Honeywell's no. T675A1540

C blower: (collector blower) chosen to produce proper flow in collector (for sizing information see chapter 9)

H blower: (house blower) delivers heat to house (for sizing information see chapter 11)

MD1: (motorized damper) shuts off collector at night; e.g., Hot Stuff's no. MD-2-14, 14″ x 14″ or equivalent

MD2: (house motorized damper) opens when heat is to be delivered to house; e.g., Hot Stuff's no. MD-2-14, 14″ x 14″ or equivalent

BD1, BD2 and BD3 (backdraft dampers) allow flow in only one direction; e.g., Hot Stuff's no. BD-1-14, 14″ × 14″ or equivalent

SRA: (solar relay A) DPDT (double pole double throw) relay with 110-volt coil, which controls C blower and MD1; e.g., W. W. Grainger's no. 5X847

SRB: (solar relay B) DPDT relay with 24-volt coil, which controls H blower (with PT and fan switch) and MD2; e.g., W. W. Grainger's no. 5X846

Transformer: (24-volt transformer) powers 24-volt components in control circuit; e.g., W. W. Grainger's no. 4X744

Switch and fuses: Fusetron's SSUs may be used with fuses appropriate to the two blowers, but we also like Buchannan's no. 352 fuse/switch blocks, used with no. 330 end block and FNM type fuses (these are small and fit handily in the big box used for controls)

Control box: e.g., Hoff's AHE, 15″ x 18″ x 4″ or equivalent

Wire: 14-ga, 600-volt stranded copper, or larger if fuse is more than 15 amp

Wire terminals: crimp-on; e.g., W. W. Grainger's no. 4X293

For Further Reference

Temple, Peter L. and Adams, Jennifer A.
Solar Heating: A Construction Manual. Radnor, Pa.: Chilton Book Co., 1980.

Electrical schematics for two-blower, four-mode systems, one-blower, three- and four-mode systems and domestic water heating in multi-mode systems are available at a nominal cost from Hot Stuff Controls, P.O. Box 306, La Jara, CO 81140. Phone: (303) 274-4069.

Checking the Control Wiring

When you've finished wiring your system, go over it carefully, without the power applied, making sure each line on the schematic is properly represented by a wire. Inspect for possible shorts, and make sure the ground wires are secured away from any power terminals in the control box.

Apply power to the system and check thoroughly for proper operation. The best way is to manipulate the inputs to the system (thermostats) so that each mode of operation is duplicated and to observe the outputs (blowers and dampers) to see if each mode functions when, and only when, it should.

In the case of the simple crawl-space or direct-delivery system, this will consist of turning the remote bulb thermostat down and up and hearing the blower go on and off (turning the collector thermostat down is equivalent to having the collector heat up). If the collector is hotter than the maximum setting on the remote bulb when this test is done (the collector shouldn't be built yet), there may be no way to check the off mode until the sun goes down.

For more complicated systems the mode logic chart can be used to check the operation of each mode. Each mode is represented on the chart by one line with the inputs from the thermostats on the left and the outputs to the blowers and dampers on the right.

Troubleshooting Control Systems

If the system operates correctly in all its modes, that's terrific and you're done with controls, but if it doesn't, the problem needs to be pinpointed and corrected.

Control wiring problems may pop the circuit breaker or may not. Let's look first at those that do. In the simple crawl-space or direct-use system, the most common problem is that the line voltage thermostat that turns on the blower is wired so that, when the collector heats up (or the thermostat is turned down), the thermostat contacts close and create a short across the power supply line. If this happens, the thermostat needs to be replaced even if it seems to work properly afterwards. The contacts are burned and will start to stick sometime soon. Another thing that can pop the breaker is over-tightening the cable clamps on metal junction boxes, causing a short. With the circuit breaker off and the blower disconnected, the short can usually be found with an ohmmeter.

In more complicated systems the popping of the circuit breaker usually means that a similar problem has occurred, and some thermostat or relay has been wired so that the closing of its contacts creates a short. It will have to be replaced. Again, with the circuit breaker off, the path of the current may be followed with an ohmmeter to locate the short.

If there are no shorts, but the control circuit fails to perform its required function, approach the circuit carefully with a voltmeter. If, for example, the blower fails to come on when it should, check to see if there is power applied to it. If not, check for power at the relay which controls the blower and at the coil of that relay, etc. In this way, working from the symptom (the right side of the logic chart and electrical schematic) to the cause (upstream toward the power supply), the problem will be located.

APPENDIX 3
WORKING WITH OFFICIALS

Building Codes

You will need a building permit to build your own collector in all urban areas. In most states it is legal for you to do all of the work yourself (carpentry, wiring and plumbing) as long as the installation is on your own home. You will still have to satisfy the local building inspector, who has essentially three concerns: fire safety in the whole system, the collector's structural integrity and safety, and contamination of your drinking water by the solar water heater.

Will the collector catch on fire? There is very little fire danger in any wood-frame collector that doesn't use a selective surface, and no danger whatsoever in all-metal designs. If all the materials you use can tolerate temperatures up to 300°F, fire won't be a problem. It may be a good idea to give the inspector a complete list of the materials you will be using, along with their temperature ranges. Many inspectors don't like wood-and-fiberglass collectors, but as long as you use a metal backing for the collector and good-quality glazing and don't use a selective surface for the absorber of a wood-frame collector, there is no fire danger.

Is there adequate structural support for the collector? The collectors we feature weigh about 6 pounds per square foot (less than the weight of 2 inches of wet snow) and don't represent much of a structural load. Vertical mounts with bottom braces are normally no problem, but roof-mounted collectors may require additional support if the rafters are smaller than 2 x 6s. For roof mounts, use 2 x 6s rather than 2 x 4s for framing, and show roof attachments in your drawings, but the dead load of a collector is nowhere near the design snow load, so no major structural analysis is necessary. Closed-in, tilt-up roof mounts have very good wind resistance so wind loading shouldn't be a problem. You may need to spend $50 to have a structural engineer approve your plans if the building inspector requires it.

Will a solar water heater contaminate your drinking water? This is a problem only in liquid collector systems. The only thing passing through the heat exchanger in an air-heating system will be air and potable water. To satisfy the inspector, you may need to show a schematic that lists all your plumbing fittings.

Once you finalize your plans, call the building inspector, tell him you will be building your own collector, and ask him what he needs to see in the way of drawings. Feel him out. Some building inspectors like do-it-yourself projects and will be helpful; others may have had bad experiences with poorly installed systems and may be uncooperative. Be prepared when you apply for your building permit. Have several good drawings and know exactly what you plan to do. The building inspector will give suggestions; you don't have to ask for them.

Although most building inspectors know little about solar heating, they know about construction practices, and you should take their advice. If, however, a building inspector tells you that you can install only a certified (factory-built) collector on your house, it's time to go to the city council meeting and raise hell.

Solar Easements

It is important to protect your access to the sun, especially in urban and suburban areas. Some states now have legal provisions for solar easements, but in most places your rights to the sun aren't protected automatically, so you need to ensure them. If shading could be a potential problem, negotiate a solar easement with your neighbor to the south, or contact your state energy office or local solar energy association to find out how to obtain one in your state.

Financing

As soon as you can, start setting aside $10 or $20 a week toward your solar installation. Purchase materials as you can afford them, keeping an eye open for bargains. If you can finish the system before winter, you'll save money on your heating bills and get a nice tax refund the following year.

You can get a home-improvement loan from most banks for a home-built installation, but this is usually unnecessary and expensive, especially when interest rates are high.

Tax Credits

More and more states are requiring collector certification in order to qualify for tax credits. This has come about by pressure from collector manufacturers and has been fueled by solar rip-offs by some shady dealers. A do-it-yourself collector can't and won't be certified. The trend toward a blanket certification requirement is one which you, the owner-builder, must resist. The truth of the matter is that do-it-yourself installations will almost always deliver heat at a better price than a commercial system (see chapter 3 on economics).

As long as we have tax credits, there is absolutely no reason that home-built systems shouldn't qualify for them. It's up to owner-builders to make sure that tax credits don't exclude their systems, which may require political involvement on the local and state level. Your local solar energy association can be a very effective lobbying force.

Resources

If you are interested in your rights as an owner-builder or are looking for tips on getting along with a hard-nosed building inspector, locate a copy of *The Owner-Builder and the Code,* by Ken Kern, Ted Kogan and Rob Thallon (Oakhurst, Calif.: Owner-Builder Publications, 1976).

For a sample solar easement and more ideas on easements and building codes, read *Adding Solar, a Denver Homeowner's Guide,* by Rachel Snyder (Denver, Colo.: Denver Solar Energy Association, 1981).

APPENDIX 4
THE CHARACTERISTICS OF AN EFFICIENT FLAT PLATE COLLECTOR

If you go shopping for a commercial solar system, one of the first things the salesman will do is proudly show you the efficiency curve for the collector he sells, hoping to convince you of the superiority of his product. This doesn't really tell you anything, however, because the efficiency of a collector has little bearing on cost-effectiveness. What really matters is the cost and durability of the system versus the amount of heat delivered from it. In this section we will examine collector efficiency to give you an understanding of what design features make a collector perform efficiently. By comparing the different design options available and how they influence efficiency, the owner-builder can design a collector that will best meet his needs. Properly designed and built collectors not only can be much more cost-effective than commercial systems, they can operate just as efficiently.

Solar air collectors convert energy from one form to another: from sunshine (radiation) to hot air (sensible) heat. In the process some of the energy gets lost. The efficiency of a collector reflects the percentage that doesn't get lost. Thus if half of the energy striking a collector in the form of sunshine is captured in the form of heated air, the efficiency of the collector is said to be 50 percent.

Efficiency can be represented by the formula:

$$\text{efficiency} = \frac{\text{output (usable heat)}}{\text{input (available solar energy)}}$$

The amount of available energy (insolation, not insulation) is expressed as the number of Btu's striking a 1-square-foot area in an hour (Btu per square foot per hour). Daily insolation is measured, with a device called a pyranometer, at many weather stations throughout the country. The amount varies a great deal with cloud cover, the time of year, the latitude and the elevation of the site. For example, the amount of insolation available at Denver on a sunny November day at noon averages 300 Btu per square foot per hour. Thus a collector operating at 50 percent efficiency under these conditions would have an output of 150 Btu per square foot per hour.

The amount of heat delivered, in this case 150 Btu per square foot per hour, is determined for air systems by multiplying the amount of air moved through the collector (in cfm) by the rise in air temperature through the collector and by the heat capacity of the outgoing air at the elevation of the installation. This is a straightforward calculation. However, we want to analyze a collector's efficiency under different conditions, and this is a bit more involved.

There are three factors which affect the efficiency of a collector operating with a given airflow: 1) the temperature of the outside air (*ambient*

Generating an Efficiency Curve

As an example, let's look at a collector that has a 50 percent efficiency when the insolation is measured (with a pyranometer) at a fairly high noontime figure of 300 Btu per square foot per hour. Assume that the ambient temperature is 30°F and the collector inlet temperature is 70°F.

$$\frac{150 \text{ Btu/ft}^2\text{/ht (delivered)}}{300 \text{ Btu/ft}^2\text{/hr (possible)}}$$

$$= 50\% \text{ efficiency}$$

$$\frac{T_i - T_a}{I} = \frac{70° - 30°}{300 \text{ Btu/ft}^2\text{/hr}}$$

$$= \frac{40}{300} = 0.133 \text{ performance parameter}$$

The same parameter can also be achieved under different conditions:

$$\frac{72° - 40°}{240 \text{ Btu/ft}^2\text{/hr}} = \frac{32}{240} = 0.133$$

Under these conditions we can expect the collector to once again operate at 50 percent efficiency and deliver 120 Btu per square foot per hour (240 x 50%).

In actual practice the collector is tested at the same flow under many different conditions,

and its efficiency is calculated for each set of conditions. Its efficiency is plotted on the Y axis, and the parameter corresponding to the test conditions is plotted on the X axis. A curve is drawn to best fit the points plotted and is usually a nearly straight line. The point where the line intercepts the Y axis (Y intercept) represents a condition in which the air entering a collector is the same temperature as the ambient air. Under these conditions the efficiency will be very high because there will be minimal heat loss out the front of the collector and the moving stream of air can pick up a large amount of heat.

Y intercept example: $T_i = 80°F$, $T_a = 80°F$
Insolation = 300 Btu/ft²/hr

$$\frac{T_i - T_a}{I} = \frac{80 - 80}{300} = \frac{0}{300} = 0$$

The measured efficiency for our example collector at these conditions is 65 percent.

NOTE: Some collector manufacturers average the collector inlet and outlet temperatures and use this average to calculate the parameter where we use only the ambient temperature. This gives a lower value to the parameter and makes the collector efficiency appear higher.

temperature), 2) the temperature of the air entering the collector, and 3) the amount of solar energy striking the collector (insolation). An efficiency curve is the standard method of expressing the interrelation of these three variables (see figure A-17).

Efficiency curves (usually straight lines) are a bit confusing at first glance since they involve so many variables. They help our understanding, however, by combining into one *performance parameter* three variables: collector inlet temperature, outside air temperature and insolation. If the performance parameter is the same for different conditions, the collector being tested will have very nearly the same measured efficiency under those different conditions.

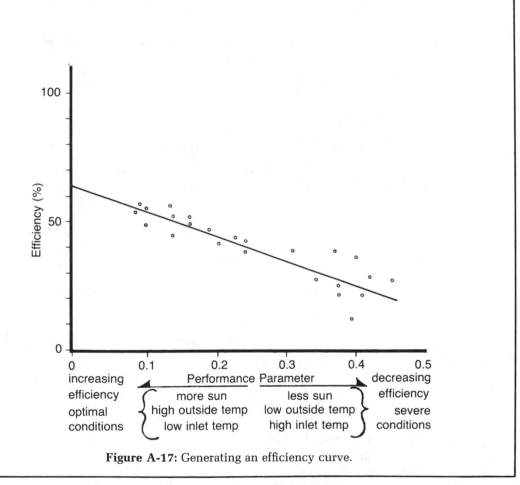

Figure A-17: Generating an efficiency curve.

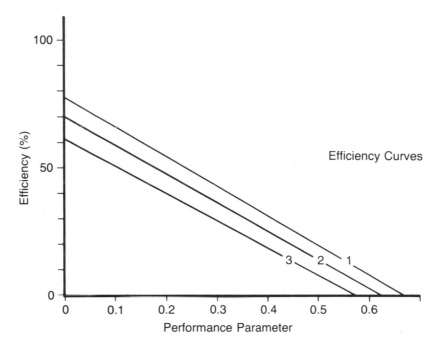

Figure A-18: These are efficiency curves for three different solar collectors. Number 1 is the curve for the site-built, roof-mounted collector on the Gonzales residence (see chapters 8 and 12). The curve was plotted from data supplied by a monitoring program of the Solar Energy Research Institute that studied site-built systems. Number 2 is the efficiency curve for a popular, liquid-medium collector tested by Lockheed according to ASHRAE Standard 93-77. Curve number 3 is taken from the product literature of a widely marketed, commercially manufactured, air-medium collector.

The formula for the performance parameter is:

$$\frac{\text{collector inlet temperature} - \text{outside air (ambient) temperature}}{\text{insolation}}$$

or

$$\text{or } \frac{T_i - T_a}{I}$$

This equation looks complicated, but there are only three things you need to keep in mind: as solar radiation (insolation) increases, efficiency increases (performance parameter decreases); as the outside air temperature rises, efficiency increases; as the collector inlet temperature falls, efficiency increases.

A collector will perform efficiently if it is tested on a warm, sunny day and is being fed cool air. The absorber will be hot because it is sunny. There will be little heat loss out the front of the collector because it is warm outside, and the cool air entering the collector can easily pick up the heat that is available inside the collector. The cooler the incoming air, the more easily it can pick up heat. A collector will, of course, operate inefficiently if it is cloudy and cold outside and if it is being fed warm air. In this case the absorber plate won't get hot, the cold outside air will rob heat from the collector, and the warm air entering the collector won't pick up what little heat is available.

Flat plate collectors in most locations should operate at about 40 percent efficiency when averaged through the course of a heating season. This 40 percent efficiency figure is quite reasonable when you consider that the average automobile is less than 15 percent efficient in converting energy. Another thing to keep in mind is that efficiency doesn't relate directly to delivery temperatures. If a collector is being fed very cold air, it will operate very efficiently, but the delivery temperatures will be quite low.

Let's look at several efficiency curves for different collectors to see how to build an efficient one.

Glazing

The glazing or transparent covering for a collector helps trap heat inside it by means of the greenhouse effect. The glazing also creates an insulating dead air space between the absorber and the glazing, which helps to prevent heat loss out the front of the collector. If a collector is double or triple glazed, there is even less heat loss out the front, but there is also less solar gain since each layer of glazing allows only about 85 to 90 percent of the solar radiation striking it to pass through. Figure A-19 shows typical efficiency curves for different glazing techniques using glass,

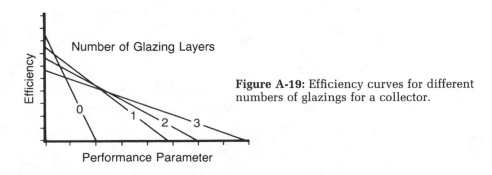

Figure A-19: Efficiency curves for different numbers of glazings for a collector.

which transmits about 85 percent of the available sunlight.

An unglazed collector lets all of the available sunlight hit the absorber and is very efficient if there is a lot of sun and it is warm outside. It will be very inefficient if it is cold outside because the unglazed collector quickly loses the heat it picks up, since it has no thermal protection and the greenhouse effect doesn't happen.

A single glazing on the collector will greatly increase its overall efficiency (the area under the efficiency curve will be increased). The glazing will screen out some of the sunlight but helps the collector operate more efficiently at lower outside temperatures. Double- and triple-glazed collectors are even more efficient at lower temperatures but block out more of the available sunlight and aren't nearly as efficient when it is warmer outside.

Cost-effectiveness is the major factor when considering double or triple glazing, and single glazing is usually adequate for flat plate installations unless the site is very cold or very windy.

The optimal size for the insulating dead air space between the absorber plate and the glazing is ¾ inch because a narrow dead air space prevents convective currents from transferring heat from the absorber to the glazing. Most collectors have a larger air space than this, however, to allow for expansion and possible bulging of the absorber plate. Larger air spaces allow for slightly more heat loss but cause only very small losses in efficiency.

The Absorber Plate

Once sunlight is inside the collector, the dark metal absorber plate converts it to heat and transfers this heat to a moving stream of air. This absorber plate should be nonreflective (capturing nearly all of the light striking it), and it should provide for a rapid transfer of heat. All of the metals traditionally used in collector construction (copper, aluminum and steel) are good at transferring this heat if they are thin (less than 0.02 inch or 20 mil). The real difference in efficiency comes from the black coating on the plate which intercepts the sunlight. The absorbers for low-cost

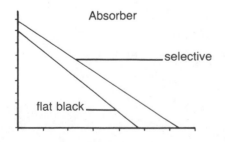

Figure A-20: Efficiency curves for selective and flat black absorber surfaces.

collectors are simply painted with high-temperature, flat black paint, but higher efficiencies can be realized if a selective absorber surface is used. A selective coating lets heat through it but not back out, and more heat passes into the air channel, especially at higher collector inlet temperatures.

Under severe-use conditions (high parameters) selective coatings increase efficiency the same way additional glazings do. Under more ideal conditions selective surfaces still help but not as much because there is less heat loss out of the front of the collector. Using a selective surface is almost always a more cost-effective approach than double glazing.

However, selective absorbers are quite expensive ($3 per square foot or more versus 30¢ to 80¢ for painted aluminum), and although they can be quite efficient at capturing and transferring heat, they may not be cost-effective for some applications.

Transferring Heat to the Airstream

In order to get rapid and efficient transfer of heat from the absorber to the moving stream of air behind it, the incoming air must be much cooler than the absorber. Heat always travels from warm objects to cold objects, and the greater the temperature differential, the more rapidly the heat will travel. If air enters a collector at a high temperature, there will be poor transfer of heat from the absorber and more losses out the glazing. The glazing is always the coldest part of a collector and will readily absorb any heat offered to it. If the glazing is hot, the collector isn't operating efficiently. Having an adequate airflow through a collector is one of the best ways to maintain the heat differential between the absorber and the airstream and ensure an efficient heat transfer.

A collector that delivers 80°F air with a very high flow will operate more efficiently and deliver more heat than an identical one that delivers 150°F air at a very low airflow. Eighty-degree air can be used effectively for slab-floor heating, but most applications require higher output tem-

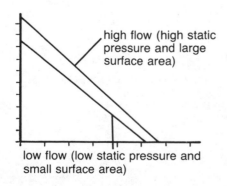

high flow (high static pressure and large surface area)

low flow (low static pressure and small surface area)

Figure A-21: Efficiency curves showing factors that influence heat transfer from the absorber.

peratures, and the collectors for them are designed to deliver 100 to 200°F air.

Another way to get a better heat transfer from the absorber plate is to create turbulence and resistance to flow in the flow cavity. Turbulent air that is forced to rub against the absorber plate will pick up more heat than an unrestricted airflow. The simplest way to accomplish this is by making the airflow cavity shallow.

Increasing the surface area of the absorber plate will also help heat transfer. Putting fins on the back of a flat absorber, bending a flat plate into an accordian shape, or using a mesh or matrix absorber will all help to increase collector efficiency. Installing and sealing these types of absorbers in an active collector is more difficult than with a simple flat plate, however, especially in do-it-yourself installations.

Figure A-21 shows efficiency curves that indicate the factors influencing heat transfer from the absorber. It is assumed that the airflow is the same for the differing static pressures and collector surface areas.

Increasing the airflow, increasing the air turbulence or increasing the surface area will all increase the efficiency under all conditions, but more so when the parameter is low. All three will require larger blowers to move air through the collector.

Insulation

Another consideration in building an efficient collector is to keep the heat in the airstream. This is done by insulating the back side and perimeter of the collector with high-temperature insulation. In flat plate collectors, insulation with an R-value of 5 to 10 is usually adequate for protecting the heat differential between the solar-heated air and the outside air. The colder it is outside, the more important insulation is in increasing efficiency. Insulation also has more value in the hot end of the collector than in the cold end.

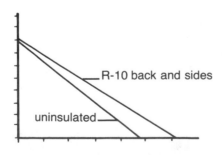

Figure A-22: Efficiency curves indicating the importance of insulation.

Reflective Gain

Increasing the amount of solar radiation striking a collector can be one of the simplest ways to increase a collector's output. After all possible obstructions to the sun in the solar window have been removed, the only way to get more sunlight to the collector site is by reflection. Locating a collector north of a reflective surface or placing reflective panels below the front of the collector can increase the amount of solar radiation received by up to 30 percent. Figure A-23 shows efficiency curves for a vertically mounted flat plate collector and for a vertical collector that is receiving 30 percent more sunshine through reflective gain.

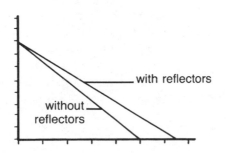

Figure A-23: Efficiency curves for collectors with and without reflectors.

Increasing the amount of insolation striking a collector will increase its *heat output* under all conditions, but it will only increase the efficiency under adverse conditions. This is a good example of how collector efficiency doesn't tell us everything about heat delivery. When the collector inlet temperature is close to the outside temperature (optimal conditions), the collector will be operating at a high efficiency no matter how much sunshine is striking it. If reflectors are used, more solar radiation will reach the collector and it will get hotter, but this additional amount of insolation will have to be figured into the efficiency calculation (heat output ÷ available sunshine). Under adverse conditions increasing the insolation will decrease the parameter and increase the efficiency. Since reflective gain will always increase heat delivery, it is a good idea to take advantage of it whenever possible.

Feeding Outside Air

Supplying your collector with outside air, which is almost always cooler than the return air from a crawl space, living area or storage, will produce very efficient operation because all collectors operate more efficiently at low collector inlet temperatures. The problem here is that the delivery temperatures may not be high enough to be usable even though the collector is operating at maximum efficiency. Feeding a collector with outside air is most effective in installations where a constant air change is required. The solar-heated air can then be directed to a furnace for further heating if it isn't warm enough for use.

Your Site and Needs

The efficiency curve for a given collector will be the same in all parts of the country, but considerations must be made to build the most efficient collector for a given site.

If you live in a place where it is cold and cloudy much of the time, as Minnesota is in the winter, your collector will usually operate at the low end (right side) of the efficiency curve. To get usable temperatures from your collector for home heating, you will need to build to an optimum design to maximize efficiency when the solar resource is marginal. Double glazing, plenty of insulation and a selective surface would all be helpful in accomplishing this. Moving a very large amount of air through this collector probably isn't a good idea. A collector at this site must be large to supply an adequate amount of heat.

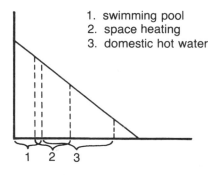

1. swimming pool
2. space heating
3. domestic hot water

Figure A-24: As shown by this illustration, different types of systems will operate at different collector efficiencies.

A collector that will be used for heating a swimming pool in a hot, sunny location like Arizona won't have to produce high temperatures in order to be useful, and it will, therefore, operate at a low inlet temperature. The collector will operate at the high end (left side) of its efficiency curve most of the time (ambient temperature may even be greater than collector inlet temperature), and it should be designed to keep this end of the curve as high as possible by using single or no glazing on the collector.

Liquid Systems

The efficiency curves for active solar systems that use liquid as the heat exchange medium will usually be slightly higher than those for air but will otherwise be quite similar because they are influenced by the same design considerations. Although liquid collectors usually operate at a bit higher efficiency than air collectors, an entire air system with storage can be more efficient than a liquid system with storage.

Convective Air Collectors

The factors affecting the efficiency of convective collectors are the same as for active collectors except for those factors that deal with heat transfer. Convective collectors must operate with an extremely low static pressure in the airflow cavity, so the convective collector we prefer uses a mesh absorber that has lots of surface area to encourage effective heat transfer. Single-glazed convective collectors with the highest efficiency rely on keeping a cool layer of air next to the glass by using a slanted mesh absorber or a doublepass (front and back) design with a flat plate. The efficiency curve for a well-built convective collector should be nearly the same as for an active collector mounted in its place.

Look and Feel

You can tell a lot about the efficiency of an operating collector by looking at it and feeling it. If there is any light reflection coming from the collector, the absorber plate isn't trapping all of the available sunlight. If the glass on an operating collector is hot to the touch, the collector is losing a lot of heat out the front. An efficient collector doesn't have any glare coming from it and its glazing is cool.

Economic Efficiency

Up to now we have been discussing efficiency in engineering terms: heat output versus heat input. We also need to consider efficiency in economic terms: useful energy output versus overall cost. It is very important to balance the two when designing a collector and to try to get the most technically efficient system possible at a reasonable price. In most residential heating applications (space heat and domestic hot water) it is a better idea to build a large, inexpensive, slightly less efficient collector than it is to build a smaller, expensive, very efficient one.

System Efficiency

A collector is only part of your solar system, and the efficiency of the entire system also needs to be considered. Good system efficiency goes hand in hand with good collector efficiency. Perhaps one of the most important factors is good heat transfer at the point of use. If the solar water preheat tank, rock storage bin or the crawl space in your system takes a lot of heat out of the solar air, the return (inlet) air to the collector will be cool, and the collector will operate more efficiently.

Proper blower and ductwork sizing are also important considerations in improving system efficiency. It is necessary to use large enough ductwork so that an efficient blower can operate the system.

Controlling heat losses will also improve system efficiency. In some systems heat losses from ductwork or storage won't make a great deal of difference. For example, if the ductwork or storage is located under the living area, any losses still produce usable wintertime heat. A poorly sealed and insulated rock box located in the garage, however, will never deliver very much usable heat no matter how efficiently the collector is operating. Likewise a solar hot water preheat tank that loses heat to the inside of a dwelling during the summer won't produce an efficient system because it makes the house uncomfortably hot or puts a load on the air conditioner.

APPENDIX 5
MONITORING YOUR SYSTEM

Monitoring your solar system can be an interesting project, but it is important to realize your limitations. The Solar Energy Research Institute (SERI) spends up to $10,000 monitoring a single system to get data that is accurate to within 10 percent. The procedures and calculations we present below are neither that expensive nor that accurate, but they should put you in the ballpark in determining your collector's performance.

If you want to know precisely how many Btu's your system is delivering, you must first find out how many pounds of air are moving through it per minute by measuring the airflow, and you must find out the temperature rise of the air as it passes through the collector. If you are also interested in determining the efficiency of the collector, you must determine how much solar energy is striking the collector at the time you measure the airflow rate and temperature rise.

To accurately measure the airflow, you will need to locate a temperature compensated *hot wire anemometer*. This device is available at major heating and cooling dealers, college labs and instrument rental places. Ask for a source at the local solar energy association. Manufacturing plants that use chemical processes all have anemometers; ask for the safety officer. OSHA (Occupational Safety and Health Administration) and NIOSH (National Institute for Occupational Safety and Health) teams all have them, too. If all else fails, electronic instrument factory representatives (consult the Yellow Pages) can steer you to one. Be sure to get instructions with the instrument since they can be rather complicated. You will have to make several readings and take the airflow measurements in a long, straight section of ductwork to get accurate results.

A hot wire anemometer senses the flow of air as its hot wire is cooled by the flow and, therefore, doesn't measure exact air speed, but speed adjusted to standard temperature and pressure (density). The mass or weight of a stream of air is calculated by multiplying the average adjusted

$$\text{speed (ft/min) x area (ft}^2\text{)} = \text{volume (ft}^3\text{/min)}$$

The air measured in this way is *standard* air, and has a density of 13.07 cubic feet per pound, so the mass flow is:

$$\text{mass flow (lbs/hr)} = \frac{\text{volume (ft}^3\text{/min) x 60 (min/hr)}}{13.07 \text{ ft}^3\text{/lb}}$$

The heat output of the collector is:

$$\text{heat output/ft}^2 \text{ (Btu/ft}^2\text{/hr)} = \text{mass flow (lbs/hr)}$$

$$\text{x specific heat of air (0.25 Btu/lb/°F) x } \frac{\text{temperature rise in collector (°F)}}{\text{area of collector (ft}^2\text{)}}$$

The best time to do all this measuring is at noon on a sunny day, when the collector is working at top efficiency. Your measurements and calculations will yield a number, say 150 Btu per square foot per hour, which is a good average figure for a well-built collector tested under these conditions. If your result is less than 100 Btu per square foot per hour on a sunny day at noon, you are having problems with your system or with your measurements and calculations.

If you want to determine the efficiency of the collector while you are finding its heat output, you will need to measure the insolation striking it, using a pyranometer. Weather stations at most large airports measure insolation. You can use their figures on a cloudless day. Ask for the present insolation value while you are doing your testing. Most weather stations measure insolation on a horizontal surface, so you will need to adjust the figure they give you, say 210 Btu per square foot per hour, to the amount of insolation striking your collector. Do this by using a protractor to make a drawing like that in figure A-25. Angle A is the angle of your collector's tilt, and angle B is the angle of the sun above the horizon at your site

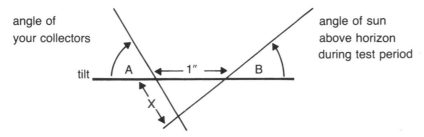

Figure A-25: This illustration shows how to determine the amount of insolation that strikes a tilted collector ($\frac{1}{x}$ x the insolation that strikes a horizontal surface).

when you measured the heat output (see the sun charts in Appendix 1). Measure the dimension X in inches with a ruler, and $\frac{1}{x}$ is your adjustment factor. For example, if X = ¾ inch, then $\frac{1}{x}$ will be 4/3, and 210 Btu/ft²/hr on the horizontal will be:

$$210 \text{ x } 4/3 = 280 \text{ Btu/ft}^2/\text{hr}$$

striking your collector during the test period.

The efficiency of the collector is calculated as the ratio of the heat output to the insolation, in our example 150 ÷ 280 = 54 percent. To really compare this to other collectors, you need to know one more thing, the air temperature outside, so that you can calculate the *performance parameter*. The parameter is

$$\frac{\text{temperature of air into collector} - \text{outside temperature}}{\text{insolation}}$$

and will be a number between 0 and 0.5 at noon. Test the collector under different conditions, and compare the parameters and calculated efficiencies to the efficiency curve given for the collectors we describe in Appendix 4.

Water-Heating Systems

You can measure the heat output and efficiency of a water-heating air collector using the previous method, but it is easier and requires less equipment simply to measure the daily heat output of the whole system. Start the day with cool water in the solar tank by opening a hot water tap for several minutes. Measure the temperature of the tank by draining a little water out. Isolate the solar tank, or close the cold supply to the solar tank by shutting the right valve, and don't use any hot water all day (be brave). Let the collector heat the tank all day. To measure the heat gathered by the system, shut off the water to the solar tank and drain all of the water into buckets, measuring the temperature and weighing the water in each bucket as you do so (subtract the weight of the bucket). The heat gathered is then:

heat gathered in tank (Btu's) = weight of water in first bucket (lbs)
x temperature rise (present temperature − morning temperature)(°F)
+ weight of water in the second bucket (lbs) x temperature rise (°F),
plus all subsequent buckets until the tank is drained

Add up the total Btu's collected, and divide this figure by the collector area to determine the heat output per square foot per day. For a sunny day this should be greater than 500 Btu per square foot per day.

One of the best ways to analyze your data is to see how much fuel your solar water heater is saving. If an 80-square-foot collector stores 56,000 Btu a day in a tank of water and you are using electricity (at 7¢ per KWH or $20.50 per million Btu) for back-up water heating, you save

$$\frac{\$20.50}{\text{MMBtu}} \text{ x } 56,000 \text{ Btu } = \$1.15$$

every sunny day.

Daily blower operating costs for this system should be less than 10¢.

Measuring Static Pressure

If you are interested in determining the exact static pressure present in your system, you will need to locate an accurate *manometer*. Look in the Yellow Pages under "Instruments" and find a company that sells or can order a Dwyer Mark II Manometer no. 25. These manometers cost only about $20 and are accurate enough for measuring the static pressure drop through collector components. For the name of a local distributor, contact Dwyer Instruments, Inc., P.O Box 373, Michigan City, IN 46360, phone: (219) 872-9141.

APPENDIX 6
CALCULATING STATIC PRESSURE

It is possible to get a more accurate calculation of the actual static pressure in your ductwork by using the information in figure A-26.

To use this chart, you need to know the desired airflow in cfm and the equivalent length of ductwork (straight runs plus transitions) for the system. For example, let's assume a system has 240 cfm and 300 feet equivalent length (square or rectangular ductwork is converted to round for convenience).

From figure A-26 we see that if we use 8-inch-round ducting, our friction loss (or static pressure, SP) will be 0.11 inch WG per 100 feet of ducting (the point on the chart where 8-inch ducting, 240 cfm, and 0.11 inch friction loss intersect). If a 10-inch-round ducting or equivalent square ducting is used, our friction loss will be 0.035 inch WG per 100 feet.

Total static pressure for the system equals:

SP per 100 ft x equivalent length
8 inches round @ 0.11 inch WG/100 ft x 300 ft = 0.33 inch WG
10 inches round @ 0.035 inch WG/100 ft x 300 ft = 0.105 inch WG

Ten-inch-diameter ductwork will be used in this system because, in order to avoid problems, the optimal static pressure in residential ductwork is almost always below 0.15 inch. Ten-inch ductwork will yield a velocity of 700 fpm (the point where velocity intersects the previously determined point). This is below the acceptable maximum of 800 fpm, so 10-inch duct is fine.

Balancing a Distribution System

The method of calculating distribution duct sizes that we use is the equal static pressure method. With this method the ductwork is sized so that the static pressures in all parallel duct lines are equal when the desired amount of air is flowing through them. If all flows encounter the same resistance, they will be correctly balanced.

For example, we will blow solar-heated air into the house from three different ducts and will want each to supply the house with a different airflow. Our three ducts, their desired flow rates and equivalent lengths are:

Friction of Air in Straight Ducts for Volumes of 10 to 2,000 Cfm

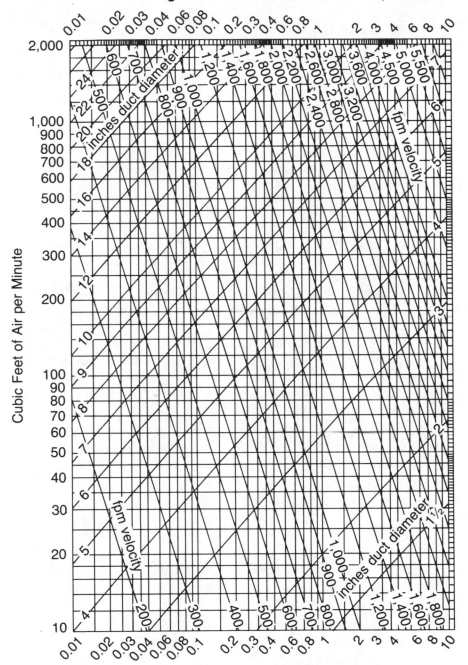

Figure A-26

SOURCE: Redrawn from American Society of Heating, Refrigerating and Air-Conditioning Engineers, Inc., *ASHRAE Handbook and Product Directory (1977 Fundamentals)*, copyright©1977, with the permission of ASHRAE, Atlanta, Ga.

Duct	Description
1	300 cfm, 70 ft equivalent length (40 ft plus 3 good elbows at 10 ft each)
2	200 cfm, 25 ft equivalent length (15 ft plus 1 good elbow at 10 ft each)
3	100 cfm, 70 ft equivalent length (30 ft plus 4 good elbows at 10 ft each)

Once again, we want the static pressure to be below 0.15 inch and the velocity to be less than 800 feet per minute.

Duct 1

- Flow rate is 300 cfm for 70 ft.
- To move 300 cfm at 800 fpm, 8.4-inch-diameter ductwork is required (the point on the chart in figure A-26 where cfm, fpm and duct diameter intersect).
- We chose 9-inch duct, which yields a static pressure of 0.09 inch WG/100 ft (the point on the chart where duct diameter, cfm and friction loss intersect).
- An SP of 0.09 inch WG/100 ft yields 0.063 inch WG/70 ft.
- The air will move at about 700 fpm under these conditions (from figure A-26).
- Both the static pressure and the velocity figures are below the acceptable maximum, so the 9-inch ductwork is a good choice.

We can now balance the other ducts to correspond to 0.063 inch static pressure for their length of run so each will carry the appropriate amount of air.

Duct 2

- Flow rate is 200 cfm for 25 ft with SP = 0.063 inch.
- 200 cfm at 800 fpm requires 6.8-inch-diameter ducting.
- We chose 7-inch, which gives 0.15 inch WG/100 ft or 0.038 inch WG/25 ft.
- A restrictor to this line will be needed to increase the static pressure to 0.063 inch, which will bring the flow rate down.

Duct 3

- Flow rate is 100 cfm for 70 ft with SP = 0.063 inch.
- 100 cfm at 800 fpm requires 4.8-inch ducting.
- We first choose 5-inch ducting, which yields 0.21 inch WG/100 ft or 0.15 inch WG/70 ft (this is above our desired SP).
- We now choose 6-inch ducting, which yields 0.09 inch WG/100 ft or 0.063 inch WG/70 ft (almost exactly the static pressure required, and no restrictions are necessary in this line).

INDEX

Page numbers in **boldface** indicate tables.